WEAK INTERACTIONS

McGRAW-HILL ADVANCED PHYSICS MONOGRAPH SERIES

COMMINS: Weak Interactions

GARRETT: Gas Lasers

JUDD: Operator Techniques in Atomic Spectroscopy

MONTGOMERY AND TIDMAN: Plasma Kinetic Theory

SPARKS: Ferromagnetic-relaxation Theory

STEELE AND VURAL: Wave Interactions in Solid State Plasmas

STIX: The Theory of Plasma Waves

UMAN: Lightning

WHITE: Quantum Theory of Magnetism

WYARD: Solid State Biophysics

McGRAW-HILL
BOOK COMPANY
New York
St. Louis
San Francisco
Düsseldorf
Johannesburg
Kuala Lumpur
London
Mexico
Montreal
New Delhi
Panama
Rio de Janeiro
Singapore
Sydney
Toronto

EUGENE D. COMMINS
Professor of Physics
University of California, Berkeley

Weak Interactions

This book was set in Times Roman.
The editors were Jack L. Farnsworth, Norma Frankel, and Shelly Langman;
the production supervisor was Ted Agrillo.
The drawings were done by Oxford Illustrators Limited.
The printer and binder was R. R. Donnelley & Sons Company.

Library of Congress Cataloging in Publication Data

Commins, Eugene D
Weak interactions.

(McGraw-Hill advanced physics monograph series)
Bibliography: p.
1. Weak interactions (Nuclear physics) I. Title.
QC794.C56 539.7'54 72-13888
ISBN 0-07-012372-1

WEAK
INTERACTIONS

234567890DODO798

CONTENTS

PREFACE

In this book, which gradually developed from lectures given at Berkeley, I attempt to describe the weak interactions in simple and concrete terms. In so doing I have been strongly influenced by the very excellent monograph "Weak Interactions of Elementary Particles" by the distinguished Soviet physicist L. Okun. This work suffers only from being slightly out of date after a decade.

The presentation here is intended for students, experimental physicists, and physicists not specializing in the field rather than for sophisticated theorists. Thus, experimental results are emphasized, while thorny and speculative theoretical topics are treated in a simplified fashion or omitted altogether. For example, nothing beyond a very elementary discussion is given on the unsolved problem of higher-order weak interactions, nor are many current algebra results presented. In short, I have attempted to write an easy and readable (but therefore incomplete) account.

For this reason the reader interested in more details will undoubtedly wish to consult other works; some of these are listed in the General Bibliography, and others are indicated in the text and listed in the References at the end of each chapter. In particular we refer to the fine encyclopedic treatise

"Theory of Weak Interactions in Particle Physics" by R. E. Marshak, Riazuddin, and C. P. Ryan (1969).

As to preparation, students should find adequate a reasonably good knowledge of the one-particle Dirac theory and the usual *Feynman rules* as expounded so well in "Relativistic Quantum Mechanics" by Bjorken and Drell (1964).

I am grateful to many people for assistance in the preparation of this book. Particularly I want to thank Professors Emilio Segrè and George Trilling, who read large portions of the manuscript carefully and offered many useful suggestions. Discussions with Professors K. Bardakci, F. Calaprice, K. Crowe, D. Dobson, J. D. Jackson, M. Halpern, R. Harris, D. H. Miller, P. G. H. Sandars, M. Suzuki, V. Telegdi, S. Treiman, and E. Wichmann and Drs. H. Gibbs and G. L. Wick were very valuable, as were conversations with many fine Berkeley students, particularly R. Cahn, D. Girvin, H. Frisch, R. McCarthy, and M. Simmons. The preparation of notes for this book began during a sabbatical leave in Rome in 1967–1968. I am indebted to Professor N. Cabibbo and others at the Rome Physics Institute for their kind hospitality and to the J. S. Guggenheim Foundation for its generosity. I gratefully acknowledge the financial support provided by the United States Atomic Energy Commission. I am also very pleased to thank Jean Atteridge for her extraordinary skill and patience in typing the manuscript.

Finally I wish to thank my wife Ulla, who gave constant loyal support. This book is dedicated to the memory of my father Saxe Commins, who would have shown me how to write it much better.

EUGENE D. COMMINS

1

INTRODUCTION TO WEAK INTERACTIONS

1.1 FERMI'S THEORY OF BETA DECAY

Our account of weak interactions begins in 1932 when the only known elementary particles were the photon, electron, proton, and the newly discovered neutron, and the only known weak interaction, nuclear beta decay, seemed utterly baffling and mysterious. Physicists believed that the energy available to the electron in beta decay must be about equal to the difference in rest energies of the initial and final nuclei: $\Delta = (M_i - M_f)c^2$, nuclear recoil being quite negligible. Yet experiments clearly demonstrated that the decay electron energy spectrum is *continuous*, with a *maximum* energy corresponding to Δ. Thus energy (and momentum) conservation appeared to have broken down. Moreover, since the electron possesses spin one-half, whereas the change in nuclear spin accompanying beta decay was known to be integral, angular momentum conservation also seemed to be violated.[1]

[1] For an excellent history of beta decay, see C. S. Wu's article in O. R. Frisch et al. (ed.), " Beiträge zur Physik und Chemie, des 20 Jahrhunderts," Friedr. Vieweg und Sohn Braunschweig, 1959.

In an attempt to rescue these fundamental conservation laws, Pauli (Pau 31, 33) proposed that a neutral half-integral spin particle of near-vanishing or zero-rest mass (later named the *neutrino* by Fermi) is emitted along with the electron in beta decay, and that it escapes detection because of its negligible interactions with matter. Shortly thereafter, directly stimulated by Pauli's radical hypothesis, Enrico Fermi presented his theory of beta decay, which contains the germ of our present theoretical ideas about all of the weak interactions (Fer 34). Fermi's ideas were formulated in close analogy to quantum electrodynamics, which had been developed several years before by Dirac.

In quantum electrodynamics, the interaction between a charged particle and the (quantized) radiation field is described by the lagrangian density

$$\mathfrak{L}_{\mathrm{EM}}(\mathbf{x}, t) = -ej_\alpha(\mathbf{x}, t)A^\alpha(\mathbf{x}, t) \tag{1.1}$$

Here $A^\alpha(\mathbf{x}, t)$ is the four-vector potential operator for the quantized electromagnetic field and $ej_\alpha(\mathbf{x}, t)$ is the electromagnetic four-vector current density of the electron. In the simplest form of the theory, which is sufficient if electrons are neither created nor destroyed, $j_\alpha(\mathbf{x}, t)$ is expressed in terms of the single-particle Dirac wave function $\psi(\mathbf{x}, t)$ of the electron:

$$j_\alpha(\mathbf{x}, t) = \bar{\psi}(\mathbf{x}, t)\gamma_\alpha\psi(\mathbf{x}, t) \tag{1.2}$$

where the γ_α are the usual 4×4 matrices and $\bar{\psi}$ is the Dirac conjugate wave function.[1] In a more formal and general statement of quantum electrodynamics that accounts for processes in which electrons are created and destroyed, $j_\alpha(\mathbf{x}, t)$ is itself an operator expressed in terms of the electron *field* $\Psi(\mathbf{x}, t)$ and its conjugate $\bar{\Psi}$:

$$j_\alpha(\mathbf{x}, t) = \bar{\Psi}(\mathbf{x}, t)\gamma_\alpha \Psi(\mathbf{x}, t)\ddagger \tag{1.3}$$

In any event, the simplest processes one can imagine in electrodynamics are the emission or absorption of a photon by a free electron, as shown in Fig. 1.1*a* and *b*. In actual fact, a free electron can neither emit nor absorb a photon if energy and linear momentum are to be conserved. Therefore the first-order matrix element for photon emission, as in Fig. 1.1*a*, is zero:

$$\int d^4x\langle p', s'; k, \hat{\varepsilon}|\mathfrak{L}_{\mathrm{EM}}(x)|p, s\rangle = 0 \tag{1.4}$$

[1] Appendix 2 contains a summary of the Dirac equation, spinor algebra, and properties of Dirac fields.

‡ Actually, in (1.3), j_α should be "normally ordered," but we may neglect this technicality here.

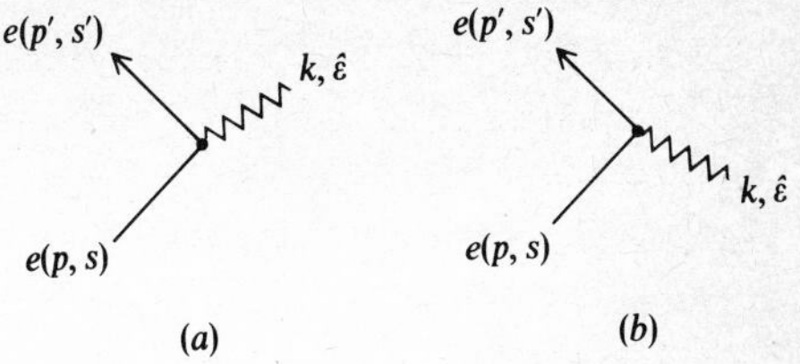

FIGURE 1.1
(*a*) Feynman diagram for emission of a photon by a free electron; (*b*) absorption of a photon by a free electron.

However, this hypothetical first-order process still serves as a convenient model for building a theory of beta decay, which, in the simplest case, is just decay of the free neutron (Fig. 1.2) to a proton, an electron, and what we know today is an antineutrino:

$$n \to pe^-\bar{\nu}_e$$

In Fermi's theory, the electron-antineutrino pair plays the role formerly assigned to the emitted photon in Fig. 1.1*a* while the transforming nucleon replaces the charged particle. Thus, in a rather obvious way the following replacements are suggested:

$$j_\alpha \to \bar{\Psi}_p \gamma_\alpha \Psi_n$$

and

$$A^\alpha \to \bar{\Psi}_e \gamma^\alpha \Psi_\nu$$

Also, the charge $-e$ of the electromagnetic lagrangian is replaced by a new constant $G/\sqrt{2}$, where G (the Fermi, or weak-interaction, coupling constant) must be determined by experiment and is found to have the value

$$G = 1.03 \times 10^{-5}\, m_p^{\ -2} \qquad (1.5)$$

where m_p is the proton mass and we employ units in which $\hbar = c = 1$.

Thus we obtain Fermi's beta decay lagrangian:

$$\mathcal{L}_\beta = \frac{G}{\sqrt{2}} \bar{\Psi}_p \gamma_\alpha \Psi_n \cdot \bar{\Psi}_e \gamma^\alpha \Psi_\nu + \text{h.c.} \qquad (1.6)$$

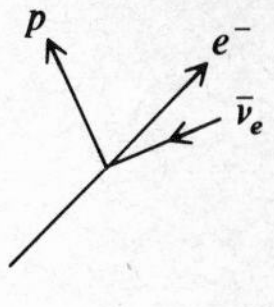

FIGURE 1.2
Feynman diagram for neutron decay.

where the hermitian-conjugate term accounts for positron emission and electron capture. When employed to calculate the beta decay transition probability in first-order perturbation theory, Fermi's lagrangian $\mathcal{L}$ gives a good account of the electron spectrum and dependence of decay rate on Δ. It also serves as a basis for all further developments in the theory of weak interactions.

1.2 THE VARIOUS WEAK INTERACTIONS

Let us now return from the 1930s to the present. In the four intervening decades, scores of new elementary particles have been discovered and nuclear beta decay is now known to be only one among many kinds of weak interactions. How shall we set about organizing all the new information and begin to reduce it to a sensible scheme? First of all, we note that each of the many elementary particles can be classified as a resonance or stable (or metastable) object. If it is a resonance, the particle decays by strong interaction and generally has such a short lifetime and correspondingly large resonance width that weak interactions are of no practical importance.

Thus, in a discussion of weak interactions we are concerned exclusively with the stable and metastable particles, which are listed in Table 1.1. Three of the particles, π^0, η^0, and Σ^0, have very short lifetimes compared with the others, although their lifetimes are much longer than any of the resonances. This is because π^0, η^0, and Σ^0 decay by the electromagnetic interaction, which is described in terms of the coupling constant α (fine structure constant):

$$\alpha = \frac{e^2}{\hbar c} = \frac{1}{137.036} \tag{1.7}$$

The remainder of the metastable particles listed in Table 1.1 are quite long-lived ($\tau \approx 10^{-10}$ s or more) and decay by the weak interaction.

At first glance it appears that these weak decays have little in common with one another. Muon decay involves only leptons (lepton $\equiv$ muon, electron, or neutrino). In other decays, such as $\pi \to \mu\nu$ or $K \to \mu\nu$, the unstable hadron[1] transforms into a final state consisting entirely of leptons. Still other decays (for example, $n \to pe^-\bar{\nu}$, $K \to \pi e\nu$ or $\pi\mu\nu$) are *semileptonic*, which means that the initial hadron decays to a final hadron plus leptons. We also have nonleptonic weak decays, such as $K^+ \to \pi^+\pi^0$ and $\Lambda^0 \to p\pi^-$. In some weak decays strangeness is conserved ($n \to pe^-\bar{\nu}_e$), while in others strangeness is violated ($K \to \mu\nu$).

[1] Hadron $\equiv$ strongly interacting particle.

The latter is in sharp contrast to the situation in strong and electromagnetic interactions, where strangeness is strictly conserved. In addition to this multitude of slow decays, the weak interaction manifests itself in the phenomena of neutrino scattering and electron and muon capture by nuclei, strangeness oscillations in the neutral K system, and parity violation in nuclear forces. Still other weak phenomena are predicted to exist but are not yet observed directly, such as electron-neutrino scattering and neutrino pair production.

In spite of their great diversity, however, it has become apparent that these weak phenomena all share common features and that they may be accounted for, at least qualitatively, by a single, universal, effective weak lagrangian. The present form of this lagrangian dates from an important paper by R. P.

Table 1.1 PROPERTIES OF STABLE AND METASTABLE PARTICLES

Particle		Spin	Mass, MeV/c^2	Mean life, s	Principal decay modes	
Photon	γ	1	0 ($<2\times10^{-21}$)	Stable		
Leptons	ν_e	$\frac{1}{2}$	0 (<60 eV/c^2)	Stable		
	ν_μ	$\frac{1}{2}$	0 (<0.6)	Stable		
	$e^\pm$	$\frac{1}{2}$	0.511	Stable ($>2\times10^{21}y$)		
	$\mu^\pm$	$\frac{1}{2}$	105.659	2.199×10^{-6}	$e\nu\bar{\nu}$	
Mesons	$\pi^\pm$	0	139.568	2.603×10^{-8}	$\mu\nu$	
	π^0	0	134.965	0.84×10^{-16}	$\gamma\gamma$	
	$K^\pm$	0	493.82	1.237×10^{-8}	$\mu\nu$	$\pi\pi^0$
					$\pi\pi^-\pi^+$	$\mu\pi^0\nu$
					$\pi\pi^0\pi^0$	$e\pi^0\nu$
	K^0, $\bar{K}^0$	0	497.76	Superpositions of $K_S{}^0$	and $K_L{}^0$	
	$K_S{}^0$			0.882×10^{-10}	$\pi^+\pi^-$	$\pi^0\pi^0$
	$K_L{}^0$			5.181×10^{-8}	$\pi^0\pi^0\pi^0$	$\pi\mu\nu$
					$\pi^+\pi^-\pi^0$	$\pi e\nu$
				CP-violating	$\pi^+\pi^-$	$\pi^0\pi^0$
	η	0	548.8	$\Gamma=2.63$ keV	$\gamma\gamma$	$3\pi^0$
					$\pi^+\pi^-\pi^0$	$\pi^0\gamma\gamma$
					$\pi^+\pi^-\gamma$	
Baryons	p	$\frac{1}{2}$	938.259	Stable ($>2\times10^{28}y$)		
	n	$\frac{1}{2}$	939.550	918	$pe^-\bar{\nu}$	
	Λ^0	$\frac{1}{2}$	1,115.60	2.51×10^{-10}	$p\pi^-$	$n\pi^0$
	Σ^+	$\frac{1}{2}$	1,189.47	0.802×10^{-10}	$p\pi^0$	$n\pi^+$
	Σ^0	$\frac{1}{2}$	1,192.48	$<1.0\times10^{-14}$	$\Lambda\gamma$	
	Σ^-	$\frac{1}{2}$	1,197.37	1.49×10^{-10}	$n\pi^-$	
	Ξ^0	$\frac{1}{2}$	1,314.7	3.03×10^{-10}	$\Lambda\pi^0$	
	Ξ^-	$\frac{1}{2}$	1,321.29	1.66×16^{-10}	$\Lambda\pi^-$	
	Ω^-	$\frac{3}{2}$	1,672.5	1.3×10^{-10}	$\Xi^0\pi^-$, $\Xi^-\pi^0$, ΛK^-	

Feynman and M. Gell–Mann (Fey 58),[1] but the essential idea goes all the way back to Fermi's theory of beta decay.

We must state at the very outset that the effective weak lagrangian is strictly phenomenological and, like Fermi's original theory, leads to sensible results for cross sections and decay rates only in the context of first-order perturbation theory. Even then, it does not yield prescriptions for calculating many of the known weak processes within its presumed range of validity (for energies ≤ 1 GeV). Moreover, just like Fermi's original theory, it surely fails when applied to processes for which the energy is approximately 300 GeV or more. Still, with all its shortcomings, the effective lagrangian represents concisely our present crude knowledge about the mysterious weak interaction. We shall therefore begin our study by writing down the lagrangian and explaining its parts.

1.3 THE EFFECTIVE WEAK LAGRANGIAN

In order to account for other weak processes as well as beta decay, we postulate the existence of the symmetric weak interaction lagrangian:

$$\mathcal{L}_W{}^S(x) = \frac{1}{2}\frac{G}{\sqrt{2}}[\mathfrak{J}^\lambda(x)\mathfrak{J}_\lambda{}^\dagger(x) + \mathfrak{J}_\lambda{}^\dagger(x)\mathfrak{J}^\lambda(x)] \tag{1.8}$$

where G is the weak coupling constant (already mentioned), $\mathfrak{J}^\lambda(x)$ is the universal weak four-current, $\mathfrak{J}_\lambda{}^\dagger(x)$ is its hermitian conjugate, and $x = (\mathbf{x}, t)$. For almost all purposes we shall find that the simpler form

$$\mathcal{L}_W(x) = \frac{G}{\sqrt{2}}\mathfrak{J}^\lambda(x)\mathfrak{J}_\lambda^\dagger(x) \tag{1.9}$$

is adequate.

The weak current $\mathfrak{J}^\lambda(x)$ consists of four parts:

$$\mathfrak{J}^\lambda(x) = j_e{}^\lambda(x) + j_\mu{}^\lambda(x) + \mathfrak{J}_0{}^\lambda(x) + \mathfrak{J}_1{}^\lambda(x) \tag{1.10}$$

The first two terms on the right-hand side of (1.10) are the *electronic* and *muonic* weak currents, respectively. They describe weak transformations of the following kinds:

$$j_e\begin{cases}\nu_e \to e^- \\ e^+ \to \bar{\nu}_e\end{cases} \tag{1.11}$$

[1] See also Sud 57, 58.

and

$$j_\mu \begin{cases} \nu_\mu & \to \mu^- \\ \mu^+ & \to \bar{\nu}_\mu \end{cases} \tag{1.12}$$

The hermitian-conjugate currents $j_e^\dagger$ and $j_\mu^\dagger$ describe the reverse transformations:

$$j_e^\dagger \begin{cases} e^- & \to \nu_e \\ \bar{\nu}_e & \to e^+ \end{cases} \tag{1.13}$$

and

$$j_\mu^\dagger \begin{cases} \mu^- & \to \nu_\mu \\ \bar{\nu}_\mu & \to \mu^+ \end{cases} \tag{1.14}$$

The spin one-half neutrinos ν_e, ν_μ and antineutrinos $\bar{\nu}_e$, $\bar{\nu}_\mu$ are assumed to be electrically neutral, upper limits on their charges as actually established by experiment being extremely small. [From charge conservation, $Q(\nu_e) < 4 \times 10^{-17}e$, and $Q(\nu_\mu) < 3 \times 10^{-5}e$, while from astrophysical arguments, $Q(\nu_e) < 10^{-13}e$ and $Q(\nu_\mu) < 10^{-13}e$. (See Ber 63.)] Thus j_e and j_μ are *charged* currents, which means that in that part of a weak process described by either term, the electric charge changes by one unit. There is no evidence for any *neutral* electronic or muonic weak current (with $\Delta Q = 0$). Quite the contrary is the case for electromagnetism; there, the current $j_{EM}{}^\lambda(x)$ is neutral: An electron possesses the same electric charge after emission or absorption of a photon as it had before. (See Fig. 1.3.)

The neutrinos ν_e, ν_μ and antineutrinos $\bar{\nu}_e$, $\bar{\nu}_\mu$ are all thought to possess zero-rest mass, upper limits on $m(\nu_e)$ and $m(\nu_\mu)$ being 60 eV/c^2 and 0.6 MeV/c^2, respectively. Also, each neutrino and antineutrino possesses a definite *handedness*, or *helicity*, which is a primary manifestation of parity violation in weak interactions (to be discussed later). Helicity is defined as the projection of the spin on the direction of motion:

$$h = \frac{\boldsymbol{\sigma} \cdot \mathbf{p}}{|\mathbf{p}|}$$

where $\boldsymbol{\sigma}$ is the Pauli spin operator and $\mathbf{p}$ is the three-momentum of the particle. One finds

$$h(\nu_e) = h(\nu_\mu) = -1$$
$$h(\bar{\nu}_e) = h(\bar{\nu}_\mu) = +1$$

The reader may thus wonder why we differentiate between ν_e and ν_μ since both possess zero charge and mass and the same helicity. The reason is that

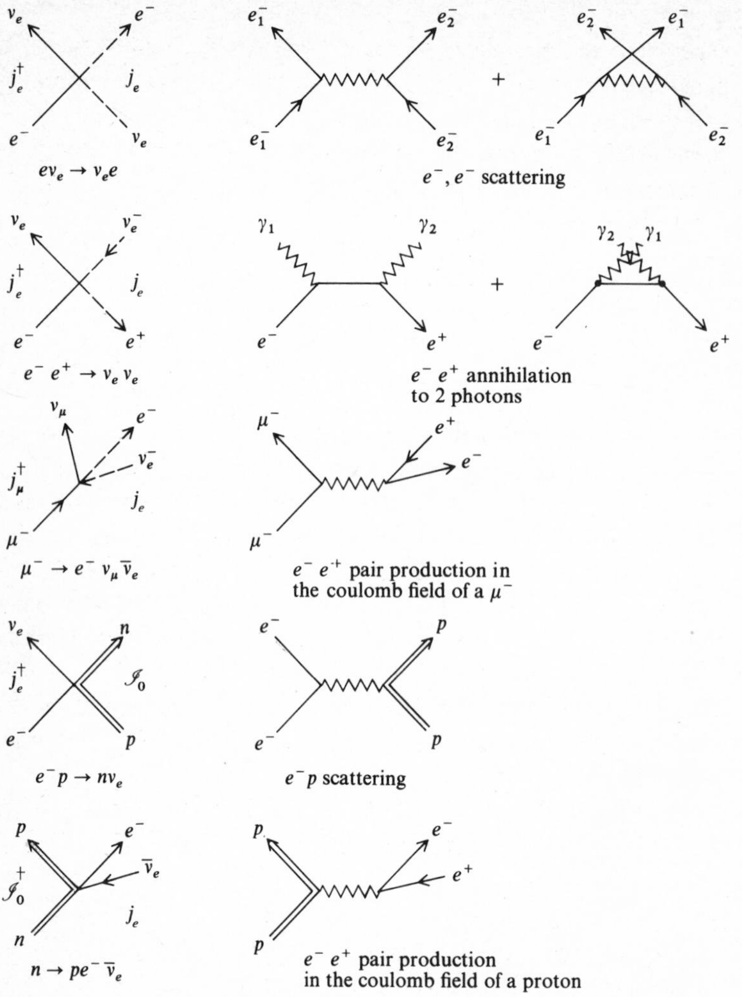

FIGURE 1.3
Schematic (Feynman) diagrams illustrating some important differences between weak and electromagnetic processes. The electromagnetic diagrams are *roughly* analogous to the corresponding weak diagrams. In each weak diagram the currents are charged and the interaction occurs "on contact": at a single point in space-time. We employ the usual convention in which time advances vertically upward, and a positron (or antineutrino) is equivalent to an electron (or neutrino) moving backward in time. (See, for example, Bjo 64.)

experiment shows conclusively that these particles are distinct. Such experiments are discussed in some detail later, but here it is appropriate to point out that the distinction between e^- and μ^-, ν_e and ν_μ, and $\bar{\nu}_e$ and $\bar{\nu}_\mu$ may be described conveniently by introducing the concept of *lepton conservation*, with electron lepton number L_e and muon lepton number L_μ. By definition,

$$\begin{aligned} L_e(e^-) &= L_e(\nu_e) = +1 \\ L_e(e^+) &= L_e(\bar{\nu}_e) = -1 \\ L_e &= 0 \qquad \text{for all other particles} \end{aligned} \tag{1.15}$$

$$\begin{aligned} L_\mu(\mu^-) &= L_\mu(\nu_\mu) = +1 \\ L_\mu(\mu^+) &= L_\mu(\bar{\nu}_\mu) = -1 \\ L_\mu &= 0 \qquad \text{for all other particles} \end{aligned} \tag{1.16}$$

According to all available experimental evidence (some of which is reviewed below and summarized in Table 1.3 later in this chapter), L_e and L_μ are separately conserved (remain invariant) in all processes. Thus, for example, muon decay $\mu^- \to e^- \bar{\nu}_e \nu_\mu$ satisfies conservation of L_e and L_μ and is an allowed process, but the transition $\mu^- \to e^- e^+ e^-$ violates the conservation law and is not observed. We note that L_e and L_μ are *additive* quantum numbers, and lepton conservation as formulated here is an additive conservation law. However, present experimental evidence is insufficient to rule out an alternative interpretation in which an ordinary lepton number L is defined such that

$$L(\nu_e) = L(e^-) = L(\nu_\mu) = L(\mu^-) = +1$$

$$L(\bar{\nu}_e) = L(e^+) = L(\bar{\nu}_\mu) = L(\mu^+) = -1$$

and a multiplicative quantum number called *muon parity* is defined, having the values -1 for μ^-, μ^+, ν_μ, $\bar{\nu}_\mu$, and $+1$ for all other particles. Such a formulation yields distinct experimental predictions, which in principle are capable of being tested. For example, a weak transformation

$$\mu^+ e^- \to e^+ \mu^-$$

is allowed by the multiplicative conservation law but forbidden by the additive conservation law. However, at this point in the development of our subject, such experimental tests have not been performed and either conservation law fits the known facts. For convenience, we shall adhere to the additive form from now on, with L_e and L_μ as expressed in Eqs. (1.15) and (1.16).

Next we turn to $\mathfrak{J}_0$, which describes the strangeness-conserving weak transformation of hadrons:

$$\begin{array}{lll} \mathfrak{J}_0 & \pi^+ \to \text{vacuum state} & \\ \mathfrak{J}_0^\dagger & \pi^- \to \text{vacuum state} & \\ \mathfrak{J}_0^\dagger & \pi^- \to \pi^0 & \\ \mathfrak{J}_0 & \pi^+ \to \pi^0 & \Delta S = 0 \\ \mathfrak{J}_0^\dagger & n \to p & \\ \mathfrak{J}_0^\dagger & \Sigma^- \to \Lambda^0 & \end{array}$$

Here ΔS quite obviously refers to the difference between the total strangeness of the final state and that of the initial state. $\mathfrak{J}_1$, the strangeness-violating portion of the hadronic weak current, corresponds to the following transformations:

$$\begin{array}{lll} \mathfrak{J}_1 & K^+ \to \text{vacuum state} & \\ \mathfrak{J}_1^\dagger & K^- \to \text{vacuum state} & \\ \mathfrak{J}_1 & K^+ \to \pi^0 & |\Delta S| = 1 \\ \mathfrak{J}_1^\dagger & \Sigma^- \to n & \\ \mathfrak{J}_1^\dagger & \Lambda^0 \to p & \end{array}$$

The currents $\mathfrak{J}_0$ and $\mathfrak{J}_1$ are also charged ($\Delta Q = -1$), so that altogether $\Delta Q = -1$ for $\mathfrak{J}$. Of course, $\Delta Q = +1$ for $\mathfrak{J}^\dagger$, and since $\mathfrak{L} \approx \mathfrak{J}\mathfrak{J}^\dagger$, electric charge is conserved for each weak interaction as a whole. Baryon number is also conserved, as in strong and electromagnetic interactions.

1.4 PARITY VIOLATION IN WEAK INTERACTIONS

Let us recall once more Fermi's original description of beta decay, embodied in the lagrangian $\mathfrak{L}_\beta$ of Eq. (1.6). Constructed as it was in close analogy to electrodynamics, $\mathfrak{L}_\beta$ is a scalar product of two (polar) four-vectors, and is thus invariant with respect to proper Lorentz transformations and spatial inversion (parity). In 1936 Gamow and Teller (Gam 36) noted that there are other ways to construct a scalar beta decay lagrangian which is linear in the fields $\overline{\Psi}_p$, Ψ_n, $\overline{\Psi}_e$, Ψ_ν and contains no derivatives of these fields.[1] Specifically, $\mathfrak{L}_\beta$ could be a linear combination of the terms

[1] Here we use the term *field* rather loosely when describing weak transformations of hadrons (i.e., protons or neutrons). In fact, it is not very useful to employ the field concept for such strongly interacting particles.

$\overline{\Psi}_p \Psi_n \overline{\Psi}_e \Psi_\nu$	(scalar × scalar, or SS)
$\overline{\Psi}_p \gamma_\lambda \Psi_n \overline{\Psi}_e \gamma^\lambda \Psi_\nu$	(vector × vector, or VV)
$\overline{\Psi}_p \sigma_{\lambda\delta} \Psi_n \overline{\Psi}_e \sigma^{\lambda\delta} \Psi_\nu$	(tensor × tensor, or TT)
$\overline{\Psi}_p \gamma_\lambda \gamma^5 \Psi_n \overline{\Psi}_e \gamma^\lambda \gamma^5 \Psi_n$	(axial vector × axial vector, or AA)
$\overline{\Psi}_p \gamma^5 \Psi_n \overline{\Psi}_e \gamma^5 \Psi_\nu$	(pseudoscalar × pseudoscalar, or PP)

The coefficients in the linear combination can only be determined by experiment, but Gamow and Teller could already observe the following: In the nonrelativistic limit for the nucleons, which is appropriate for beta decay, since the momentum transfer to the leptons is so small (approximately 1 MeV), the pseudoscalar term vanishes and the following nuclear-spin-selection rules are obeyed:

For SS, VV	$\Delta J = 0$	
For AA, TT	$\Delta J = 0, \pm 1$	but $J_i = 0 \nleftrightarrow J_f = 0$

Therefore, while Fermi's description might be appropriate for $\Delta J = 0$ transitions, some AA and/or TT amplitude must be present to account for $|\Delta J| = 1$ beta decays.

Naturally, all of this was predicated on the assumption, universally made at the time, that $\mathfrak{L}_\beta$ is in fact a true scalar, i.e., parity-invariant. However, in 1956, Lee and Yang (Lee 56) discovered that parity is actually violated in weak interactions. They were first led to this finding by consideration of the $\tau\theta$ puzzle, which concerned itself with the question: How can the decays

$$\theta \qquad K^+ \to \pi^+\pi^0 \tag{1.17}$$

$$\tau \qquad K^+ \to \pi^+\pi^+\pi^- \tag{1.18}$$

both originate from the same initial state when the parity of the $\pi^+\pi^0$ final state in (1.17) is even whereas the parity of the $\pi^+\pi^+\pi^-$ final state in (1.18) is odd?[1]

[1] It is easy to see why $P(\pi^+\pi^0) = +1$, $P(\pi^+\pi^+\pi^-) = -1$. For, since the spin $J(K^+)$ satisfies $J(K^+) = 0$, we must have $J(\pi^+\pi^0) = J(\pi^+\pi^+\pi^-) = 0$ by conservation of angular momentum. Therefore the space wave function of $\pi^+\pi^0$ has $l = 0$ orbital angular momentum, and so the parity is

$$P(\pi^+\pi^-) = (-1)^2(-1)^l = (-1)^0 = +1 \tag{i}$$

Here the factor $(-1)^2$ comes from the odd intrinsic parity of each pion. As for $\pi^+\pi^+\pi^-$, if the total angular momentum is J, then J may be regarded as the vector sum of the angular momentum L of the $\pi^+\pi^-$ about its center of mass plus the angular momentum l of π^- about the three-pion center of mass. For $J = 0$ we have $L = l$, and the total parity of the $\pi^+\pi^+\pi^-$ system is

$$P(\pi^+\pi^+\pi^-) = (-1)^3(-1)^L(-1)^l = (-1)(-1)^{2l} = -1 \tag{ii}$$

Comparing (i) and (ii) it is evident that parity must be violated in the $K^+ \to \pi^+\pi^0$ and $K^+ \to \pi^+\pi^+\pi^-$ decays.

Lee and Yang eventually recognized that invariance with respect to space inversion had always been taken for granted but had never before been tested in weak interactions. This perception immediately led them to suggest a variety of experimental tests for parity conservation in nuclear beta decay, muon decay, etc. There ensued a period of intense experimentation and theoretical activity (1956 to 1958) culminating in the present opinion, which regards each weak current as composed of a polar vector portion and an axial vector portion. Thus, in the weak lagrangian, scalar products (vector × vector, axial × axial) *and* pseudoscalar products (vector × axial) occur. Thus the so-called V − A lagrangian as a whole is of mixed parity and is capable of describing parity-violating weak processes.

For example, let us consider the leptonic weak currents j_e and j_μ. All experimental results are consistent with the forms

$$j_e^{\lambda}(x) = \overline{\Psi}_e(x)\gamma^{\lambda}(1 + \gamma^5)\Psi_{\nu e}(x) \qquad (1.19)$$

$$j_\mu^{\lambda}(x) = \overline{\Psi}_\mu(x)\gamma^{\lambda}(1 + \gamma^5)\Psi_{\nu\mu}(x) \qquad (1.20)$$

As is discussed in detail in Appendix 2, one finds that in the limit of zero mass the operator $\frac{1}{2}(1 + \gamma^5)$ projects out the *left-handed*, or negative helicity, state of any Dirac spinor immediately to the right of it; similarly, the operator $\frac{1}{2}(1 - \gamma^5)$ is a *right-handed*, or positive helicity, projection operator. Thus, Eqs. (1.19) and (1.20) contain the statement that all (zero-mass) neutrinos are left-handed (and all antineutrinos are right-handed). This is frequently referred to as the *two-component neutrino* theory. Furthermore, since $a = \frac{1}{2}(1 + \gamma^5)$ and $\bar{a} = \frac{1}{2}(1 - \gamma^5)$ are projection operators,

$$a^2 = a$$

$$\bar{a}^2 = \bar{a}$$

$$a\bar{a} = \bar{a}a = 0$$

we have

$$\overline{\Psi}_l\gamma^{\lambda}a\Psi_\nu = \overline{\Psi}_l\gamma^{\lambda}a^2\Psi_\nu$$

and since

$$\gamma^{\lambda}\gamma^5 = -\gamma^5\gamma^{\lambda}$$

we obtain

$$\begin{aligned}\overline{\Psi}_l\gamma^{\lambda}a\Psi_\nu &= \overline{\Psi}_l\bar{a}\gamma^{\lambda}a\Psi_\nu \\ &= (\overline{a\Psi_l})\gamma^{\lambda}a\Psi_\nu\end{aligned}$$

Thus Eqs. (1.19) and (1.20) contain the additional information that *to the extent to which we can approximate their masses as zero*, e^- and μ^- participate in weak interactions only through their left-handed components (and e^+ and μ^+ only through their right-handed components).

As for the hadronic weak currents, it is not possible to write explicit forms for them in terms of field operators [such as (1.19) and (1.20), for example] because of the complications of strong interactions. Still, we know a good deal about the matrix elements of these hadronic weak currents between specific states, and, in particular, we know that $\mathfrak{J}_0$ and $\mathfrak{J}_1$ can each be expressed as the sum of a vector and an axial vector portion:

$$\mathfrak{J}_0{}^\lambda = \mathcal{V}_0{}^\lambda + \mathcal{A}_0{}^\lambda \tag{1.21}$$

$$\mathfrak{J}_1{}^\lambda = \mathcal{V}_1{}^\lambda + \mathcal{A}_1{}^\lambda \tag{1.22}$$

1.5 CLASSIFICATION OF WEAK PROCESSES

We may now undertake to classify all weak processes in a convenient way by writing

$$\begin{aligned} \mathcal{L}_W = \frac{G}{\sqrt{2}} (& j_e^\dagger j_e + j_e^\dagger j_\mu + j_e^\dagger \mathfrak{J}_0 + j_e^\dagger \mathfrak{J}_1 \\ & + j_\mu^\dagger j_e + j_\mu^\dagger j_\mu + j_\mu^\dagger \mathfrak{J}_0 + j_\mu^\dagger \mathfrak{J}_1 \\ & + \mathfrak{J}_0^\dagger j_e + \mathfrak{J}_0^\dagger j_\mu + \mathfrak{J}_0^\dagger \mathfrak{J}_0 + \mathfrak{J}_0^\dagger \mathfrak{J}_1 \\ & + \mathfrak{J}_1^\dagger j_e + \mathfrak{J}_1^\dagger j_\mu + \mathfrak{J}_1^\dagger \mathfrak{J}_0 + \mathfrak{J}_1^\dagger \mathfrak{J}_1) \end{aligned} \tag{1.23}$$

The expression on the right-hand side of (1.23) is suggestive of a *matrix* in which the rows are labeled $j_e^\dagger, j_\mu^\dagger, \ldots$, and the columns are labeled $j_e, j_\mu, \ldots$. In Table 1.2, the matrix entries are the weak processes corresponding to the various terms of Eq. (1.23).

In the remainder of this chapter let us survey the entries in Table 1.2. This will give us an opportunity to become acquainted in a general way with the various weak processes before discussing them in detail.

1.5.1 Purely Leptonic Interactions (Entries 1 to 4)

1 Box 1 contains *electron-neutrino* scattering,

$$e^- + \nu_e \rightarrow \nu_e + e^- \tag{1.24}$$

$$e^- + \bar{\nu}_e \rightarrow \bar{\nu}_e + e^- \tag{1.25}$$

and neutrino pair production,

$$e^+ + e^- \to \nu_e + \bar{\nu}_e \tag{1.26}$$

None of these processes has been observed directly. However, their existence is predicted by the lagrangian, and the cross sections can be calculated explicitly. The results show in a very simple way that our effective lagrangian breaks down at high energies (300 GeV) since otherwise the cross sections would become larger than the limit imposed by unitarity.

Neutrino pair production is thought to play a dominant role as an energy-loss mechanism in the last evolutionary stages of very hot, dense stars. Direct laboratory observations of $\bar{\nu}_e e$ scattering may become possible within the next few years.

2 Boxes 2 and 3 contain *muon decay:*

$$\mu^\pm \to e^\pm + \nu_e(\bar{\nu}_e) + \bar{\nu}_\mu(\nu_\mu)$$

This is the only purely leptonic weak process that has been observed. [There are no other known modes of muon decay. The transition $\mu^\pm \to e^\pm \gamma$, $\mu \to 3e$,

Table 1.2

	j_e	j_μ	$\mathfrak{J}_0$	$\mathfrak{J}_1$
$j_e^\dagger$	1 $e^- + \nu_e \to \nu_e + e^-$ $e^- + \bar{\nu}_e \to \bar{\nu}_e + e^-$ $e^+ + e^- \to \nu_e + \bar{\nu}_e$ etc.	2 $\mu^+ \to e^+ + \nu_e + \bar{\nu}_\mu$	5 $(Z, N) \to (Z-1, N+1)e^+\nu_e$ $pp \to de^+\nu_e$ $\bar{\nu}_e p \to ne^+$ $\pi^+ \to e^+\nu_e$ $\pi^+ \to \pi^0 e^+\nu_e$	9 $K^+ \to e^+\nu_e$ $K^+ \to \pi^0 e^+\nu_e$ $K_L{}^0 \to \pi^- e^+\nu_e$
$j_\mu^\dagger$	3 $\mu^- \to e^- + \bar{\nu}_e + \nu_\mu$	4 $\mu^- + \nu_\mu \to \nu_\mu + \mu^-$ $\mu^- + \bar{\nu}_\mu \to \bar{\nu}_\mu + \mu^-$ $\mu^+ + \mu^- \to \nu_\mu + \bar{\nu}_\mu$ etc.	7 $\pi^+ \to \mu^+\nu_\mu$ $\mu^- p \to n\nu_\mu$ $\bar{\nu}_\mu p \to n\mu^+$	11 $K^+ \to \mu^+\nu_\mu$ $K^+ \to \pi^0\mu^+\nu_\mu$ $K_L{}^0 \to \pi^-\mu^+\nu_\mu$ etc.
$\mathfrak{J}_0^\dagger$	6 $n \to pe^-\bar{\nu}_e$ $\nu_e + n \to pe^-$ $\pi^- \to e^-\bar{\nu}_e$	8 $\pi^- \to \mu^-\bar{\nu}_\mu$ $\nu_\mu n \to p\mu^-$	15 $n + p \to p + n$	13 $\Sigma^+ \to n\pi^+$ $\Sigma^+ \to p\pi^0$ $K \to 2\pi, 3\pi$
$\mathfrak{J}_1^\dagger$	10 $K^- \to e^-\bar{\nu}_e$ $\Lambda^0 \to pe^-\bar{\nu}_e$ $\Sigma^- \to ne^-\bar{\nu}_e$ $K_L{}^0 \to \pi^+ e^-\bar{\nu}_e$ $\Xi \to \Lambda e^-\bar{\nu}_e$	12 $K^- \to \mu^-\bar{\nu}_\mu$ $K_L{}^0 \to \pi^+\mu^-\bar{\nu}_\mu$ etc.	14 $\Lambda^0 \to p\pi^-$ $\Sigma^- \to n\pi^-$ $K \to 2\pi, 3\pi$	16 $n + p \to p + n$ etc.

or $\mu \to e\gamma\gamma$ would indicate a breakdown of L_e and L_μ conservation. Presumably, if ν_e and ν_μ were not distinct, the ν and $\bar{\nu}$ in μ decay could be virtual and combine to form a γ ray. Experimentally (Par 64) one finds

$$\frac{W(\mu \to e\gamma)}{W(\mu \to e\nu\bar{\nu})} < 2 \times 10^{-8} \tag{1.27}$$

in accord with lepton conservation and distinct e and μ neutrinos.]

If we assume the leptonic weak currents take the V-A forms (1.19) and (1.20), then the resulting theoretical transition probabilities for $\mu^\pm$ decay, when modified by radiative corrections, are in excellent agreement with existing precise experimental results. Unfortunately, it remains possible to account for these experimental results with more general forms for the leptonic weak currents. This may be explained by the fact that when we consider the most general possible muon decay amplitude which is linear in the lepton fields, contains no lepton field derivatives, is invariant under proper Lorentz transformations, and obeys lepton conservation, we find that such an amplitude contains 19 real parameters. On the other hand, all possible experiments on muon decay in which the two neutrinos are not observed can determine only six relations between these parameters.

In muon decay, two observable effects are characteristic of parity violation. They are longitudinal polarization of the emitted e^- or e^+ and an asymmetry in electron emission with respect to the muon-spin axis in the decay of polarized muons. The average electron kinetic energy in muon decay is so large that it is a reasonable approximation to neglect the electron-rest mass for most purposes; thus, the VA law leads us to expect helicities of -1 for e^- in μ^- decay and $+1$ for e^+ in μ^+ decay, and the experiments confirm these expectations. Furthermore, the asymmetries in μ^- and μ^+ decay are predicted and observed to be of opposite sign. This shows that in muon decay, not only *parity* (P), but also *charge-conjugation* (C) invariance is violated.[1] By definition, under charge conjugation the signs of all charges on a particle are reversed (electric, baryonic, leptonic, etc.) but its spin and momentum are left unchanged.

3 Box 4 contains *muon-neutrino scattering* and *neutrino pair production.* The cross sections for these processes are calculated by the same rules that apply to $e\nu_e$ scattering, and the results are similar. Muon-neutrino scattering would be extremely difficult to observe in the laboratory, and at this time the possibilities for successful experiments are remote.

[1] Charge conjugation should really be called *particle-antiparticle* conjugation.

1.5.2 Strangeness-conserving Semileptonic Interactions (Entries 5 to 8)

1 Boxes 5 and 6 of Table 1.2 contain *neutron beta decay*

$$n \to pe^- \bar{\nu}_e \tag{1.28}$$

and the related process of nuclear *β^- decay*

$$(Z, N) \to (Z + 1, N - 1) + e^- + \bar{\nu}_e \tag{1.29}$$

β^+ decay

$$(Z', N') \to (Z' - 1, N' + 1) + e^+ + \nu_e \tag{1.30}$$

orbital electron capture

$$e^- + (Z', N') \to (Z' - 1, N' + 1) + \nu_e \tag{1.31}$$

inverse beta decay

$$\nu_e + n \to e^- + p \tag{1.32}$$

$$\bar{\nu}_e + p \to n + e^+ \tag{1.33}$$

and the *pp reaction*

$$p + p \to de^+ \nu_e \tag{1.34}$$

Of course, beta decay was the first known weak process; it has been studied intensively for decades, and until about fifteen years ago it yielded most of the fundamental information about weak interactions. We have already noted Pauli's hypothesis of the neutrino, Fermi's theory of beta decay, and Gamow and Teller's generalization of that theory. Orbital electron capture was first discussed in detail by Yukawa and Sakata (Yuk 35) and first observed by Alvarez (Alv 37). The electron- (anti-) neutrino $\bar{\nu}_e$ was first observed directly by Reines and Cowan (Rei 56) in inverse beta decay. Although the notion that parity is violated in weak interactions first suggested itself to Lee and Yang through their preoccupation with the $\tau\theta$ puzzle, the first direct demonstration of their new ideas was an observation of the asymmetry in beta emission from polarized Co^{60} in 1956 (Wu 57). Subsequently, many experiments on beta asymmetries, polarizations, and angular correlations, together with a detailed analysis of beta decay coupling constants and comparison with those of muon decay, led to the present explicit forms (1.19) and (1.20) for the leptonic weak currents and the understanding of the β decay matrix elements of the strangeness-conserving hadronic current. This culminated in the *conserved vector current hypothesis* (Fey 58). Conservation of electron-lepton number L_e was verified by sensitive experiments which place an upper limit on the probability for *neutrinoless* double beta decay:[1]

$$(Z, N) \rightarrow (Z + 2, N - 2) + 2e^- \qquad (1.35)$$

In the field of astrophysics, the cross section for the proton-proton reaction (1.34) was calculated by Bethe and Critchfield (Bet 38) on the basis of the theory of Gamow and Teller. Bethe's work led to the contemporary view of thermonuclear reactions in stellar interiors. At present, much effort is devoted to detecting the low-energy neutrinos (ν_e) which should be emitted from the core of the sun during certain of these thermonuclear reactions. These neutrinos must be detected by inverse electron capture [Eq. (1.32)]:

$$\nu_e + (Z, N) \rightarrow (Z + 1, N - 1) + e^-$$

Thus, the study of nuclear beta decay has led to general understanding of the weak interactions as well as to the development of other fields. It is quite possible, however, that little of a fundamental nature will be forthcoming from beta decay in the future. The reason is simply that the energies and momentum transfers in beta decay are extremely small, whereas most contemporary unsolved problems in the area of weak interactions arise in the high-energy domain.

2 Boxes 5 to 8 also contain the *decay modes of the charged pions.* The principal mode is $\pi_{\mu 2}$:

$$\pi^\pm \rightarrow \mu^\pm \nu_\mu(\bar{\nu}_\mu) \qquad (1.36)$$

Observation of the asymmetry in $e^\pm$ emission from decay of muons brought to rest without depolarization shows that the muons (and therefore also the muon-neutrinos) are polarized in $\pi_{\mu 2}$ decay. This is an especially clear and simple demonstration of P and C violation (see Fig. 1.4).

Besides $\pi_{\mu 2}$, there are two rare decay modes of great importance:

$$\pi_{e2}: \quad \pi^\pm \rightarrow e^\pm \nu_e(\bar{\nu}_e) \qquad \text{BR} = 1.24 \times 10^{-4} \qquad (1.37)$$

$$\pi_{e3}: \quad \pi^\pm \rightarrow \pi^0 e^\pm \nu_e(\bar{\nu}_e) \qquad \text{BR} = 1.01 \pm 0.09 \times 10^{-8} \qquad (1.38)$$

[1] Such a process may be viewed as occurring in two steps:

$$\begin{aligned} (Z, N) &\rightarrow (Z + 1, N - 1) + e^- + \bar{\nu}_e(\nu_e) \\ \nu_e(\bar{\nu}_e) + (Z + 1, N - 1) &\rightarrow (Z + 2, N - 2) + e^- \end{aligned} \qquad \text{(i)}$$

If L_e is conserved, the first step must be accompanied by emission of $\bar{\nu}_e$, not ν_e, whereas the second step must occur by absorption of ν_e but not $\bar{\nu}_e$. Thus, L_e conservation forbids the occurrence of double beta decay.

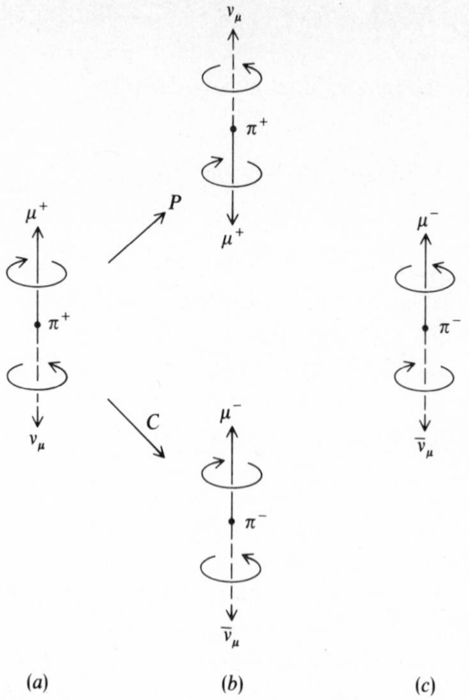

FIGURE 1.4
Schematic diagrams of $\pi_{\mu 2}$ decay illustrating P and C violation. (*a*) Observed; (*b*) not observed; (*c*) observed.

The observed π_{e2} branching ratio (BR) is in accord with a calculation based on an axial vector matrix element for the hadronic weak current and indicates that the other possible type of π_{l2} coupling (pseudoscalar) is absent. The observed π_{e3} branching ratio (DePom 68, Bac 65) constitutes impressive evidence in favor of the *isotriplet vector hypothesis* (conserved vector current) according to which the vector ($\mathcal{V}_0$) portion of $\mathfrak{J}_0$, $\mathcal{V}_0^\dagger$, and the isovector component of the hadronic electromagnetic current form the I^-, I^+, and I_3 components, respectively, of a single isovector current (Fey 58).

In observed $\pi^+ \rightarrow \mu^+ \nu_\mu$ decay (Fig. 1.4*a*), the μ^+ has helicity $h(\mu^+) = -1$. Since $J(\pi^+) = 0$, this implies that $h(\nu_\mu) = -1$, in accord with theoretical expectations. A spatial inversion, or parity (P) transformation, results in reversal of μ^+, ν_μ momenta but leaves spins invariant. Thus under P, the μ^+ helicity is reversed. This is *never* observed (Fig. 1.4*b*). Similarly, a charge conjugation C leaves spins and momenta unchanged, producing a μ^- with negative helicity. This is *never* observed. In fact, in actual π^- decay (Fig. 1.4*c*), the μ^- has helicity $h(\mu^-) = +1$.

3 Boxes 7 and 8 of Table 1.2 also contain the reactions

$$\nu_\mu + n \to p + \mu^- \tag{1.39}$$

$$\bar{\nu}_\mu + p \to n + \mu^+ \tag{1.40}$$

and negative muon capture,

$$\mu^- + p \to n + \nu_\mu \tag{1.41}$$

Reactions (1.39) and (1.40) are of great interest in connection with the question of lepton conservation and the distinction between ν_e and ν_μ. One observes that the neutrinos emerging from the decays $\pi^+ \to \mu^+$ are capable of inducing reactions of the form (1.39) but *not* reactions of the form (1.40). This is consistent with conservation of L_μ. Moreover, neutrinos generated in $\pi \to \mu$ decay induce *neither* of the reactions

$$\nu_e + n \to p + e^-$$

$$\bar{\nu}_e + p \to n + e^+$$

This is clear-cut evidence of the distinction between ν_e and ν_μ and shows why we are compelled to use two separate lepton numbers L_e and L_μ.

One may ask whether lepton conservation has any independent meaning whatsoever! After all, in each test of lepton conservation—whether it be double beta decay (L_e) or $\pi^+ \to \mu^+\nu$, $\bar{\nu} + p \to n + \mu^+(L_\mu)$—one looks for a double process which may be symbolized by the expressions

$$A \to B + \nu$$

$$\bar{\nu} + C \to D$$

If one fails to observe the double process, one claims it is because lepton conservation forbids it. But could not the double process be equally well forbidden by the fact that zero-mass neutrinos always have helicity $h(\nu) = -1$, whereas $h(\bar{\nu}) = +1$? If A always emits left-handed neutrinos, whereas C is only capable of absorbing right-handed ones, then the double process could not occur.

Indeed, *if* $m_{\nu_e} = m_{\nu_\mu} = 0$ is strictly true, and *if* neutrinos are correctly described by a two-component theory, then lepton conservation has no independent meaning (although, of course, we must continue to distinguish between ν_e and ν_μ). However, neutrino helicity measurements have very limited accuracy, and in the foreseeable future it will probably be impossible to determine neutrino helicities to better than 2 to 5 percent precision at the very best. Moreover, it is impossible to establish that the neutrino mass is strictly zero.

Table 1.3 TESTS OF LEPTON CONSERVATION AND RELATED MATTERS

Test	Experimental results	Theoretical remarks	References
1. Conservation of L_e ($\nu_e \neq \bar{\nu}_e$):	Double beta decay		
	$T_{1/2}(\beta\beta_0, Ca^{48}) > 1.6 \times 10^{21} y$	$T_{1/2}(\beta\beta_0, Ca^{48}) = 5 \times 10^{15 \pm 2} y$ $T_{1/2}(\beta\beta_2, Ca^{48}) = 10^{21 \pm 2.5} y$	Bar 67
	$T_{1/2}(\beta\beta_2, Te^{130}) = 10^{21.34 \pm 0.12} y$	$T_{1/2}(\beta\beta_2, Te^{130}) = 10^{22.5 \pm 2.0} y$ ($\beta\beta_0$ theoretical lifetimes calculated for maximum violation of L_e conservation)	Kir 68
2. Conservation of L_μ ($\nu_\mu \neq \bar{\nu}_\mu$):	$\pi^+ \to \mu^+ + \nu_\mu, \nu_\mu + A \to A' + \mu^-$ (or μ^+?) Violation $< 3.8 \times 10^{-3}$ (90% confidence)	[two positive muons would violate L_μ conservation]	
3. L_μ and L_e are distinct ($\nu_\mu \neq \nu_e$):	(*a*) Investigation of type of charged lepton produced in collision of ν_μ with neutrons: $\nu_\mu \to \mu^- + \cdots, \nu_\mu \to e^- \cdots$ $\sigma_e = (0.011 \pm 0.010)\sigma_\mu$ (5,000 events)		Bie 64
	(*b*) $\mu^+ \not\to e^+ + \gamma$ $R = \dfrac{W(\mu^+ \to e^+ + \gamma)}{W(\mu^+ \to e^+ + \nu_e + \bar{\nu}_\mu)} < 2 \times 10^{-8}$	$R_{theo} \approx \dfrac{\alpha}{2\pi} \varepsilon^2$ where ε is the relative amplitude of the interaction that does not conserve leptonic charge	Par 64

[The present upper limit on $m(\nu_\mu)$ is about one electron mass!] Thus, from a purely operational point of view, lepton conservation is a useful concept, independent of any theoretical statements about neutrino helicities. For convenience we summarize the various tests of lepton conservation in Table 1.3.

The reactions

$$\nu_l + n \to p + l^-$$
$$\bar{\nu}_l + p \to n + l^+$$

(where l refers to μ or e) are of great importance in still another way. With the development of high-energy neutrino beams at the largest accelerators, it becomes possible to study in detail the cross sections for these and other neutrino-nucleon reactions at very large energies. This offers a practically unique means for investigating the structure of the hadronic weak currents at high energies. Similarly, in *negative muon capture* on nuclei (unlike orbital electron capture), the lepton mass is so large that the matrix element of the hadronic weak current $\mathfrak{J}_0$, taken between initial and final nuclear states, contains momentum transfer-dependent terms of considerable importance. In particular, μ^- capture is useful for investigating the so-called "induced pseudoscalar" coupling, which can be predicted on the basis of the Goldberger–Treiman relation (Gol 58).

1.5.3 Leptonic and Semileptonic Strangeness-violating Interactions (Entries 9 to 12)

Continuing our discussion of Table 1.2, with the strangeness-violating interactions we encounter a bewildering assortment of decays involving charged and neutral kaons and hyperons. We seek some fundamental clarifying principle, which should somehow relate $\mathfrak{J}_0$ to $\mathfrak{J}_1$ and utilize the basic hadron classification scheme SU_3. The key to a preliminary understanding is *Cabibbo's hypothesis* (Cab 63), which states that the hadronic weak current $\mathfrak{J}_0 + \mathfrak{J}_1$ can be expressed as follows:

$$\begin{aligned} \mathfrak{J}_{\text{had}} &= \mathfrak{J}_0 + \mathfrak{J}_1 = (\mathcal{V}_0 + \mathcal{A}_0) + (\mathcal{V}_1 + \mathcal{A}_1) \\ \mathcal{V}_0 + \mathcal{V}_1 &= \cos\Theta(j_1 - ij_2) + \sin\Theta(j_4 - ij_5) \\ \mathcal{A}_0 + \mathcal{A}_1 &= \cos\Theta(g_1 - ig_2) + \sin\Theta(g_4 - ig_5) \end{aligned} \tag{1.42}$$

where the j's and g's are members of two distinct SU_3 octets of current operators.[1] The number Θ, the *Cabibbo angle*, is a new physical constant whose value cannot

[1] See Chap. 7.

be calculated until the relationship between the weak and strong interactions is better understood than at present. Empirically,

$$\Theta \approx 15^\circ$$
$$\sin\Theta = 0.26$$
$$\cos\Theta = 0.97$$
$$\tan\Theta = 0.27$$

The Cabibbo hypothesis ties together many otherwise fragmentary and mysterious empirical peculiarities of the weak interactions. Let us merely enumerate some of these, saving explanations for later chapters.

1 Over a decade ago it was noticed that the *vector* coupling strengths for muon decay and nuclear beta decay are essentially equal in spite of the fact that the muon is a lepton whereas the nucleon participates in strong interactions. This made clear that the $\Delta S = 0$ weak vector coupling of hadrons is *unrenormalized* by strong interactions and led to the invention of the *CVC hypothesis*. Actually, however, very careful measurements of pure-vector beta decays ($O^{14} \to N^{14*}$, and so on) lead to the conclusion that the vector coupling strength for these decays is not *precisely* the same as that for the muon, but slightly less (0.975 times as much). Radiative corrections are not sufficient to account for this discrepancy, but the Cabibbo formulation does. It predicts that the coupling constant entering muon decay is G, whereas that in vector beta decays ought to be $G \cos\Theta$.

2 The decays K_{e2}, $K_{\mu 2}$ are similar to π_{e2}, $\pi_{\mu 2}$, respectively, but they proceed at relatively slow rates: When *phase space* factors have been accounted for, one still finds that the *kaon decay coupling constant* f_K is only 0.27 times the *pion decay coupling constant* f_π. According to Cabibbo's theory, the ratio of these quantities should be equal to $\tan\Theta = 0.27$.

3 *The* $\Delta S = \Delta Q$ *rule.* This rule applies to strangeness-violating semileptonic decays and states that the change in hadron strangeness is equal to the change in hadron electric charge. The rule is implicitly contained in the form given by Cabibbo for the hadronic weak current [Eq. (1.42)]. As examples of the $\Delta S = \Delta Q$ rule, we note that decays $\Sigma^- \to n e^- \nu_e$ are allowed and observed, but decays $\Sigma^+ \to n e^+ \nu_e$ are forbidden[1]. Also, $K^+ \to \pi^+ \pi^- e^+ \nu_e$ occurs, but $K^+ \to \pi^+ \pi^+ e^- \bar{\nu}_e$ does not.

4 *The* $|\Delta I| = \frac{1}{2}$ *rule.* This empirical rule applies to all strangeness-violating weak decays. It states that the initial and final isotopic spins of hadron states in a $|\Delta S| = 1$ weak decay differ by $\frac{1}{2}$. The form assumed by

[1] Actually there are a very small number of observed events $\Sigma^+ \to n e^+ \nu_e$.

Cabibbo for the hadronic current [Eqs. (1.42)] automatically guarantees that the $|\Delta I| = \frac{1}{2}$ rule is obeyed for leptonic and semileptonic decays since $\mathfrak{J}_1$ transforms as a member of an isodoublet under isospin transformations (while $\mathfrak{J}_0$ transforms as a member of an isotriplet). It is *not* obvious why the $|\Delta I| = \frac{1}{2}$ rule is generally obeyed in nonleptonic decays (see boxes 13 and 14 of Table 1.2, since, according to the vector model, a coupling of $\mathfrak{J}_1$ with $\mathfrak{J}_0{}^\dagger$ could give $|\Delta I| = \frac{1}{2}$ *and* $|\Delta I| = \frac{3}{2}$ amplitudes. In fact, in at least one nonleptonic kaon decay, $K^+ \to \pi^+\pi^0$, the decay amplitude must be $|\Delta I| = \frac{3}{2}$ (although the decay rate here is relatively slow). Tentative explanations have been offered for the general suppression of $|\Delta I| = \frac{3}{2}$ amplitudes in nonleptonic decays, but the situation cannot yet be regarded as satisfactory.

5 *The decay rates for baryon → baryon + leptons.* Cabibbo's hypothesis enables one to calculate the relative strengths of axial vector and vector amplitudes for $\Delta S = 0$ and $\Delta S \neq 0$ baryon semileptonic decays and clarifies the relationship between these amplitudes and the SU_3 theory of baryon-baryon-meson trilinear couplings.

1.5.4 Nonleptonic Strangeness-violating Decays (Entries 13 and 14)

In this category of Table 1.2 we include the nonleptonic decays of hyperons (for example, $\Lambda^0 \to p\pi^-$) and the pionic decay modes of the K mesons ($K \to 2\pi, 3\pi$). Unfortunately, we yet lack the means to calculate the absolute rates for any of these decays in terms of G, Θ, or indeed any fundamental constants. However, the nonleptonic hyperon decays are worth discussing as examples of C and P violation and as tests of the empirical selection rule $|\Delta I| = \frac{1}{2}$.

The decays of neutral kaons are of special interest because of the peculiar and unique *particle-mixture* properties of the $K^0\bar{K}^0$ system. The particles K^0 and $\bar{K}^0$ are charge conjugates of one another and possess definite strangeness eigenvalues (+1 and −1, respectively). Thus, it is appropriate to think in terms of the states $|K^0\rangle$ and $|\bar{K}^0\rangle$, which describe these particles at rest, while discussing the strangeness-conserving strong and electromagnetic interactions. Yet K^0 and $\bar{K}^0$ do not possess definite lifetimes for weak decay because the weak interaction does not conserve strangeness! Thus, in discussing the decays, it is appropriate to consider the states $|K_L{}^0\rangle$ and $K_S{}^0\rangle$ [each of which is a distinct linear combination of $|K^0\rangle$ and $|\bar{K}^0\rangle$ (Gel 53)[1]. The states $|K_L{}^0\rangle$ and $|K_S{}^0\rangle$ do have definite lifetimes for weak decay, but they do not have definite strangeness.

The short-lived $K_S{}^0$ decays in only two significant modes: $\pi^+\pi^-$ and $\pi^0\pi^0$. Each of these final states has CP eigenvalue +1. On the other hand, there are

[1] See also Nak 53.

many known modes of decay for the long-lived component $K_L{}^0$, including the fully allowed decay $K_L{}^0 \to \pi^+\pi^-\pi^0$, in which the final state is (predominantly) an eigenstate of CP with eigenvalue -1.‡ For years it was generally believed that CP invariance was valid for all interactions (even though C and P are separately violated in the weak interactions) and that $K_S{}^0$ and $K_L{}^0$ were eigenstates of CP with eigenvalues $+1$ and -1, respectively. Thus it was believed that the decays $K_L{}^0 \to 2\pi^0$ and $K_L{}^0 \to \pi^+\pi^-$ are strictly forbidden. However, all of this was changed in 1964 with the observation that such decays $K_L{}^0 \to 2\pi$ actually occur with small but finite probability (Chr 64). Thus CP invariance is broken.

What is the basic mechanism for CP violation in $K_L{}^0$ decay? This is one of the outstanding problems of contemporary physics, and, in spite of enormous experimental and theoretical efforts since 1964, it remains unsolved. However, the possibility exists that CP violation is caused by a new *superweak* interaction obeying the selection rule $|\Delta S| = 2$.

1.5.5 Parity-violating Nuclear Forces (Entries 15 and 16)

We complete our discussion of Table 1.2 with a consideration of terms in the lagrangian of the form $\mathfrak{J}_0{}^\dagger\mathfrak{J}_0 + \mathfrak{J}_0\mathfrak{J}_0{}^\dagger$ and $\mathfrak{J}_1{}^\dagger\mathfrak{J}_1 + \mathfrak{J}_1\mathfrak{J}_1{}^\dagger$. Such terms lead to the expectation of a weak component in the interaction potential for nucleon-nucleon scattering (a weak nuclear force) which must coexist with the ordinary potential of strong and electromagnetic interactions. Since the weak interaction violates parity, the total nuclear hamiltonian is no longer inversion-symmetric when we include such weak terms, and in general its eigenstates are not truly eigenstates of parity. Thus we may expect such effects as the emission by unpolarized nuclei of γ rays with a slight degree of circular polarization and fore-aft asymmetry in the emission of γ rays from polarized nuclei. Recently such effects have been observed.

Thus we conclude our preliminary and qualitative survey of the weak interactions. We turn next to detailed calculations and comparisons with experiments.

‡ The angular momentum of the $\pi^+\pi^-\pi^0$ final state is zero. Thus the parity of $\pi^+\pi^-\pi^0$ is $(-1)^3(-1)^L(-1)^L = -1$ where

$$\begin{aligned} L &= \text{orbital angular momentum of } \pi^+\pi^- \text{ about its center of mass} \\ &= \text{orbital angular momentum of } \pi^0 \text{ about } \pi^+\pi^-\pi^0 \text{ center of mass} \end{aligned}$$

The C eigenvalue of $\pi^+\pi^-\pi^0$ is $(-1)^L$ since under C, $\pi^+ \to \pi^-$, $\pi^- \to \pi^+$, $\pi^0 \to \pi^0$. Thus

$$\mathrm{CP}(\pi^+\pi^-\pi^0) = (-1)^{L+1}$$

Now L can be 0, 1, 2, ..., but $L = 0$ is strongly favored because of centrifugal-barrier effects.

REFERENCES

Alv 37	Alvarez, L.: *Phys. Rev.*, **52**:134(L) (1937).
Bac 65	Bacastow, R. B., C. Chesquière, C. Wiegand, and R. Larsen: *Phys. Rev.*, **139B**:407 (1965).
Bar 67	Bardin, R. K., P. J. Gollon, J. D. Ullman, and C. S. Wu: *Phys. Letters*, **26B**:112 (1967).
Ber 63	Bernstein, J., G. Feinberg, and M. Ruderman: *Phys. Rev.*, **132**:1227 (1963).
Bet 38	Bethe, H. A., and C. L. Critchfield: *Phys. Rev.*, **54**:248 (1938).
Bie 64	Bienlein, J., et al.: *Phys. Letters*, **13**:80 (1964).
Bjo 64	Bjorken, J. D., and S. D. Drell: "Relativistic Quantum Mechanics," McGraw-Hill, New York, 1964.
Boo 67	Booth, Johnson, Williams, and Wormald: *Phys. Letters*, **26B**:39 (1967).
Bor 67	Borer, K., et al.: *Phys. Letters*, **29B**:614 (1969).
Cab 63	Cabibbo, N.: *Phys. Rev. Letters*, **10**:531 (1963).
Chr 64	Christenson, J. H., J. W. Cronin, V. L. Fitch, and R. Turlay: *Phys. Rev. Letters*, **13**:138 (1964).
DePom 68	DePommier, P., Du Clos, Heintze, and Kleinknecht: *Nucl.* Phys., **B4**:189 (1968).
Fer 34	Fermi, E.: *Z. Physik.*, **88**:161 (1934).
Fey 58	Feynman, R. P., and M. Gell–Mann: *Phys. Rev.*, **109**:193 (1958).
Gam 36	Gamow, G., and E. Teller: *Phys. Rev.*, **49**:895 (1936).
Gel 53	Gell–Mann, M.: *Phys. Rev.*, **92**:833 (1953).
Gol 58	Goldberger, M. L., and S. B. Treiman: *Phys. Rev.*, **110**:1178 (1958); *ibid.*, **111**:354 (1958).
Kir 68	Kirsten, T., W. Gentner, and O. A. Schaeffer: *Phys. Rev. Letters*, **20**:1300 (1968).
Lee 56	Lee, T. D., and C. N. Yang: *Phys. Rev.*, **104**:254 (1956).
Nak 53	Nakano, T., and K. Nishijima: *Progr. Theoret. Phys., Osaka*, **10**:581 (1953).
Par 64	Parker, S., et al.: *Phys. Rev.*, **133B**:768 (1964).
Pau 31	Pauli, W.: 1931 correspondence with Hahn and Meitner.
Pau 33	Pauli, W.: *Proc. Solvay Congr.*, 324 (1933).
Rei 56	Reines, F., and C. L. Cowan, Jr.: *Science*, **124**:103 (1956). [See also *Phys. Rev.*, **113**:273 (1959).]
Sud 57	Sudarshan, E. C. G., and R. E. Marshak: *Proc. Padua-Venice Conf. Fundamental Particles, Padua*, Società Italiana di Fisica, Bologna, Italy, v–14 (1957).
Sud 58	Sudarshan, E. C. G., and R. E. Marshak: *Phys. Rev.*, **109**:1860 (1958).
Wu 57	Wu, C. S., E. Ambler, R. W. Hayward, D. D. Hoppes, and R. F. Hudson: *Phys. Rev.*, **105**:1413 (1957).
Yuk 35	Yukawa, H., and S. Sakata: *Proc. Phys. Soc. Japan*, **17**:48 (1935).

2

MUON DECAY

2.1 V−A AMPLITUDE FOR MUON DECAY; GENERAL FORMULA FOR THE TRANSITION PROBABILITY

Among all purely leptonic weak interactions, the decays

$$\mu^- \to e^- + \bar{\nu}_e + \nu_\mu$$

and

$$\mu^+ \to e^+ + \nu_e + \bar{\nu}_\mu$$

are of special importance since they have been observed in great detail with accurate experiments. In this chapter we shall derive the transition probability per unit time for muon decay according to the V-A law and compare the results with observations. It will be seen that the agreement is very satisfactory, but that a more general transition probability including the possibility of S, T, and/or P couplings as strong as approximately 30 percent of the V and A couplings can also account for the experimental results. In this chapter we also include a discussion of several important applications of parity violation in muon decay, namely, observations of the muon magnetic moment and $g-2$ and the hyperfine structure (hfs) of muonium.

We begin by considering a physical system which makes a weak transition from some initial state i to some final state f. In general, this may be described by means of the S-matrix element S_{fi}:

$$S_{fi} = \delta_{f,i} + i\mathcal{T}_{fi}$$

where $\mathcal{T}_{fi}$ is the matrix element of the transition operator $\mathcal{T}$.

$$\mathcal{T}_{fi} = \int \langle f | \mathcal{L}_w(x)\, d^4x | i \rangle \tag{2.1}$$

Here, $\mathcal{L}_w(x)$ is the weak lagrangian density defined in Chap. 1, and $|i\rangle$ and $|f\rangle$ are initial and final plane-wave states, respectively. For example, in μ^- decay, we would have

$$|i\rangle = |\mu^-(p_\mu, s_\mu)\rangle$$
$$|f\rangle = |e^-(p_e, s_e), \bar{\nu}_e(q_1, s_1), \nu_\mu(q_2, s_2)\rangle$$

where p_μ, p_e, q_1, and q_2 are the four-momenta of the muon, electron, $\bar{\nu}_e$, and ν_μ, respectively, and s_μ, s_e, s_1, and s_2 are their four-polarizations. Thus, for muon decay, (2.1) would become

$$\mathcal{T}_{fi}(\mu^- \to e^- \bar{\nu}_e \nu_\mu) = \frac{G}{\sqrt{2}} \int \langle e^-(p_e, s_e), \bar{\nu}_e(q_1, s_1),$$
$$\nu_\mu(q_2, s_2) | j_{e\alpha}(x) j_\mu^{\alpha\dagger}(x) | \mu^-(p_\mu, s_\mu)\rangle\, d^4x \tag{2.2}$$

where

$$j_{e\alpha}(x) = \overline{\Psi}_e(x)\gamma_\alpha(1 + \gamma^5)\Psi_{\nu e}(x) \tag{2.3}$$

is the electronic weak current expressed in terms of e- and ν_e-field operators and

$$\begin{aligned}[j_\mu{}^\alpha(x)]^\dagger &= [\overline{\Psi}_\mu(x)\gamma^\alpha(1 + \gamma^5)\Psi_{\nu\mu}(x)]^\dagger \\ &= \Psi_{\nu\mu}{}^\dagger(x)(1 + \gamma^5)\gamma^{\alpha\dagger}\gamma^0\Psi_\mu(x) \\ &= \overline{\Psi}_{\nu\mu}(x)\gamma^0\gamma^{\alpha\dagger}\gamma^0(1 + \gamma^5)\Psi_\mu(x) \\ &= \overline{\Psi}_{\nu\mu}(x)\gamma^\alpha(1 + \gamma^5)\Psi_\mu(x)\end{aligned} \tag{2.4}$$

is a similar expression for the muonic weak current. Inserting (2.3) and (2.4) in (2.2), we obtain

$$\mathcal{T}_{fi}(\mu^- \to e^- \bar{\nu}_e \nu_\mu)$$
$$= \frac{G}{\sqrt{2}} \int d^4x \langle e^-(p_e, s_e), \bar{\nu}_e(q_1, s_1) | \overline{\Psi}_e(x)\gamma_\alpha(1 + \gamma^5)\Psi_{\nu_e}(x) | 0\rangle$$
$$\times \langle \nu_\mu(q_2, s_2) | \overline{\Psi}_{\nu\mu}(x)\gamma^\alpha(1 + \gamma^5)\Psi_\mu(x) | \mu^-(p_\mu, s_\mu)\rangle$$

Next we consider the effect of the field operators on the indicated states. For the muon, for example, the field operator[1] is

$$\Psi_\mu(x) = \frac{1}{\sqrt{V}} \sum_{\pm s} \int d^3\mathbf{p} \frac{1}{\sqrt{2E}} [d^\dagger(p, s)v_\mu(p, s)e^{ip\cdot x} + b(p, s)u_\mu(p, s)e^{-ip\cdot x}]$$

where $u_\mu(p, s)$ and $v_\mu(p, s)$ are single-particle and single-antiparticle Dirac spinors, respectively, for the muon; that is, u_μ corresponds to μ^-, and v_μ corresponds to μ^+. Also, V is a large arbitrary volume. In general, the u's and v's are normalized as follows[1]:

$$u^\dagger u = 2E$$

$$v^\dagger v = 2E$$

Also, $E = +\sqrt{|\mathbf{p}|^2 + m_\mu{}^2}$, and in the above formula for $\Psi_\mu(x)$, $b(p, s)$ and $d^\dagger(p, s)$ are annihilation and creation operators, respectively, for μ^- and μ^+. Applying $\Psi_\mu(x)$ to $|\mu^-(p_\mu, s_\mu)\rangle$, we obtain

$$\Psi_\mu(x)|\mu^-(p_\mu, s_\mu)\rangle = \frac{1}{\sqrt{2E_\mu V}} e^{-ip_\mu\cdot x}u_\mu(p_\mu, s_\mu)|0\rangle$$

where $E_\mu = +\sqrt{|\mathbf{p}_\mu|^2 + m_\mu{}^2}$.

Carrying out similar operations for the other particles participating in muon decay, we arrive at

$$\mathcal{T}_{fi}(\mu^- \to e^-\bar{\nu}_e\nu_\mu) = \frac{G}{\sqrt{2}} \int d^4x[\bar{u}_e(p_e, s_e)\gamma_\alpha(1 + \gamma^5)v_1(q_1, s_1)]$$

$$\times [\bar{u}_2(q_2, s_2)\gamma^\alpha(1 + \gamma^5)u_\mu(p_\mu, s_\mu)]$$

$$\times \frac{e^{ip_e\cdot x}}{\sqrt{2E_e V}} \frac{e^{iq_1\cdot x}}{\sqrt{2\omega_1 V}} \frac{e^{iq_2\cdot x}}{\sqrt{2\omega_2 V}} \frac{e^{-ip_\mu\cdot x}}{\sqrt{2E_\mu V}}$$

where ω_1 and ω_2 are the energies of $\bar{\nu}_e$ and ν_μ, respectively, and

$$E_e = \sqrt{|\mathbf{p}_e|^2 + m_e{}^2}$$

Integrating, we obtain

$$\mathcal{T}_{fi}(\mu^- \to e^-\bar{\nu}_e\nu_\mu) = \frac{(2\pi)^4\delta^4(p_e + q_1 + q_2 - p_\mu)}{\sqrt{2E_e V 2\omega_1 V 2\omega_2 V 2E_\mu V}} \mathcal{M}(\mu^- \to e^-\bar{\nu}_e\nu_\mu) \tag{2.5}$$

where

$$\mathcal{M}(\mu^- \to e^-\bar{\nu}_e\nu_\mu) = \frac{G}{\sqrt{2}} \bar{u}_e\gamma_\alpha(1 + \gamma^5)v_1\bar{u}_2\gamma^\alpha(1 + \gamma^5)u_\mu \tag{2.6}$$

[1] See Appendix 2.

is the *amplitude* for weak decay. Equation (2.5) is only a special case of the general formula

$$\mathcal{T}_{fi} = \frac{(2\pi)^4\,\delta^4(p_f - p_i)\,\mathcal{M}_{fi}}{\sqrt{\prod_i(2E_iV)\prod_f(2E_fV)}} \tag{2.7}$$

where p_i and p_f are the initial and final four-momenta, respectively, and $\mathcal{M}_{fi}$ is a general amplitude.

The transition probability in the general case is

$$|\mathcal{T}_{fi}|^2 = \frac{[(2\pi)^4\,\delta^4(p_f - p_i)]^2}{\prod_i(2E_iV)\prod_f(2E_fV)}\,|\mathcal{M}_{fi}|^2 \tag{2.8}$$

In order to translate Eq. (2.8) into a workable expression we first write

$$\begin{aligned}[(2\pi)^4\,\delta^4(p_f - p_i)]^2 &= (2\pi)^4\,\delta^4(p_f - p_i)(2\pi)^4\,\delta^4(0)\\ &= \lim_{V,T\to\infty}(2\pi)^4\,\delta^4(p_f - p_i)VT\end{aligned}$$

where VT is the four-dimensional volume element (space × time). Also, instead of considering the transition probability to a single state f, we are interested in the transition probability *per unit time* dW to a group of final states f. Therefore we must divide by T and multiply by the phase-space factor

$$\Phi_f = \prod_f\left[\frac{d^3\mathbf{p}_f\cdot V}{(2\pi)^3}\right] \tag{2.9}$$

and we obtain the differential transition probability per unit time dW:

$$dW = \frac{(2\pi)^4\,\delta^4(p_f - p_i)V\,|\mathcal{M}_{fi}|^2}{\prod_i(2E_i\,V)\prod_f(2E_f\,V)}\prod_f\left[\frac{d^3\mathbf{p}_f\cdot V}{(2\pi)^3}\right]$$

or

$$dW = \frac{(2\pi)^4\,\delta^4(p_f - p_i)V}{\prod_i(2E_i\,V)}\,|\mathcal{M}_{fi}|^2\prod_f\frac{d^3\mathbf{p}_f}{(2\pi)^3(2E_f)} \tag{2.10}$$

For the case of decay of a single particle, formula (2.10) reduces to

$$dW = \frac{(2\pi)^4\,\delta^4(p_f - p_i)}{2E_i}\prod_f\frac{d^3\mathbf{p}_f}{(2\pi)^3(2E_f)}\,|\mathcal{M}_{fi}|^2 \tag{2.11}$$

which is independent of the volume, as, of course, it should be. Equation (2.11) is a general formula which will be used repeatedly in this and the following chapters.

2.2 CALCULATION OF MUON DECAY

Our next task is to calculate $|\mathcal{M}|^2$, where $\mathcal{M}$ is given by Eq. (2.6). We have

$$|\mathcal{M}|^2 = \frac{G^2}{2}[\bar{u}_e \gamma_\alpha(1+\gamma^5)v_1 \bar{u}_2 \gamma^\alpha(1+\gamma^5)u_\mu][\bar{u}_e \gamma_\beta(1+\gamma^5)v_1 \bar{u}_2 \gamma^\beta(1+\gamma^5)u_\mu]^*$$

$$= \frac{G^2}{2}[\bar{u}_e \gamma_\alpha(1+\gamma^5)v_1 \bar{v}_1 \gamma_\beta(1+\gamma^5)u_e][\bar{u}_2 \gamma^\alpha(1+\gamma^5)u_\mu \bar{u}_\mu \gamma^\beta(1+\gamma^5)u_2] \tag{2.12}$$

At this point we remind ourselves of several essentials concerning the single-particle and single-antiparticle spinors $u(p, s)$ and $v(p, s)$.[1] Such spinors associated with mass m satisfy the Dirac equations

$$(\gamma_\alpha p^\alpha - m)u(p, s) \equiv (\not{p} - m)u(p, s) = 0 \tag{2.13}$$

and

$$(\not{p} + m)v(\not{p}, s) = 0 \tag{2.14}$$

The Dirac-conjugate spinors satisfy

$$\bar{u}(p, s)(\not{p} - m) = 0$$

and

$$\bar{v}(p, s)(\not{p} + m) = 0$$

The normalization we have chosen leads to

$$\bar{u}u = +2m$$

and

$$\bar{v}v = -2m$$

From this it follows quite easily that

$$u\bar{u} = (\not{p} + m)\left(\frac{1 - \gamma^5 \not{s}}{2}\right) \tag{2.15}$$

and

$$v\bar{v} = (\not{p} - m)\left(\frac{1 - \gamma^5 \not{s}}{2}\right) \tag{2.16}$$

As for the four-polarization s, if the particle possesses finite mass m, then one can find a frame in which the particle is at rest and for which s becomes

$$s = (0, \hat{\mathbf{s}}) \tag{2.17}$$

[1] Appendix 2 contains a more detailed summary of the Dirac equation, spinor algebra, and Dirac fields.

where $\hat{\mathbf{s}}$ is a unit vector denoting the three-spin in the rest frame. From Eq. (2.17) we have $p \cdot s = 0$, $s \cdot s = -1$. Making a Lorentz transformation to a new frame in which the particle has three-momentum $\mathbf{p}$, we find

$$s_0 = \frac{\mathbf{p}}{m} \cdot \hat{\mathbf{s}} \tag{2.18}$$

$$\mathbf{s} = \hat{\mathbf{s}} + \frac{\mathbf{p} \cdot \hat{\mathbf{s}}\mathbf{p}}{m(E + m)} \tag{2.19}$$

For a neutrino with zero mass there is no rest frame and we must redefine s, as follows:

$$s_\nu = \left(1, -\frac{\mathbf{p}}{|p|}\right) \tag{2.20}$$

$$s_{\bar{\nu}} = \left(1, +\frac{\mathbf{p}}{|p|}\right) \tag{2.21}$$

Note that $s_\nu{}^2 = s_{\bar{\nu}}{}^2 = 0$. For neutrinos, Eqs. (2.15) and (2.16) simplify to

$$u\bar{u} = v\bar{v} = \not{p} \tag{2.22}$$

Now let us return to $|\mathcal{M}|^2$ as given by Eq. (2.12). The bracketed factors on the right-hand side of (2.12) can be rewritten conveniently as traces of certain 4×4 matrices. For example,

$$\begin{aligned}
[\bar{u}_e \gamma_\alpha(1 + \gamma^5)v_1\bar{v}_1\gamma_\beta(1 + \gamma^5)u_e] &\equiv T_1 \\
&= \{\bar{u}_{em}[\gamma_\alpha(1 + \gamma^5)]_{mn}(v_1\bar{v}_1)_{nl}[\gamma_\beta(1 + \gamma^5)]_{lr}\, u_{er}\} \\
&= \{(u_e\bar{u}_e)_{rm}[\gamma_\alpha(1 + \gamma^5)]_{mn}(v_1\bar{v}_1)_{nl}[\gamma_\beta(1 + \gamma^5)]_{lr}\} \\
&= \operatorname{tr}[\Lambda_e \gamma_\alpha(1 + \gamma^5)\Lambda_1\gamma_\beta(1 + \gamma^5)]
\end{aligned}$$

and similarly

$$\begin{aligned}
&[\bar{u}_2\gamma^\alpha(1 + \gamma^5)u_\mu\bar{u}_\mu\gamma^\beta(1 + \gamma^5)u_2] \\
&\qquad = \operatorname{tr}[\Lambda_2\gamma^\alpha(1 + \gamma^5)\Lambda_\mu\gamma^\beta(1 + \gamma^5)] \equiv T_2
\end{aligned}$$

where the projection operators Λ are given by

$$\begin{aligned}
\Lambda_1 &= \not{q}_1 \\
\Lambda_2 &= \not{q}_2 \\
\Lambda_e &= \tfrac{1}{2}(\not{p}_e + m_e)(1 - \gamma^5\not{s}_e) \\
\Lambda_\mu &= \tfrac{1}{2}(\not{p}_\mu + m_\mu)(1 - \gamma^5\not{s}_\mu)
\end{aligned}$$

and m_e, m_μ denote electron and muon masses, respectively.

We proceed to evaluate the traces T_1 and T_2 in $|\mathcal{M}|^2 = \frac{G^2}{2} T_1 T_2$:

$$
\begin{aligned}
T_1 &= \mathrm{tr}[\Lambda_e \gamma_\alpha(1+\gamma^5)\Lambda_1\gamma_\beta(1+\gamma^5)] \\
&= \tfrac{1}{2}\,\mathrm{tr}[(\not p_e + m_e)(1-\gamma^5 \not s_e)\gamma_\alpha(1+\gamma^5)\not q_1\gamma_\beta(1+\gamma^5)] \\
&= \mathrm{tr}[(\not p_e - m_e\gamma^5\not s_e)\gamma_\alpha\not q_1\gamma_\beta(1+\gamma^5)] \\
&= \mathrm{tr}[(\not p_e - m_e\not s_e)\gamma_\alpha\not q_1\gamma_\beta(1+\gamma^5)] \qquad (2.23)
\end{aligned}
$$

Here we use the facts that the trace of a product of an odd number of γ matrices is zero, that

$$\gamma^5(1+\gamma^5) = (1+\gamma^5)$$
$$(1+\gamma^5)^2 = 2(1+\gamma^5)$$

and that

$$\gamma_\beta\gamma^5 = -\gamma^5\gamma_\beta$$

for all γ_β. Similarly, we have

$$T_2 = \mathrm{tr}[\not q_2\gamma^\alpha(\not p_\mu - m\not s_\mu)\gamma^\beta(1+\gamma^5)] \qquad (2.24)$$

Now we employ the result that[1]

$$
\begin{aligned}
\mathrm{tr}[\gamma_\rho\gamma_\alpha\gamma_\sigma\gamma_\beta(1+\gamma^5)] &= 4\chi_{\rho\sigma\alpha\beta} \\
&= 4g_{\rho\alpha}g_{\sigma\beta} + 4g_{\alpha\sigma}g_{\rho\beta} - 4g_{\rho\sigma}g_{\alpha\beta} - 4i\varepsilon_{\rho\alpha\sigma\beta}
\end{aligned}
$$

to write

$$
\begin{aligned}
|\mathcal{M}|^2 &= \frac{G^2}{2} T_1 T_2 \\
&= 8G^2(p_e - m_e s_e)^\rho q_1{}^\sigma(p_\mu - m_\mu s_\mu)_\phi q_{2\theta}\chi_{\rho\sigma\alpha\beta}\chi^{\theta\phi\alpha\beta} \qquad (2.25)
\end{aligned}
$$

Since

$$\chi_{\rho\sigma\alpha\beta}\chi^{\theta\phi\alpha\beta} = 4g_\rho{}^\theta g_\sigma{}^\phi$$

Eq. (2.25) becomes

$$|\mathcal{M}|^2 = 32G^2(p_e - m_e s_e)\cdot q_2(p_\mu - m_\mu s_\mu)\cdot q_1 \qquad (2.26)$$

The neutrinos are undetected in all experiments to date in muon decay. Thus, we must integrate dW over the neutrino momenta in order to obtain an expression which can be compared with experimental results. For this purpose

[1] See Appendix 2.

we insert (2.26) in the transition rate dW [Eq. (2.11)] and integrate over $d^3\mathbf{q}_1, d^3\mathbf{q}_2$:

$$dW = \frac{2G^2\, d^3\mathbf{p}_e}{(2\pi)^5 E_e E_\mu} (p_e - m_e s_e)^\alpha (p_\mu - m_\mu s_\mu)^\beta I_{\alpha\beta} \tag{2.27}$$

where
$$I_{\alpha\beta} = \iint \frac{\delta^4(q_1 + q_2 - q) q_{2\alpha} q_{1\beta}\, d^3\mathbf{q}_1\, d^3\mathbf{q}_2}{\omega_1 \omega_2} \tag{2.28}$$

In Eq. (2.28) we have used the symbol q, denoting *four-momentum transfer from muon to electron:*

$$q = p_\mu - p_e = q_1 + q_2$$

Our immediate task is to evaluate the second-rank tensor $I_{\alpha\beta}$. We try to express it as a linear combination of $q^2 g_{\alpha\beta}$ and $q_\alpha q_\beta$:

$$I_{\alpha\beta} = A q^2 g_{\alpha\beta} + B q_\alpha q_\beta \tag{2.29}$$

with numerical coefficients A and B to be determined. To find A and B we multiply both sides of (2.29) with $g^{\alpha\beta}$ and then $q^\alpha q^\beta$ to obtain

$$g^{\alpha\beta} I_{\alpha\beta} = 4Aq^2 + Bq^2 \tag{2.30}$$

$$q^\alpha q^\beta I_{\alpha\beta} = Aq^4 + Bq^4 \tag{2.31}$$

From (2.28) it is also evident that

$$g^{\alpha\beta} I_{\alpha\beta} = \iint \frac{\delta^4(q_1 + q_2 - q) q_1 \cdot q_2}{\omega_1 \omega_2}\, d^3\mathbf{q}_1\, d^3\mathbf{q}_2 \tag{2.32}$$

$$q^\alpha q^\beta I_{\alpha\beta} = \iint \frac{\delta^4(q_1 + q_2 - q)(q_1 \cdot q_2)^2}{\omega_1 \omega_2}\, d^3\mathbf{q}_1\, d^3\mathbf{q}_2 \tag{2.33}$$

Since $g^{\alpha\beta} I_{\alpha\beta}$ and $q^\alpha q^\beta I_{\alpha\beta}$ are invariants, we are at liberty to choose any coordinate system in which to evaluate the integrals (2.32) and (2.33). For convenience we choose that system for which $\mathbf{q}_1 = -\mathbf{q}_2$ [thus for which $q = (q_0, 0)$]. Now,

$$\int \delta^3(\mathbf{q}_1 + \mathbf{q}_2 - \mathbf{q})\, d^3\mathbf{q}_2 = 1$$

Thus (2.32) and (2.33) become, respectively,

$$g^{\alpha\beta} I_{\alpha\beta} = 8\pi \int \omega_1{}^2\, d\omega_1\, \delta(2\omega_1 - q_0) = \pi q^2 \tag{2.34}$$

$$q^\alpha q^\beta I_{\alpha\beta} = 16\pi \int \omega_1{}^4\, d\omega_1\, \delta(2\omega_1 - q_0) = \frac{\pi}{2} q^4 \tag{2.35}$$

Now, inserting (2.34) and (2.35) into (2.30) and (2.31), we find

$$g^{\alpha\beta}I_{\alpha\beta} = \pi q^2 = 4Aq^2 + Bq^2$$

$$q^\alpha q^\beta I_{\alpha\beta} = \frac{\pi q^4}{2} = Aq^4 + Bq^4$$

which yield $A = \pi/6$ and $B = \pi/3$. Thus finally we have

$$I_{\alpha\beta} = \frac{\pi}{6}(q^2 g_{\alpha\beta} + 2q_\alpha q_\beta) \tag{2.36}$$

Now, inserting this into the expression (2.27) for dW, we obtain

$$dW = \frac{\pi}{3}\frac{G^2}{(2\pi)^5}\frac{d^3\mathbf{p}_e}{E_\mu E_e}(p_e - m_e s_e)^\alpha(p_\mu - m_\mu s_\mu)^\beta(q^2 g_{\alpha\beta} + 2q_\alpha q_\beta) \tag{2.37}$$

In order to complete the calculation we now go to the muon rest frame, where

$$q = (q_0, \mathbf{q})$$

$$q_0 = m_\mu - E_e$$

and

$$\mathbf{q} = -\mathbf{p}_e$$

Furthermore,

$$s_\mu = (0, \hat{\mathbf{s}}_\mu)$$

and

$$s_{e0} = \frac{\mathbf{p}\cdot\hat{\mathbf{s}}_e}{m_e}$$

$$\mathbf{s} = \hat{\mathbf{s}}_e + \frac{(\mathbf{p}\cdot\hat{\mathbf{s}}_e)\mathbf{p}}{m_e(E_e + m_e)}$$

where $\hat{\mathbf{s}}_e$ is the e^- polarization in the electron rest frame. With these substitutions, (2.37) becomes

$$\begin{aligned} dW = \frac{\pi G^2}{3(2\pi)^5 m_\mu}\, d\Omega_e\, p_e\, dE_e \Bigg\{ &(m_\mu{}^2 + E_e{}^2 - 2m_\mu E_e - \mathbf{p}_e{}^2) \\ &\times \left[(E_e - \mathbf{p}_e\cdot\hat{\mathbf{s}}_e)m_\mu + m_\mu\left(\mathbf{p}_e - m_e\hat{\mathbf{s}}_e - \frac{\mathbf{p}_e\cdot\hat{\mathbf{s}}_e}{E_e + m_e}\mathbf{p}_e\right)\cdot\hat{\mathbf{s}}_\mu\right] \\ &+ 2\left[(E_e - \mathbf{p}_e\cdot\hat{\mathbf{s}}_e)(m_\mu - E_e) + \left(\mathbf{p}_e - m_e\hat{\mathbf{s}}_e - \frac{\mathbf{p}_e\cdot\hat{\mathbf{s}}_e}{E_e + m_e}\mathbf{p}_e\right)\cdot\mathbf{p}_e\right] \\ &\times [m_\mu{}^2 - m_\mu E_e - m_\mu\hat{\mathbf{s}}_\mu\cdot\mathbf{p}_e]\Bigg\} \end{aligned} \tag{2.38}$$

Since $m_\mu = 206 m_e$, we can safely neglect the electron rest mass in first approximation. Thus we have

$$\mathbf{p}_e = E_e \hat{\mathbf{n}}$$

where $\hat{\mathbf{n}}$ is a unit vector in the direction of electron motion in the muon rest frame. Therefore dW becomes

$$dW = \frac{\pi G^2}{3(2\pi)^5 m_\mu} d\Omega_e p_e \, dE_e (1 - \hat{\mathbf{n}} \cdot \hat{\mathbf{s}}_e) E_e m_\mu^2 [(3m_\mu - 4E_e) + (m_\mu - 4E_e)\hat{\mathbf{n}} \cdot \hat{\mathbf{s}}_\mu] \tag{2.39}$$

The maximum electron energy and momentum are, respectively,

$$E_{e,\max} = \frac{m_\mu^2 + m_e^2}{2m_\mu} \qquad \text{and} \qquad \left|\mathbf{p}_{e,\max}\right| = \frac{m_\mu^2 - m_e^2}{2m_\mu}$$

and these values are taken when the two neutrinos are emitted in one direction and the electron in the opposite direction. For $m_e = 0$ the maximum electron energy reduces to

$$E_{e,\max} = \frac{m_\mu}{2}$$

Let us write

$$\varepsilon = E_e / E_{e,\max}$$

so that the entire electron energy range is spanned by the interval $0 < \varepsilon < 1$. Now, substituting $\varepsilon m_\mu/2$ for E_e and choosing the muon spin axis along the $+z$ direction in polar coordinates

$$\hat{\mathbf{n}} \cdot \hat{\mathbf{s}}_\mu = \cos\theta$$

we obtain

$$\boxed{dW = \frac{G^2 m_\mu^5}{3 \times 2^6 \pi^3} [2\varepsilon^2(3 - 2\varepsilon)] \left[1 + \left(\frac{1 - 2\varepsilon}{3 - 2\varepsilon}\right)\cos\theta\right] \left[\frac{1 - \hat{\mathbf{n}} \cdot \hat{\mathbf{s}}_e}{2}\right] d\varepsilon \frac{\sin\theta \, d\theta \, d\phi}{4\pi}} \tag{2.40}$$

On the right-hand side of (2.40) the first factor in brackets is the normalized electron energy spectrum $n(\varepsilon) = 2\varepsilon^2(3 - 2\varepsilon)$:

$$\int_0^1 n(\varepsilon) \, d\varepsilon = 1$$

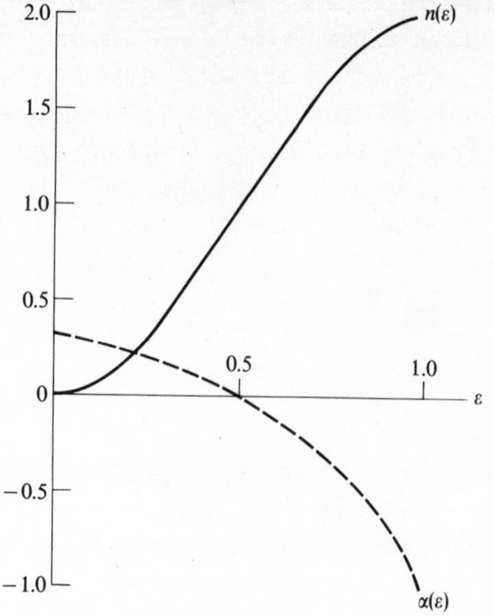

FIGURE 2.1
Normalized electron spectrum $n(\varepsilon)$ and asymmetry $\alpha = (1 - 2\varepsilon)/(3 - 2\varepsilon)$ for μ^- decay, as predicted by the VA theory, in the approximation $m_e = 0$.

The function $n(\varepsilon)$ is represented by the solid curve in Fig. 2.1. The second factor in brackets represents the asymmetry in electron emission with respect to the muon polarization axis; the asymmetry coefficient $\alpha = (1 - 2\varepsilon)/(3 - 2\varepsilon)$ is also plotted in Fig. 2.1 (dashed curve). The third factor in brackets represents the electron helicity. We see that $h(e^-) = -1$ independent of electron energy, in accord with the VA theory and the approximation $m_e = 0$. Taking into

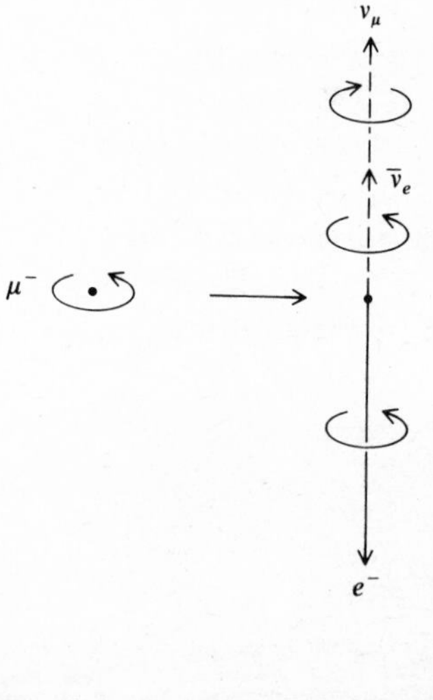

FIGURE 2.2
Schematic diagram illustrating μ^- decay for $\varepsilon = 1$, where neutrinos are emitted with equal momentum, with opposite spins in upward direction: $h(\bar{\nu}_e) = +1$, $h(\nu_\mu) = -1$. Electron is emitted in downward direction, with spin in same direction as muon spin: $h(e^-) = -1$, $\alpha = -1$.

account that $h(e^-) = h(\nu_\mu) = -1$, $h(\bar{\nu}_e) = +1$, it is easy to see why the asymmetry coefficient is $\alpha = -1$ for maximum electron energy $\varepsilon = 1$ (Fig. 2.2).

We obtain the total transition rate for unpolarized muons by averaging over muon spin, summing over electron spin, integrating over electron energy, and integrating over the solid angle. In these operations each bracketed factor of Eq. (2.40) contributes unity, and so the final result for the total transition rate for unpolarized muons, unmodified by radiative corrections, is

$$W = \frac{G^2 m_\mu{}^5}{3 \times 2^6 \pi^3} \tag{2.41}$$

2.3 DECAY OF μ^+; VIOLATION OF C INVARIANCE

In order to calculate the transition probability per unit time for μ^- decay, we employed the term $G/\sqrt{2}\, j_{e\lambda} j_\mu{}^{\dagger\lambda}$ in the weak lagrangian. For μ^+ decay, the relevant term in the lagrangian is instead

$$\frac{G}{\sqrt{2}} j_{\mu\lambda} j_e{}^{\dagger\lambda} = \overline{\Psi}_{\nu e} \gamma^\lambda (1 + \gamma^5) \Psi_e \overline{\Psi}_\mu \gamma_\lambda (1 + \gamma^5) \Psi_{\nu\mu}$$

which leads to an amplitude

$$\mathcal{M}_+ \equiv \mathcal{M}(\mu^+ \to e^+ \nu_e \bar{\nu}_\mu) = \frac{G}{\sqrt{2}} \bar{u}_{\nu e} \gamma^\lambda (1 + \gamma^5) v_e \bar{v}_\mu \gamma_\lambda (1 + \gamma^5) v_{\nu\mu}$$

Thus we find

$$\begin{aligned} |\mathcal{M}_+|^2 &= \frac{G^2}{2} \operatorname{tr}[\not{q}_1 \gamma^\lambda (1 + \gamma^5)(\not{p}_e + m_e \not{s}_e) \gamma^\sigma (1 + \gamma^5)] \\ &\quad \times \operatorname{tr}[(\not{p}_\mu + m_\mu \not{s}_\mu) \gamma_\lambda (1 + \gamma^5) \not{q}_2 \gamma_\sigma (1 + \gamma^5)] \\ &= 32 G^2 (p_e + m_e s_e) \cdot q_2 (p_\mu + m_\mu s_\mu) \cdot q_1 \end{aligned} \tag{2.42}$$

Comparing this expression with the analogous formula (2.26) for μ^- decay, we see that the terms in s_e and s_μ differ in sign in the two cases. Physically this leads to changes in sign of the helicity $h(e^+) = +1$ and the asymmetry parameter for polarized muon decay $\alpha(e^+) = (2\varepsilon - 1)/(3 - 2\varepsilon)$. Thus the VA law violates C as well as P invariance.

Formally, we can see that the leptonic portion of the weak lagrangian is C-noninvariant, as follows:

$$\mathcal{L}_{\text{lep}} = \frac{G}{\sqrt{2}} (j_{e\lambda} j_\mu{}^{\lambda\dagger} + j_{\mu\lambda} j_e{}^{\lambda\dagger})$$

$$\hat{C} \mathcal{L}_{\text{lep}} \hat{C}^{-1} = \frac{G}{\sqrt{2}} [\hat{C} j_e \hat{C}^{-1} \hat{C} j_\mu{}^\dagger \hat{C}^{-1} + \hat{C} j_\mu \hat{C}^{-1} \hat{C} j_e{}^\dagger \hat{C}^{-1}]$$

Now, $$\hat{C}\Psi\hat{C}^{-1} = i\gamma^2\gamma^0\tilde{\bar{\Psi}}$$

and $$\hat{C}\bar{\Psi}\hat{C}^{-1} = i\tilde{\Psi}\gamma^2\gamma^0$$

(see Appendix 3). Thus,

$$\begin{aligned}
\hat{C}\mathcal{L}_{\text{lep}}\hat{C}^{-1} &= \frac{G}{\sqrt{2}}[\tilde{\Psi}_e\gamma^2\gamma^0\gamma_\lambda(1+\gamma^5)\gamma^2\gamma^0\tilde{\bar{\Psi}}_{\nu e}\tilde{\Psi}_{\nu\mu}\gamma^2\gamma^0\gamma^\lambda(1+\gamma^5)\gamma^2\gamma^0\tilde{\bar{\Psi}}_\mu \\
&\quad + \tilde{\Psi}_\mu\gamma^2\gamma^0\gamma_\lambda(1+\gamma^5)\gamma^2\gamma^0\tilde{\bar{\Psi}}_{\nu\mu}\tilde{\Psi}_{\nu e}\gamma^2\gamma^0\gamma^\lambda(1+\gamma^5)\gamma^2\gamma^0\tilde{\bar{\Psi}}_e] \\
&= \frac{G}{\sqrt{2}}[\bar{\Psi}_{\nu e}\gamma^0\gamma^2(1+\gamma^5)\tilde{\gamma}_\lambda\gamma^0\gamma^2\Psi_e\bar{\Psi}_\mu\gamma^0\gamma^2(1+\gamma^5)\tilde{\gamma}^\lambda\gamma^0\gamma^2\Psi_{\nu\mu} \\
&\quad + \bar{\Psi}_{\nu\mu}\gamma^0\gamma^2(1+\gamma^5)\tilde{\gamma}_\lambda\gamma^0\gamma^2\Psi_\mu\bar{\Psi}_e\gamma^0\gamma^2(1+\gamma^5)\tilde{\gamma}^\lambda\gamma^0\gamma^2\Psi_{\nu e}] \\
&= \frac{G}{\sqrt{2}}[\bar{\boldsymbol{\Psi}}_{\nu e}\gamma_\lambda(1-\gamma^5)\boldsymbol{\Psi}_e\bar{\boldsymbol{\Psi}}_\mu\gamma^\lambda(1-\gamma^5)\boldsymbol{\Psi}_{\nu\mu}] + \cdots
\end{aligned}$$

(In this calculation we used the relations $\tilde{\gamma}^5 = \gamma^5$, $\tilde{\gamma}^0 = \gamma^0$, $\tilde{\gamma}^2 = \gamma^2$, $\tilde{\gamma}^1 = -\gamma^1$, $\tilde{\gamma}^3 = -\tilde{\gamma}^3$, and $\gamma^\lambda\gamma^5 = -\gamma^5\gamma^\lambda$.) We see that $\hat{C}\mathcal{L}_{\text{lep}}\hat{C}^{-1}$ differs from $\mathcal{L}_{\text{lep}}$ since it includes $(1-\gamma^5)$ instead of $(1+\gamma^5)$. Thus, $\mathcal{L}_{\text{lep}}$ is manifestly C-noninvariant.

2.4 A MORE GENERAL DECAY AMPLITUDE

How well does the V-A transition probability agree with experiment? Or, to put it another way, if we write down a very general muon decay amplitude, to what extent do experimental results restrict it to the V-A form? We continue to assume that the new amplitude is invariant under proper Lorentz transformations and linear in the four lepton spinors (or their Dirac conjugates), and that it contains no derivatives of these spinors. Also we assume lepton conservation, for which there is good evidence.[1] Then the most general amplitude we can write is

$$\mathcal{M} = \frac{G}{\sqrt{2}}\sum_i \bar{u}_e(A_i + A_i'\gamma^5)O_i v_1 \bar{u}_2 O_i u_\mu \tag{2.43}$$

where the O_i are the usual Dirac matrices

$$\{O_S = 1, O_V = \gamma^\alpha, O_T = \sigma^{\alpha\beta}, O_A = i\gamma^\alpha\gamma^5, O_P = \gamma^5\}$$

and the index i in (2.43) runs through S, V, T, A, P.

The coupling constants A_i and A_i' must be determined by experiment. If parity were conserved, $\mathcal{M}$ would include only "even" coefficients A_i or "odd"

[1] See Table 1.3. Without the assumption of lepton conservation, the amplitude is considerably more complicated. See, e.g., Pau 57.

coefficients A_i', but not both. Time reversal (T) invariance demands that all the A_i and A_i' be relatively real.[1] However, it is not necessary to assume T invariance a priori, and so there are 10 complex numbers A_i, A_i', or 20 real parameters in Eq. (2.43). However, an overall phase is arbitrary, and so we are left with 19 real parameters to be determined by experiment.

It would appear that the amplitude of (2.43) is not unique because we can still satisfy the stated assumptions by writing a new amplitude in which two of the lepton spinors are permuted (for example, v_1 and u_μ):

$$\mathcal{M} = \frac{G}{\sqrt{2}} \sum \bar{u}_e(C_i + C_i'\gamma^5)O_i u_\mu \bar{u}_2 O_i v_1 \tag{2.44}$$

However, it is easy to show[2] that the coefficients C_i, C_i' may be expressed as linear combinations of the coefficients A_i, A_i', so that by permuting v_1 and u_μ

[1] See Appendix 3 and Chap. 4.

[2] To show that the C_i and C_i' may be expressed as linear combinations of the A_i, A_i', we write the 16 matrices in the form

$$\Gamma^x \begin{cases} \gamma^0,\ i\gamma^k \\ \sigma^{jk},\ \sigma^{0k} \\ i\gamma^0\gamma^5,\ \gamma^k\gamma^5 \\ 1,\ \gamma^5 \end{cases} \quad j, k = 1, 2, 3$$

The Γ^x are so chosen that $\sum_{x=1}^{16} \Gamma_{\alpha\beta}{}^x \Gamma_{\rho\delta}{}^x = 4\delta_{\alpha\delta}\delta_{\rho\beta}$, as can be shown easily. Now consider a term from (2.44) such as

$$\begin{aligned} \bar{\psi}_e \Gamma^x \psi_\mu \bar{\psi}_2 \Gamma^x \psi_1 &= (\bar{\psi}_e \Gamma^x)_\alpha \delta_{\alpha\delta}(\psi_\mu)_\delta (\bar{\psi}_2 \Gamma^x)_\rho \delta_{\rho\beta} \psi_{1\beta} \\ &= \tfrac{1}{4} \sum_y (\bar{\psi}_e \Gamma^x)_\alpha \Gamma_{\alpha\beta}{}^y \psi_{1\beta} (\bar{\psi}_2 \Gamma^x)_\rho \Gamma_{\rho\delta}{}^y \psi_{\mu\delta} \\ &= \tfrac{1}{4} \sum_y \bar{\psi}_e \Gamma^x \Gamma^y \psi_1 \bar{\psi}_2 \Gamma^x \Gamma^y \psi_\mu \end{aligned}$$

However, the product $\Gamma^x\Gamma^y$ is just another Γ matrix. Thus, each term $\bar{\psi}_e \Gamma^x \psi_\mu \bar{\psi}_2 \Gamma^x \psi_1$ can be expressed as a linear combination of the

$$\bar{\psi}_e \Gamma^x \psi_1 \bar{\psi}_2 \Gamma^x \psi_\mu$$

Explicitly, on interchanging u_μ and v_1, we find

$$\begin{aligned} C_S &= \tfrac{1}{4}(A_S + 4A_V + 6A_T + 4A_A + A_P) \\ C_V &= \tfrac{1}{4}(A_S - 2A_V \qquad\quad + 2A_A - A_P) \\ C_T &= \tfrac{1}{4}(A_S \qquad\quad - 2A_T \qquad\quad + A_P) \\ C_A &= \tfrac{1}{4}(A_S + 2A_V \qquad\quad - 2A_A - A_P) \\ C_P &= \tfrac{1}{4}(A_S - 4A_V + 6A_T - 4A_A + A_P) \end{aligned}$$

Similar expressions hold for the primed coefficients. In the particular case where $A_S = A_{S'} = A_T = A_{T'} = A_P = A_{P'} = 0$, we obtain

$$\begin{aligned} C_V &= \tfrac{1}{2}(A_A - A_V) = -C_A \\ C_V' &= \tfrac{1}{2}(A_A' - A_V') = -C_A' \end{aligned}$$

For the VA law, $A_V = -A_A = +1$ and $A_V' = -A_A' = -1$. In this case, interchange of u_μ and v_1 results only in a change of sign in amplitude.

we merely rename the coupling constants (Fie 37). In what follows we shall always employ the so-called charge-retention ordering of Eq. (2.44) since it leads to expressions in the coupling constants which are relatively simple.

2.5 TRANSITION PROBABILITIES IN THE GENERAL CASE

We shall now compare the differential transition probability derived from the general decay amplitude (2.44) with experiments. It is convenient to divide the experiments into two categories: first, those in which the electron helicity is observed in the decay of unpolarized muons; and second, those in which the electron spectrum, including the asymmetry, is observed in the decay of polarized or unpolarized muons. In the helicity experiments electrons are observed over the entire energy spectrum, corrections for finite electron mass are unimportant on account of the limited accuracy of the measurements (Kuz 61), and radiative corrections are also negligible. Measurements of the electron spectrum, at least at the low-energy end, must take into account the finite electron mass, and radiative corrections are important.

First we present an expression derived from (2.44) which is useful for discussing the helicity. Let us define the quantities

$$\begin{aligned} a_{ij} &= C_i^* C_j + C_i'^* C_j' \\ b_{ij} &= C_i^* C_j' + C_i'^* C_j \end{aligned} \qquad i, j = 1, \ldots, 5\ (S, V, T, A, P) \tag{2.45}$$

Carrying out the calculation of the transition probability averaged over μ-spin states and integrated over electron energies, and assuming $m_e = 0$, we find

$$dW = \frac{G^2 m_\mu{}^5}{3 \times 2^{10}\, \pi^3} \frac{\sin\theta\, d\theta\, d\phi}{4\pi} K[1 + h\hat{\mathbf{n}} \cdot \hat{\mathbf{s}}_e] \tag{2.46}$$

where

$$K = a_{11} + a_{55} + 4(a_{22} + a_{44}) + 6a_{33} \tag{2.47}$$

and

$$h = \frac{1}{K}[b_{11} + b_{55} + 4(b_{22} + b_{44}) + 6b_{33}] \tag{2.48}$$

The helicity of negatons from μ^- decay is observed to be[1]

$$h_{\text{exp}}(e^-) = -0.89 \pm 0.28 \tag{2.49}$$

[1] See Sec. 2.7 for a brief description of experiments.

while in the decay of μ^+ the observed positron helicity is

$$h_{\text{exp}}(e^+) = 1.03 \pm 0.13 \qquad (2.50)$$

Since opposite e^+ and e^- helicities are to be expected, we may combine the data of (2.49) and (2.50) to obtain

$$|h| = 1.00 \pm 0.14 \qquad (2.51)$$

The condition $h(e^-) = -1$ would imply from (2.47) and (2.48) that

$$(b_{11} + b_{55}) + 4(b_{22} + b_{44}) + 6b_{33} + (a_{11} + a_{55}) + 4(a_{22} + a_{44}) + 6a_{33} = 0$$

or

$$|C_S + C_S'|^2 + |C_P + C_P'|^2 + 4|C_V + C_V'|^2 + 4|C_A + C_A'|^2 + 6|C_T + C_T'|^2 = 0$$

In other words, $h(e^-) = -1$ implies

$$C_i' = -C_i \qquad \text{for all } i \qquad (2.52)$$

Next we turn to the electron spectrum in the decay of polarized μ^-. Here we take into account finite electron mass and radiative corrections (see Sec. 2.6 for details). The differential transition probability summed over final electron spin for the decay of polarized μ^- is found to be

$$dW = \frac{K}{16} \frac{G^2 m_\mu{}^5}{3 \times 2^6 \pi^3} \frac{\sin\theta \, d\theta \, d\phi}{4\pi}$$

$$\times \left\{ \frac{1 + h(x)}{1 + 4\eta(m_e/m_\mu)} \left[12(x-1) + \frac{4}{3}\rho(8x-6) + 24 \frac{m_e}{m_\mu} \frac{(1-x)}{x} \eta \right] \right.$$

$$\left. + \xi \cos\theta \left[4(1-x) + \frac{4}{3}\delta(8x-6) + \frac{\alpha}{2\pi} \frac{g(x)}{x^2} \right] \right\} x^2 \, dx \qquad (2.53)$$

Here, $x = |\mathbf{p}_e/\mathbf{p}_{e,\max}|$, while $h(x)$† and $g(x)$ are radiative corrections (to be discussed later) and the *Michel* parameters ρ, η, ξ,†† and δ (Mic 50, Bou 57) are given by the expressions

$$K\rho = 3a_{22} + 3a_{44} + 6a_{33} \qquad (2.54)$$

$$K\eta = a_{11} - a_{55} + 2(a_{22} - a_{44}) \qquad (2.55)$$

$$K\xi = [6(b_{15} + b_{51}) - 4(b_{24} + b_{42}) - 14b_{33}] \qquad (2.56^2)$$

$$K\delta = -\frac{1}{\xi}[3(b_{24} + b_{42}) + 6b_{33}] \qquad (2.57)$$

† $h(x)$, the radiative correction, should not be confused with h, the helicity.

†† There is considerable confusion in the literature over the definition of the sign of ξ. The sign is chosen here to give $\xi = -1$ for the VA law with $C_V = -C_V' = -C_A = C_A'$.

What is the significance of these parameters? For the moment, let us neglect m_e/m_μ and radiative corrections. We see that the terms in ρ and δ give zero when integrated over the whole spectrum. These quantities characterize the shape of the isotropic and anisotropic spectra at the high-energy end. The integrated asymmetry parameter $\bar{a}$ in $W \approx (1 + \bar{a} \cos \theta)$ is $\bar{a} = \xi/3$. Finally, η characterizes the low-energy end of the isotropic spectrum.

If it were true that $C_S = C_S' = C_T = C_T' = C_P = C_P' = 0$, Eqs. (2.47) and (2.54) to (2.57) would reduce to

$$K = 4(a_{22} + a_{44}) \tag{2.58}$$

$$\rho = \tfrac{3}{4} \tag{2.59}$$

$$\eta = \frac{2(a_{22} - a_{44})}{4(a_{22} + a_{44})} = \frac{1}{2} \frac{|C_V|^2 - |C_A|^2}{|C_V|^2 + |C_V|^2} \tag{2.60}$$

$$\xi = -\frac{b_{24} + b_{42}}{a_{22} + a_{44}} \tag{2.61}$$

and

$$\delta = \tfrac{3}{4} \tag{2.62}$$

If, in addition, we impose the condition $C_i = -C_i'$ from the helicity results, ξ reduces to

$$\xi = \frac{C_V C_A^* + C_A^* C_V}{|C_V|^2 + |C_A|^2} \tag{2.63}$$

Moreover, if the VA law holds, with $C_A = -C_V$, we have

$$K = 16|C_V|^2 \qquad \rho = \delta = \tfrac{3}{4} \qquad \eta = 0 \qquad \xi = -1 \tag{2.64}$$

Thus, for the VA law, (2.53) with $C_V = 1$ reduces to our original transition probability expression (2.40) in the limit of zero electron mass and negligible radiative corrections, as indeed it should.

2.6 RADIATIVE CORRECTIONS TO MUON DECAY

The radiative corrections to muon decay have been analyzed by Behrends, Finkelstein, and Sirlin (Beh 56), Berman (Ber 58), Kinoshita and Sirlin (Kin 57, Kin 59), Berman and Sirlin (Ber 62), and Kuznetsov (Kuz 60, Kuz 61). In Beh 56 a general four-component parity-conserving interaction was assumed, but since the discovery of parity violation, all the other calculations assumed the VA theory as a starting point.[1] The reason for this is that, in the case of STP

[1] See, however, a more general treatment by Florescu and Kamei (Flo 68).

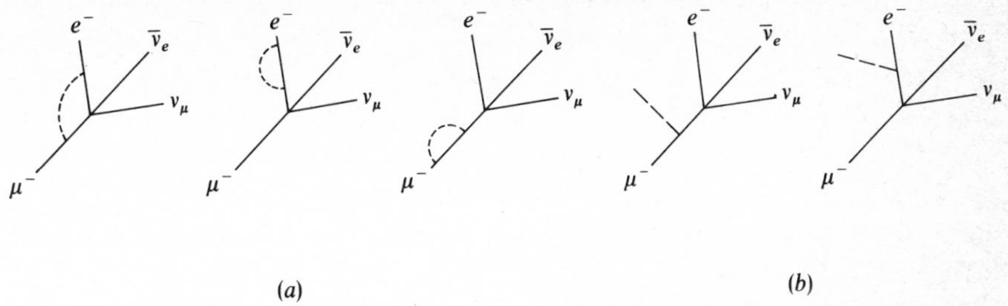

FIGURE 2.3
Lowest-order radiative corrections to μ decay. (*a*) Virtual photon diagrams; (*b*) inner bremsstrahlung diagrams.

couplings, the radiative corrections depend on an arbitrary cutoff and are thus ambiguous.

To order α, the diagrams which contribute to the electromagnetic correction to μ decay are shown in Fig. 2.3. The inner *bremsstrahlung* is very important for the low-energy part of the electron spectrum, while the virtual photon diagrams are relatively more important at higher energies. In quantum electrodynamics the same type of radiative corrections are incurred (for example, for e^- or μ^- interacting with an external field).[1] There, the infrared divergences cancel when all diagrams are taken into account, while the ultraviolet divergences cancel by virtue of Ward's identity.

In the present case of radiative corrections to μ decay, the infrared divergences also cancel, but in dealing with the ultraviolet divergences, the possibility arises of an ambiguity in the *regulator* procedure[2] because of the difference in masses of the initial and final charged lepton. However, as shown by Berman and Sirlin (Ber 62), the appropriate choice of gauge eliminates these difficulties and leads to an unambiguous result in agreement with earlier calculations. Thus one finds that for the VA law the functions $h(x)$ and $g(x)$ of (2.53) are given by the formulas

$$h(x) = \frac{1}{3 - 2x} \frac{\alpha}{2\pi} f(x)$$

[1] See, for example, Bjo 64, p. 124.

[2] In order to deal in a consistent way with internal and external photon lines, it is necessary to assign a small finite mass to the photon and then take the limit as this mass goes to zero. This is called the *regulator* procedure.

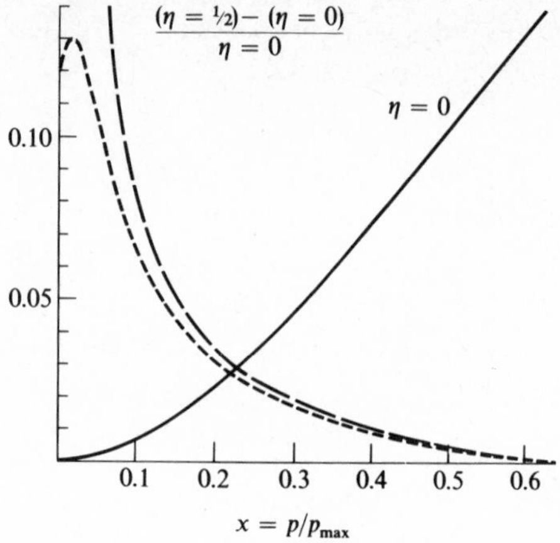

FIGURE 2.4
Sensitivity of the decay spectrum to η. Solid line shows μ^+ decay spectrum for $\eta = 0$. Long-dashed line shows fractional change in spectrum before radiative corrections when η is varied from 0 to $\frac{1}{2}$. Short-dashed line shows same change after radiative corrections. [*From S. E. Derenzo, Phys. Rev.,* ***181:*** *1854 (1969).*]

with

$$f(x) = (6 - 4x)R(x) + (6 - 6x)\ln x + \frac{1-x}{3x^2}[(5 + 17x - 34x^2)(w + \ln x) - 22x + 34x^2]$$

$$g(x) = (2 - 4x)R(x) + (2 - 6x)\ln x - \frac{1-x}{3x^2}\left[(1 + x + 34x^2)(w + \ln x) + 3 - 7x - 32x^2 + \frac{4(1-x)^2}{x}\ln(1-x)\right]$$

where

$$R(x) = 2\sum_{n=1}^{\infty}\frac{x^n}{n^2} - \frac{\pi^2}{3} - 2 + w\left[\frac{3}{2} + 2\ln\frac{(1-x)}{x}\right] - \ln x(2\ln x - 1) + (3\ln x - 1) - \frac{1}{x}\ln(1-x)$$

and where $w = \ln m_\mu/m_e = 5.332$.

The radiative correction to the overall decay rate may be calculated with sufficient accuracy by setting $m_e = 0$, whereupon the algebraic complexity is reduced considerably. One finds

$$\tau^{-1} = W^{\text{corr}} = \left(\frac{G^2 m_\mu{}^5}{192\pi^3}\right)\left[1 - \frac{\alpha}{2\pi}\left(\pi^2 - \frac{25}{4}\right)\right] \tag{2.65}$$

where the factor $G^2 m_\mu{}^5/192\pi^3$ is the uncorrected inverse lifetime [Eq. (2.41)]. Using Eq. (2.65), one obtains a precise value of G from measured values of τ and m_μ. In cgs units

$$G = 1.43506 \pm 0.00026 \times 10^{-49} \text{ erg-cm}^3 \qquad (2.66)$$

The electron mass *cannot* be neglected in radiative corrections to the low-energy end of the electron spectrum; these are large and must be taken into account in measurements of η. Recently, a useful approximate formula for these corrections has been found by Grotch (Gro 68). Figure 2.4 shows the effect of radiative corrections on the variation of η from 0 to $\frac{1}{2}$ in the low-energy spectrum.

2.7 EXPERIMENTAL DETERMINATION OF MUON DECAY PARAMETERS

2.7.1 The Lifetime τ

In all modern observations of $\tau(\mu^+)$ or $\tau(\mu^-)$, one measures the time interval between the arrival of a stopping muon in a suitable target and the emission of the ensuing decay electron, with the aid of sophisticated timing circuitry. Since negative muons can be captured by target nuclei, a correction must be applied for this competing process in determination of $\tau(\mu^-)$. For example, in an experiment performed at Columbia University (Mey 63) muons are stopped in a liquid H_2 target containing less than 1 ppm of O_2 and less than 1 part in 10^9 of other impurities. Thus, the small capture rate in H_2 is the only significant correction to the free μ^- decay rate, and it was measured separately. The lifetimes $\tau(\mu^+)$ and $\tau(\mu^-)$ were observed with the same apparatus and under essentially the same conditions of rate and geometry, with the result

$$R = \frac{\tau(\mu^-)}{\tau(\mu^+)} = 1.000 \pm 0.001 \qquad (2.67)$$

consistent with CPT invariance.[1] The best value of τ, obtained by averaging the results of this and other experiments (Lun 62, Eck 63, Far 62), is

$$\tau = (2.199 \pm 0.001) \times 10^{-6} \text{ s} \qquad (2.68)$$

2.7.2 The Helicities $h(e^-)$ and $h(e^+)$

The quantity $h(e^-)$ has been determined by observation of Møller (e^-e^-) scattering of negatons incident on polarized target electrons in a magnetized iron foil (Sch 67). The cross sections for this process and the related process of

[1] See Sec. 11.14.

Bhabha (e^+e^-) scattering are spin-dependent (Bin 57). The helicity $h(e^+)$ ha. been observed, not only by Bhabha scattering [which gives the most precise results (Duc 64)], but also by e^+e^- annihilation in a magnetized iron target (Buh 63) [the cross section for which is spin-dependent (Pag 62, McM 60)] and also by circular polarization of *bremsstrahlung* from longitudinally polarized e^+ (Blo 64). The results, quoted earlier, are $h(e^-) = -0.89 \pm 0.28$ and $h(e^+) = 1.03 \pm 0.14$.

2.7.3 Momentum Spectrum of Electrons in Decay of Unpolarized Muons

2.7.3.1 The high-momentum end of the spectrum: ρ The parameter ρ has been the object of many experimental observations (Pla 60, She 67, Fry 68). A very precise result was obtained at Columbia (Bar 65, Peo 66) with an apparatus

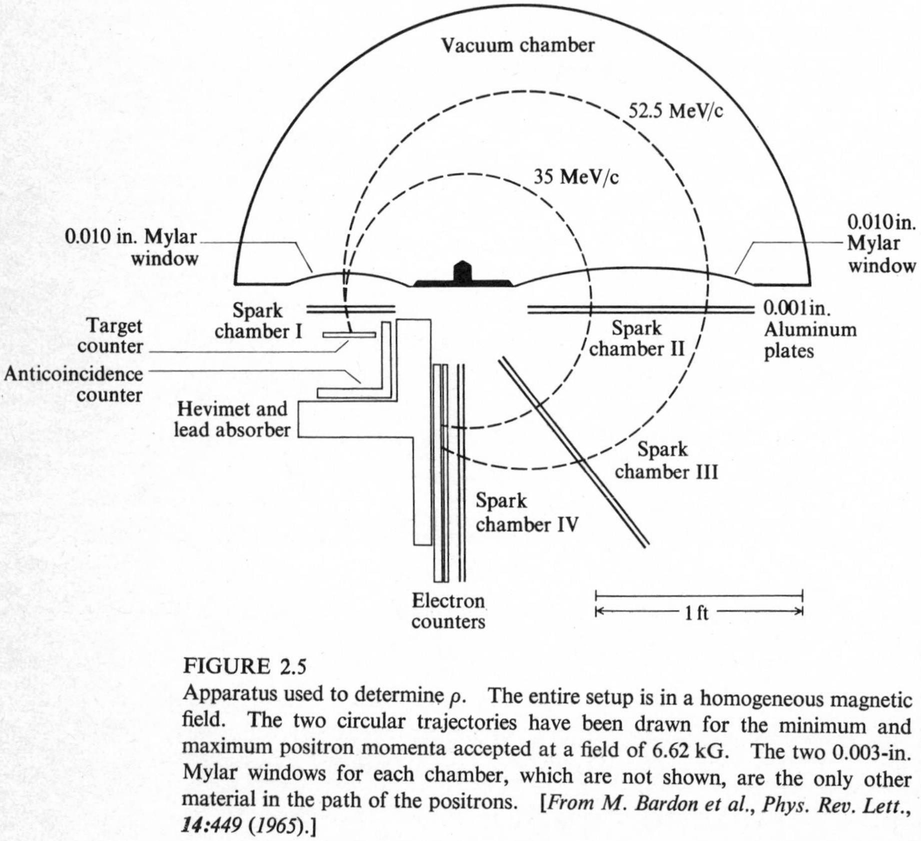

FIGURE 2.5
Apparatus used to determine ρ. The entire setup is in a homogeneous magnetic field. The two circular trajectories have been drawn for the minimum and maximum positron momenta accepted at a field of 6.62 kG. The two 0.003-in. Mylar windows for each chamber, which are not shown, are the only other material in the path of the positrons. [*From M. Bardon et al., Phys. Rev. Lett.*, ***14**:449* (*1965*).]

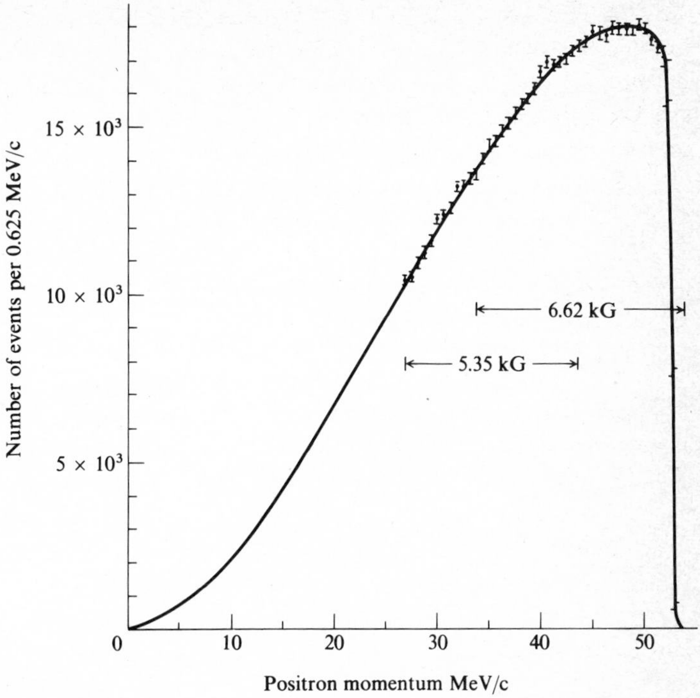

FIGURE 2.6
Results of experiment to determine ρ. Experimental points are plotted, together with a theoretical curve for $\rho = \frac{3}{4}$ corrected for radiative effects and ionization loss. Compare with Fig. 2.1. [*From M. Bardon et al., Phys. Rev. Lett.*, ***14**:449* (*1965*).]

illustrated in Fig. 2.5, which functions as follows: Positive pions incident along the magnetic field direction stop in a thin plastic counter. About 80 percent of the decay muons also stop and decay in this target. Approximately 5 percent of the resulting positrons reach electron counters. A fast coincidence between target and electron counters falling within a suitable gate initiated by the stopping π^+ triggers the spark chambers. The observed decay spectrum is shown in Fig. 2.6. The best result for ρ, based on this and other observations, is

$$\rho = 0.7518 \pm 0.0026 \qquad (2.69)$$

2.7.3.2 The low-momentum end of the spectrum: η The parameter η has been measured as follows (Der 69): A bubble chamber in a strong magnetic field is exposed to a beam of stopping π^+ and μ^+ from the University of Chicago synchrocyclotron. Over 400,000 exposures were taken, including 2,070,000 μ^+ decays. Events with e^+ tracks of sufficiently large radius of curvature to ensure momentum less than 7.1 MeV/c were selected. These momentum measurements are carefully calibrated with internal conversion electron tracks of known momentum from a radioactive source placed inside the chamber. The result is

$$\eta = -0.12 \pm 0.21 \qquad (2.70)$$

2.7.4 Asymmetry in the Decay of Polarized Muons

2.7.4.1 The integrated asymmetry $a = \xi/3$ The most accurate determination of ξ has been made in the Soviet Union (Akh 68). Positrons are observed from μ^+ stopping in a photoemulsion immersed in a magnetic field of 140,000 G oriented parallel to the muon spin. A positive muon stopping in the emulsion captures an electron to form a muonium atom. The intense magnetic field is used to decouple the hyperfine interaction in muonium and thus prevent muon depolarization. An emulsion is especially suitable for such measurements because it is sensitive to very low e^+ energies and this yields an undistorted spectrum. On the basis of 190,000 events, a value

$$\bar{a}(e^+) = +0.325 \pm 0.005$$

is obtained. This result, when combined with other less precise measurements, yields the value

$$\xi = -0.973 \pm 0.014 \qquad (2.71)$$

2.7.4.2 The asymmetry shape factor δ In a recent experiment to measure δ, a variation of the well-known and important Garwin–Lederman method (Gar 57, Swa 58) is employed. In this variation, positive muons stop in a non-depolarizing target immersed in a magnetic field B perpendicular to the muon spin. The muons then precess in the plane normal to B at the Larmor frequency $\omega = \mu B/\hbar = g(e/2m_\mu c)B$. If a fixed counter is located in this plane and gated on by the arrival of the muon, then the distribution of times for positron events will be proportional to $(\frac{1}{2}\pi\tau)e^{-t/\tau}(1 - A\cos\omega t)$, where A is the experimental asymmetry and τ is the muon mean life.

In the measurement of δ (Fry 68), μ^+ with polarization $\mathcal{P}$ are stopped in the magnetic field of a wire spark-chamber spectrometer, which is used both for momentum analysis and precession. The parameters δ and $\mathcal{P}\xi$ are determined by fitting predicted asymmetry functions to experimental results. The best value of δ obtained from this and other measurements is

$$\delta = 0.7540 \pm 0.0085 \qquad (2.72)$$

2.8 SUMMARY AND CONCLUSIONS ON MUON DECAY

Table 2.1 presents the various muon decay parameters and compares them with predictions of the VA theory. Obviously, the agreement is very satisfactory. However, there is considerable latitude in the choice of coupling constants for the more general amplitude (2.44). Specifically, it is found (Der 69) that the following range of values is consistent with experiment:

$$a_{11} = |C_S|^2 + |C_S'|^2 \le 0.33a_{22} \qquad a_{22} = |C_V|^2 + |C_V'|^2$$

$$a_{55} = |C_P|^2 + |C_P'|^2 \le 0.33a_{22}$$

$$0.76a_{22} \le a_{44} \le 1.20a_{22} \qquad a_{44} = |C_A|^2 + |C_A'|^2$$

$$a_{33} = |C_T|^2 + |C_T'|^2 \le 0.28a_{22}$$

$$\phi_{A,V} \equiv \operatorname{Arg}\frac{C_A}{C_V} = 180° \pm 15° \tag{2.73}$$

We see that experiments cannot exclude appreciable S, T, and P couplings. Moreover, there is considerable latitude in the ratio C_A/C_V. Unfortunately, there is little likelihood of a drastic improvement in this situation until electron-neutrino coincidence observations of muon decay become possible (Jar 66).

Table 2.1 **PROPERTIES OF THE MUON AND DECAY PARAMETERS**

Muon mass	$m_\mu = 105.659 \pm 0.002$ MeV/c^2
Ratio of muon mass to electron mass	$m_\mu/m_e = 206.7683(6)$
Muon-neutrino mass	$m_{\nu\mu} < 0.6$ MeV/c^2
Electron-neutrino mass	$m_{\nu\mu} < 60$ eV/c^2
Maximum electron energy	$E_{\max} = \dfrac{m_\mu^2 + m_e^2}{2m_\mu} = 52.827$ MeV
Muon mean life	$\tau = (2.199 \pm 0.001) \times 10^{-6}$ s

Michel parameters	Experimental values	VA values
ρ	0.7518 ± 0.0026	$\frac{3}{4}$
η	-0.12 ± 0.21	0
ξ	-0.973 ± 0.014	-1
δ	0.7540 ± 0.0085	$\frac{3}{4}$
$h(e^-)$	$(-)1.00 \pm 0.13$	-1

2.9 A DIGRESSION: THE MUON MAGNETIC MOMENT, $g-2$, AND MUONIUM

Although the topics to be discussed in this section do not really constitute *weak interaction* physics, they represent interesting and important applications of parity violation. Each of the experiments to be described utilizes the asymmetry in electron emission which occurs in the decay of polarized muons.

2.9.1 The Muon Magnetic Moment and Anomalous g Value

According to Dirac's theory of relativistic quantum mechanics, the magnetic moment of the electron or muon is given by the formula

$$\mu_l = 2\frac{e\hbar}{2m_l c} = g\frac{e\hbar}{2m_l c} \tag{2.74}$$

that is, $g = 2$. (Here m_l is the lepton mass.) Actually, g is not exactly 2, either for the electron or the muon, because of radiative corrections. Instead, defining $a = \frac{1}{2}(g-2)$, we have to order α^3:

$$a_{e,\text{theo}} = \frac{\alpha}{2\pi} - 0.32848\frac{\alpha^2}{\pi^2} + [0.26(5) + 0.13(\text{est})]\frac{\alpha^3}{\pi^3} \tag{2.75}$$

and

$$a_{\mu,\text{theo}} - a_{e,\text{theo}} = 1.09426\frac{\alpha^2}{\pi^2} + [20.3 \pm 1.3]\frac{\alpha^3}{\pi^3} \tag{2.76}$$

The anomalies a_e and a_μ are of the most fundamental interest in quantum electrodynamics. In Eq. (2.75) the first two terms are the well-known second- and fourth-order corrections (Sch 48, 49; Pet 57; Som 57, 58). The various contributions to the sixth-order (α^3/π^3) term in (2.75) have been examined by a number of workers (Ald 69, 70; Lau 69). The anomalies a_μ and a_e are not equal because vacuum polarization contributions to the α^2/π^2 and α^3/π^3 terms differ for muon and electron (Suu 57; Pet 57, 58; Ele 66). The anomaly a_e has been determined by very refined experiments (Wes 70) to be

$$a_{e,\,\text{exp}} = \frac{\alpha}{2\pi} - 0.32848\frac{\alpha^2}{\pi^2} + (0.54 \pm 0.58)\frac{\alpha^3}{\pi^3} \tag{2.77}$$

This result is in good agreement with the theoretical value for a_e. A discussion of the experimental determination of a_μ is given below.

The *magnetic moment* of μ^+ has been determined by an experiment of the Garwin–Lederman type performed by Hague et al. (Hag 70). As noted earlier, the time variation between the muon-arrival and decay-electron emission in such an experiment is proportional to $1 - A\cos(g_\mu eB/2m_\mu c)t$. The magnetic field B is measured by proton resonance, and so determination of μ_μ reduces to a measurement of the frequency ratio ω_μ/ω_p, where $\omega_p = \mu_{\text{proton}} B/\hbar$. The chief problem in an experiment of this kind is the diamagnetic correction, which may depend on the chemical environment and is not necessarily the same for μ^+ and the proton a priori (Rud 66). In the Hague experiment, observations were made in three separate liquids: NaOH solution, distilled water, and methylene cyanide [$CH_2(CH)_2$]. The results agree to high precision, showing that, in fact, the diamagnetic shielding correction for μ^+ *is* essentially the same as for protons in water (approximately 26 ppm). The result

$$\frac{\mu_\mu}{\mu_p} = 3.183347(9) \qquad (2.8 \text{ ppm}) \tag{2.78}$$

is obtained. Knowledge of μ_μ by itself is not enough to obtain g_μ with sufficient accuracy to permit meaningful comparison with the theoretical formula (2.76) because in Eq. (2.74) the most uncertain quantity is the muon mass. In fact, an independent determination of $g - 2$ (to be described below), together with the magnetic moment determination, gives the most accurate value of the muon mass, which we express in units of the electron mass as follows:

$$\frac{m_\mu}{m_e} = 206.7683(6) \tag{2.79}$$

The anomaly $a = (g - 2)/2$ is determined for μ^+ and μ^- by the following elegant method (Bai 68): Muons of momentum 1.28 GeV/c are stored in a weak-focusing circular ring magnet (see Fig. 2.7). The muons are formed by allowing an external proton beam to strike a target inside the ring. Thus pions are created, some of which revolve in the ring and decay in flight, giving rise to some muons which acquire stable orbits and are stored for the duration of their lifetimes. The muons have appreciable longitudinal polarization when formed, but the spin turns $1 + (e\bar{B}/m_\mu c)a$ times as fast as the momentum vector in the magnetic field. Thus after a time T the angle between spin and momentum is given by

$$\omega_a T = a \frac{e\bar{B}}{m_\mu c} T$$

The muon precession difference frequency ω_a is recorded by observing the asymmetry in decay electrons. An experimental curve is shown in Fig. 2.8.

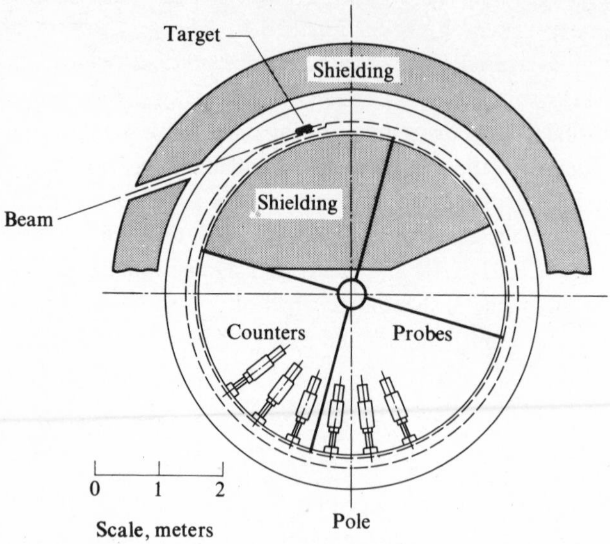

FIGURE 2.7
Plan of 5-m-diameter ring magnet. The proton beam enters through a hole in the yoke and hits a target in the magnetic field. Muons created by $\pi\mu$ decay are stored. The counters detect electrons from μe decay. [*From J. W. Bailey et al., Phys. Lett.*, ***28B****:287 (1968)*.]

The results of this experiment are (Bai 68)

$$a(\mu^+) = (11{,}657.5 \pm 7.1) \times 10^{-7} \tag{2.80}$$

$$a(\mu^-) = (11{,}662.5 \pm 2.4) \times 10^{-7} \tag{2.81}$$

$$a(\mu^-) - a(\mu^+) = (5.0 \pm 7.5) \times 10^{-7} \tag{2.82}$$

Equation (2.82) is in agreement with CPT invariance. Results (2.80) and (2.81) are in good agreement with the theoretical formula (2.76) and thus provide remarkable confirmation of quantum electrodynamics.

2.9.2 The Hyperfine Structure of Muonium

Muonium is a hydrogenic atom (μ^+e^-) formed in the laboratory by allowing a beam of μ^+ to stop in a gas, e.g., argon. The reaction

$$\mu^+ + \mathrm{Ar} \rightarrow (\mu^+e^-) + \mathrm{Ar}^+$$

has a maximum cross section for μ^+ kinetic energies around 200 eV. The muonium atoms quickly come to thermal equilibrium with the surrounding gas

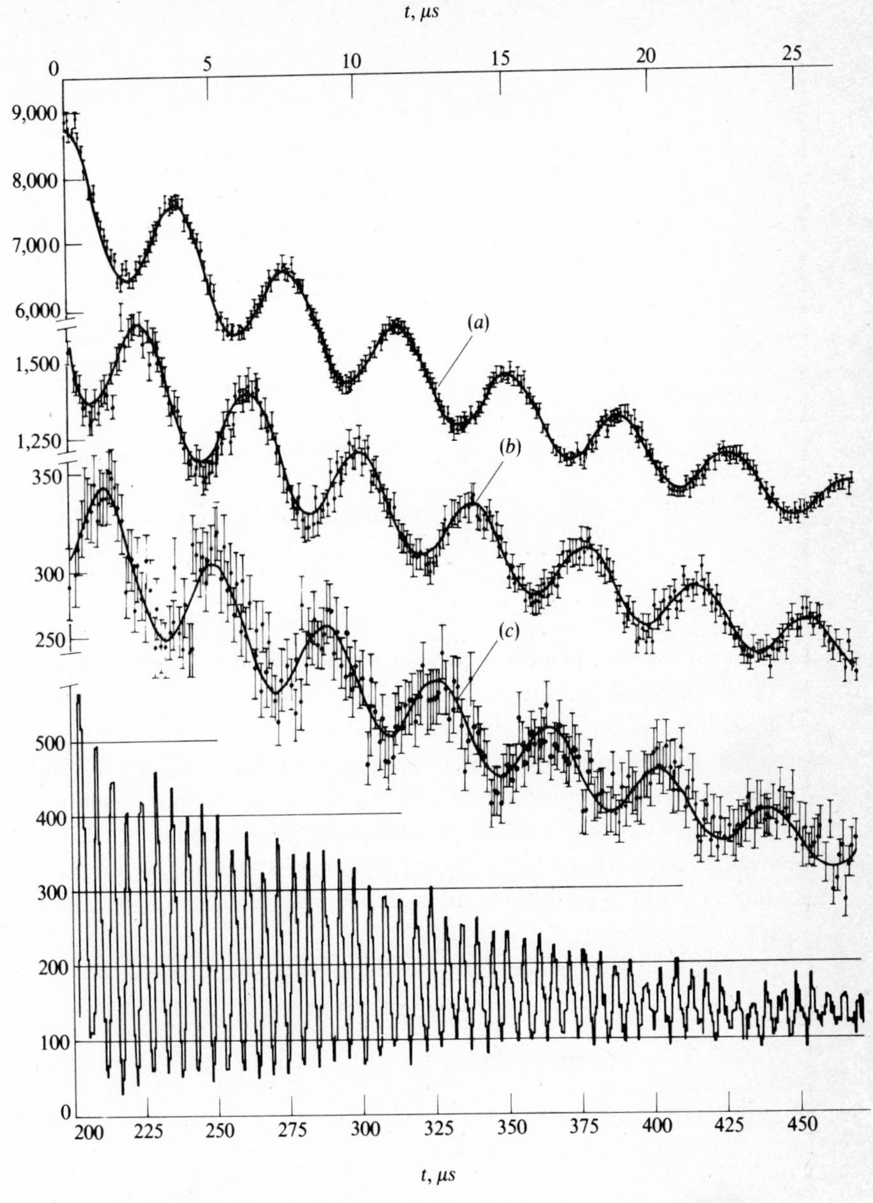

FIGURE 2.8
Distribution of decay-electron events as a function of time. Lower curve shows rotation frequency of muon at early time. (*a*), (*b*), and (*c*) Late time data, 20 to 130 μs, showing $(g-2)$-precession. Data are fitted from 21 to 190 μs. (*a*) 20 to 45 μs; (*b*) 65 to 90 μs; (*c*) 105 to 130 μs. [*From J. W. Bailey et al., Phys. Lett.*, ***28B**:287 (1968)*.]

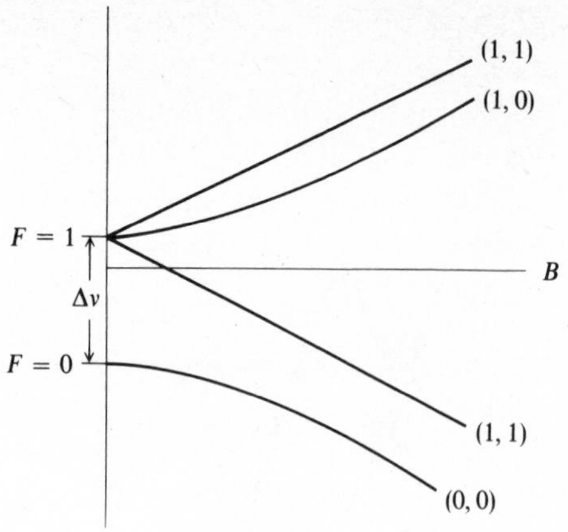

FIGURE 2.9
Breit–Rabi diagram for hfs of muonium in $1^2S_{1/2}$ ground state.

(except that there is negligible spin depolarization). Since the rate for the process

$$\mu^+ e^- \to \bar{\nu}_\mu \nu_e$$

in the $1^2S_{1/2}$ state of muonium is negligible, a muonium atom exists for the lifetime of the muon itself: 2×10^{-6} s.

The Zeeman effect of the hyperfine structure (hfs) of muonium in the $1^2S_{1/2}$ ground state is given by the *Breit–Rabi* formula (Ram 56a, Bre 31) and is shown in the well-known diagram (Fig. 2.9). Since the incoming muons are polarized, the four Zeeman sublevels are unequally populated and a net transition between levels (1, 1) and (0, 0), for example, can be induced by application of a suitable rf field. Transitions are detected by a change in the electron asymmetry in muon decay.

Pioneering work with muonium was done by Hughes and collaborators (see, e.g., Hug 66, 70), who recently obtained the result

$$\Delta\nu(0) = 4{,}463.310 \pm 0.011 \text{ MHz} \qquad \text{(Cra 71)} \qquad (2.83)$$

for the zero field hfs splitting. In a remarkably precise and elegant experiment, Telegdi and coworkers achieved the result

$$\Delta\nu(0) = 4{,}463.30133 \pm 0.00395 \text{ MHz} \qquad \text{(Fav 71)} \qquad (2.84)$$

by a method which may be described as follows: We consider muons with an initial polarization $P_z(0) = 1$ which form muonium in the $F = 1$, $M = 1$ state at zero external static magnetic field. A microwave magnetic field B_1 oscillating in the x direction with frequency ω close to the hyperfine resonance frequency $\omega_0 = 2\pi\,\Delta\nu(0)$ induces transitions to the $F = 0$, $M_F = 0$ state, and from it back to both $|M_F| = 1$ states. Thus $P_z(t)$ varies with time, being initially at the value $+1$ and going through zero at a time τ when the populations of both $M_F = \pm 1$ states are equal. If B_1 is switched off at τ, then $\mathbf{P}$ continues to oscillate in the x, y plane along $\mathbf{P}(\tau)$ with frequency ω_0 because one has prepared a coherent superposition of $F = 1$ and $F = 0$ states. When the microwave field is switched on again at a later time T, $\mathbf{P}$ is rotated to $P_z = -1$ only if $\mathbf{B}_1$ and $\mathbf{P}$ are in the same phase relation to each other at $T + \tau$ as they were at τ. This occurs at resonance, i.e., when $\omega = \omega_0$. This method achieves very narrow resonance linewidths and thus leads to extremely precise results for $\Delta\nu$, as indicated.

Why is $\Delta\nu(0)$ interesting? The reason is that the muon is a lepton, and so the theoretical formula for $\Delta\nu$ contains no uncertain terms arising from strong interaction effects and thus may be compared directly with experiment to yield a precise value of the fine structure constant α. By contrast, consider the case of ordinary hydrogen: p^+e^-. There, $\Delta\nu$ has been measured to twelve significant figures, but the theoretical formula for $\Delta\nu$ contains a troublesome and uncertain proton polarizability correction factor $(1 + \delta_p)$.

The theoretical formula for $\Delta\nu(\mu^+e^-)$ can be expressed as an expansion in α and m_e/m_μ. Including terms of order α^3, we find

$$\Delta\nu_{\text{theo}}(\mu^+e^-) = \left[\frac{16}{3}\alpha^2 c R_\infty \frac{\mu_\mu}{\mu_p}\frac{\mu_p}{\mu_e}\left(1 + \frac{\alpha}{2\pi} - 0.328\frac{\alpha^2}{\pi^2}\right)\right]\left[1 + \left(\frac{m_e}{m_\mu}\right)\right]^{-3}$$
$$\times\left[1 + \frac{3}{2}\alpha^2 + a_e + \varepsilon_1 + \varepsilon_2 + \varepsilon_3 - \delta_\mu\right] Hz \qquad (2.85)$$

where μ_μ, μ_e, and μ_p are the spin magnetic moments of the muon, electron, and proton, respectively; the first factor in brackets on the right-hand side is the well-known Fermi formula for S-state hfs (Ram 56b, Fer 30); the next factor is a reduced mass correction; and the last factor contains relativistic, virtual radiative, and relativistic recoil corrections.

Inserting the best values of various constants, including μ_μ/μ_p as determined by the experiment of Hague et al. (Eq. 2.78), and using the experimental result (2.84) for $\Delta\nu$, we find

$$\alpha^{-1} = 137.03638(19) \qquad (1.7 \text{ ppm}) \qquad (2.86)$$

which is in good agreement with the result

$$\alpha^{-1} = 137.03610(22) \qquad (2.87)$$

obtained from observation of $2e/h$ by the Josephson effect (Fin 70). Actually, employing the experimental values for $\Delta\nu(\mu^+e^-)$, $\Delta\nu(p^+e^-)$, and μ_μ/μ_p, one can compute the proton polarizability correction δ_p. This turns out to be (Fav 71)

$$\delta_p = -(4 \pm 4) \text{ ppm}$$

which is small and agrees with recent theoretical estimates.

PROBLEMS

1 Consider the matrices

$$\Gamma^x = \begin{cases} \gamma^0,\ i\gamma^k \\ \sigma^{jk},\ i\sigma^{0k} \\ i\gamma^0\gamma^5,\ \gamma^k\gamma^5 \\ 1,\ \gamma^5 \end{cases} \qquad j, k = 1, 2, 3$$

Prove that $\sum_{x=1}^{16} \Gamma_{\alpha\beta}{}^x \Gamma_{\rho\delta}{}^x = 4\delta_{\alpha\delta}\,\delta_{\rho\beta}$.

2 Consider the $1^2S_{1/2}$ state of muonium. Calculate the branching ratio for the processes

$$\mu^+ + e^- \to \nu_e + \bar{\nu}_\mu$$

$$\mu^+ \to e^+ + \nu_e + \bar{\nu}_\mu$$

3 In the text, an electron energy spectrum was derived for muon decay, assuming the $V - A$ law and zero electron mass:

$$n(\varepsilon)d\varepsilon = 2\varepsilon^2(3 - 2\varepsilon)\, d\varepsilon$$

Derive the correction to $n(\varepsilon)$ arising from finite electron mass to lowest nonvanishing order in m_e/m_μ, where m_e and m_μ are electron and muon masses, respectively.

REFERENCES

Akh 68 Akhmanov, V. V., et al.: *Sov. J. Nucl. Phys.*, **6**:230 (1968).

Ald 69 Aldins, J., T. Kinoshita, S. J. Brodsky, and A. J. Dufner, *Phys. Rev. Letters*, **23**:441 (1969).

Ald 70 Aldins, J., S. J. Brodsky, A. J. Dufner, and T. Kinoshita: *Phys. Rev.*, **1**:2378 (1970).

Bai 68 Bailey, J., W. Bartl, G. von Bochman, R. C. A. Brown, F. J. M. Farley, H. Jostlein, E. Picasso, and R. W. Williams: *Phys. Letters*, **28B**:287 (1968).

Bar 65 Bardon, M., P. Norton, A. M. Sachs, and J. Lee–Franzini: *Phys. Rev.*, **14**:449 (1965).

Beh 56 Behrends, R. E., R. J. Finkelstein, and A. Sirlin: *Phys. Rev.*, **101**:866 (1956).

Ber 58 Berman, S. M.: *Phys. Rev.*, **112**:267 (1958).

Ber 62 Berman, S. M., and A. Sirlin: *Ann. Phys.*, **20**:20 (1962).

Bin 57 Bincer, A. M.: *Phys. Rev.*, **107**:1434 (1957).

Bjo 64 Bjorken, J. D., and S. D. Drell: "Relativistic Quantum Mechanics," McGraw-Hill, New York, 1964.

Blo 64 Bloom, S., L. A. Dick, L. Feuvrais, G. R. Henry, P. C. Macq, and M. Spighel: *Phys. Letters*, **8**:87 (1964).

Bou 57 Bouchiat, C., and L. Michel: *Phys. Rev.*, **106**:170 (1957).

Bre 31 Breit, G., and I. I. Rabi: *Phys. Rev.*, **38**:2082 (1931).

Buh 63 Buhler, A., et al.: *Phys. Letters*, **7**:368 (1963).

Cra 71 Crane, T., D. Casperson, P. Crane, P. Egan, V. W. Hughes, R. Stambaugh, P. A. Thompson, and K. zu Putlitz: *Phys. Rev. Letters*, **27**:474 (1971).

Der 69 Derenzo, S. E.: *Phys. Rev.*, **181**:1854 (1969).

Duc 64 Duclos, J., J. Heintze, A. DeRujula, and V. Soergel: *Phys. Letters*. **9**:62 (1964).

Eck 63 Eckhause, M., and T. A. Filippas: *Phys. Rev.*, **132**:442 (1963).

Ele 66 Elend, H. H.: *Phys. Letters*, **20**:682 (1966); *ibid.*, **21**:720(E) (1966).

Far 62 Farley, F., Massam, Muller, and A. Zichichi: *Proc. Internatl. Conf. Elementary Particle Physics., CERN*, 415 (1962).

Fav 71 Favart, D., P. McIntyre, D. Stowell, V. Telegdi, R. DeVoe, and R. Swanson: *Phys. Rev. Letters*, **27**:1336 (1971).

Fer 30 Fermi, E.: *Z. Physik.*, **60**:320 (1930).

Fie 37 Fierz, M.: *Z. Physik.*, **104**:553 (1937).

Fin 70 Finnegan, T. F., A. Denenstein, and D. N. Langenberg: *Phys. Rev. Letters*, **24**:738 (1970).

Flo 68 Florescu, V., and O. Kamei: *Nuovo Cimento*, **56**:967 (1968).

Fry 68 Fryberger, D.: *Phys. Rev.*, **166**:1379 (1968).

Gar 57 Garwin, R. L., L. M. Lederman, and M. Weinreich: *Phys. Rev.*, **105**:1415 (1957).

Gro 68 Grotch, H.: *Phys. Rev.*, **168**:1872 (1968).

Hag 70 Hague, J. F., J. E. Rothberg, A. Schenck, D. L. Williams, R. W. Williams, K. K. Young, and K. M. Crowe: *Phys. Rev. Letters*, **25**:628 (1970).

Hug 66 Hughes, V. W.: *Ann. Rev. Nucl. Sci.*, **16**:445 (1966).

Hug 70 Hughes, V. W., D. W. McColm, K. Ziock, and R. Prepost: *Phys. Rev.*, **1A**:595 (1970).

Jar 66 Jarlskog, C.: *Nucl. Phys.*, **75**:659 (1966).

Kin 57 Kinoshita, T., and A. Sirlin: *Phys. Rev.*, **106**:110 (1957); *ibid.*, **107**:593 (1957); *ibid.*, **107**:638 (1957); *ibid.*, **108**:844 (1957).

Kin 59 Kinoshita, T., and A. Sirlin: *Phys. Rev.*, **113**:1652 (1959).

Kuz 60 Kuznetsov, V. P.: *Soviet Phys. JEPT*, **10**:989 (1960).

Kuz 61 Kuznetsov, V. P.: *Soviet Phys. JEPT*, **12**:1202 (1961).

Lau 69 Lautrup, B. E., and E. de Rafael: *Nuovo Cimento*, **A64a**:322 (1969).

Lun 62 Lundy, R. A.: *Phys. Rev.*, **125**:1686 (1962).

McM 60 McMaster, W. M.: *Nuovo Cimento*, **17**:395 (1960).

Mey 63 Meyer, S. L., E. W. Anderson, E. Bleser, L. M. Lederman, J. L. Rosen, J. Rothberg, and T. T. Wang: *Phys. Rev.*, **132**:2693 (1963).

Mic 50 Michel, L.: *Proc. Phys. Soc.* (*London*), **A63**:514 (1950).

Pag 62 Page, I. A.: *Ann. Rev. Nucl. Sci.*, **12**:43 (1962).

Pau 57 Pauli, W.: *Nuovo Cimento*, **6**:204, 1957.

Peo 66 Peoples, J.: *Columbia University*, *Nevis Cycl. Rep.* 147, 1966.

Pet 57 Petermann, A.: *Phys. Rev.*, **105**:1931 (1957).

Pet 58 Petermann, A.: *Fortschr. Physik.*, **6**:505 (1958).

Pla 60 Plano, R.: *Phys. Rev.*, **119**:1400 (1960).

Ram 56a Ramsey, N. F.: "Molecular Beams," p. 104, Oxford, New York, 1956.

Ram 56b Ramsey, N. F.: "Molecular Beams," p. 267, Oxford, New York, 1956.

Rud 66 Ruderman, M. A.: *Phys. Rev. Letters*, **17**:794 (1966).

Sch 48 Schwinger, J.: *Phys. Rev.*, **73**:416 (1948).

Sch 49 Schwinger, J.: *Phys. Rev.*, **76**:790 (1949).

Sch 67 Schwartz, D. M.: *Phys. Rev.*, **162**:1306 (1967).

She 67 Sherwood, B. A.: *Phys. Rev.*, **156**:1475 (1967).

Som 57 Sommerfield, C. M.: *Phys. Rev.*, **107**:328 (1957).

Som 58 Sommerfield, C. M.: *Ann. Phys.* (*N.Y.*), **5**:26 (1958).

Suu 57 Suura, H., and E. H. Wichmann: *Phys. Rev.*, **105**:1930 (1957).

Swa 58 Swanson, R. A.: *Phys. Rev.*, **112**:580 (1958).

Wes 70 Wesley, J. C., and A. Rich: *Phys. Rev. Letters*, **24**:1320 (1970).

3

NEUTRINO-ELECTRON SCATTERING AND RELATED PROCESSES; THE INTERMEDIATE VECTOR BOSON

3.1 AMPLITUDES FOR NEUTRINO-LEPTON SCATTERING AND NEUTRINO PAIR PRODUCTION

We now consider the leptonic weak interactions

$$\nu_\mu + e^- \to \nu_e + \mu^- \tag{3.1}$$

$$\nu_e + e^- \to e^- + \nu_e \tag{3.2}$$

$$\bar{\nu}_e + e^- \to e^- + \bar{\nu}_e \tag{3.3}$$

$$e^- + e^+ \to \nu_e + \bar{\nu}_e \tag{3.4}$$

and

$$\nu_\mu + \mu^- \to \mu^- + \nu_\mu \tag{3.5}$$

Although none of these reactions has yet been observed directly, each is of considerable interest since the phenomenological VA lagrangian predicts their existence and yields values for their cross sections. On the experimental side, it now appears barely feasible, if extremely difficult, to observe reaction $\nu_\mu e^- \to \nu_e \mu^-$ (3.1) with high-energy ν_μ beams available at the largest accelerators (see Chap. 13). From this point of view, reaction $\nu_e e^- \to e^- \nu_e$ (3.2) is much more difficult to observe since the ν_e component in an accelerator neutrino beam is far less intense than the ν_μ component. However, $\bar{\nu}_e e^- \to e^- \bar{\nu}_e$ scattering is another matter; here, $\bar{\nu}_e$ beams of large flux, if low energy, are available from U^{235} fission reactors, and at the time of writing experiments are almost sensitive enough to detect this reaction if it is truly described by the standard theory.

Neutrino pair production $e^- e^+ \to \nu_e \bar{\nu}_e$ (3.4) is thought to play a leading role as an energy-loss mechanism in hot stars at very late stages of evolution, and there seems to be some positive, if not yet decisive, evidence for the existence of this reaction from astrophysical data (see Chap. 14). Finally, the reaction $\nu_\mu \mu^- \to \mu^- \nu_\mu$ (3.5) hardly could be observed in the laboratory, but its existence might be inferred from a study of the related reaction of muon pair production by neutrinos in a coulomb field:

$$\nu_\mu + Z \to \mu^+ + \mu^- + \nu_\mu + Z \tag{3.6}$$

In this chapter our first task is to see how $|\mathcal{M}|^2$ for each of the reactions (3.1) to (3.5) may be obtained from $|\mathcal{M}|^2$ for μ^- decay [Eq. (2.26)]:

$$|\mathcal{M}|^2_{\mu^- \to e^- \bar{\nu}_e \nu_\mu} = 32G^2(p_e - m_e s_e)\cdot q_{\nu_\mu}(p_\mu - m_\mu s_\mu)\cdot q_{\bar{\nu}_e}$$

When averaged over initial spins and summed over final spins, this formula becomes

?

$$|\bar{\mathcal{M}}|^2_{\mu^- \to e^- \bar{\nu}_e \nu_\mu} = 64G^2 p_\mu \cdot q_{\bar{\nu}_e} p_e \cdot q_{\nu_\mu} \tag{3.7}$$

Here, we recall that ν_μ and $\bar{\nu}_e$ are completely polarized.

Muon decay is represented by Fig. 3.1*a*. We may transform that figure by changing the outgoing $\bar{\nu}_e$ line to an incoming ν_e line, whereupon we obtain Fig. 3.1*b*, which represents the process $\nu_e \mu^- \to e^- \nu_\mu$. Under such a transformation the only modification of $|\bar{\mathcal{M}}|^2$ is the replacement of $q_{\bar{\nu}_e}$ by q_{ν_e}:

$$|\bar{\mathcal{M}}|^2_{\nu_e \mu^- \to e^- \nu_\mu} = 64G^2 p_\mu \cdot q_{\nu_e} p_e \cdot q_{\nu_\mu}$$

For convenience in what follows, we change the notation at this point and label incoming and outgoing particles with subscripts 1 and 2, respectively. Also, from now on, we refer to the four-momentum of a charged lepton (e or μ) with symbol p, while q denotes the four-momentum of ν or $\bar{\nu}$. Thus, $|\bar{\mathcal{M}}|^2$ for $\nu_e \mu^- \to e^- \nu_\mu$ becomes

$$|\bar{\mathcal{M}}|^2_{\nu_e \mu^- \to e^- \nu_\mu} = 64G^2 p_1 \cdot q_1 p_2 \cdot q_2 \tag{3.8}$$

Because of $e\mu$ universality, exactly the same expression holds for the reactions $\nu_e e^- \to e^- \nu_e$, $\nu_\mu e^- \to \mu^- \nu_e$, or $\nu_\mu \mu^- \to \mu^- \nu_\mu$:

$$|\bar{\mathcal{M}}|^2 = 64G^2 p_1 \cdot q_1 p_2 \cdot q_2 \tag{3.9}$$

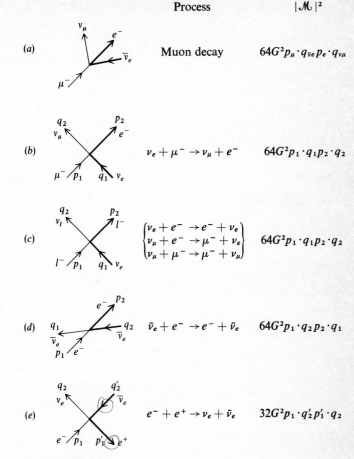

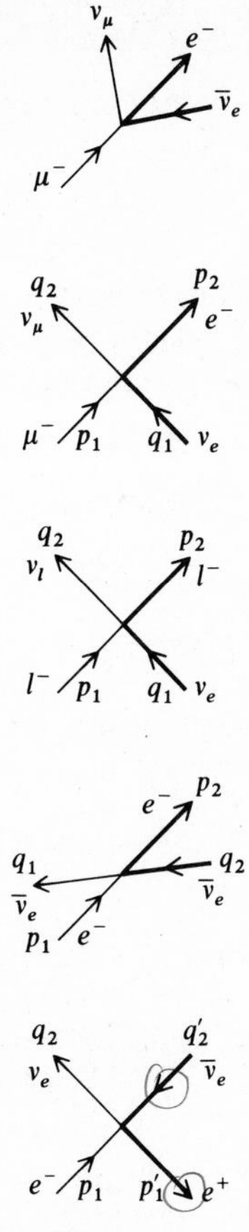

	Process	$\lvert\mathcal{M}\rvert^2$
(a)	Muon decay	$64G^2 p_\mu \cdot q_{\bar{\nu} e} p_e \cdot q_{\nu\mu}$
(b)	$\nu_e + \mu^- \to \nu_\mu + e^-$	$64G^2 p_1 \cdot q_1 p_2 \cdot q_2$
(c)	$\begin{Bmatrix} \nu_e + e^- \to e^- + \nu_e \\ \nu_\mu + e^- \to \mu^- + \nu_e \\ \nu_\mu + \mu^- \to \mu^- + \nu_\mu \end{Bmatrix}$	$64G^2 p_1 \cdot q_1 p_2 \cdot q_2$
(d)	$\bar{\nu}_e + e^- \to e^- + \bar{\nu}_e$	$64G^2 p_1 \cdot q_2 p_2 \cdot q_1$
(e)	$e^- + e^+ \to \nu_e + \bar{\nu}_e$	$32G^2 p_1 \cdot q_2' p_1' \cdot q_2$

FIGURE 3.1
Feynman diagrams for related purely leptonic weak interactions.

In order to obtain expressions for $\bar{\nu}_e e^- \to e^- \bar{\nu}_e$ scattering and $e^- e^+ \to \nu_e \bar{\nu}_e$, we transform Fig. 3.1*c* to Fig. 3.1*d* by the replacements

$$\nu \text{ incoming} \to \bar{\nu} \text{ outgoing}$$

$$\nu \text{ outgoing} \to \bar{\nu} \text{ incoming}$$

which yields

$$|\bar{\mathcal{M}}|^2 = 64G^2 p_1 \cdot q_2\, p_2 \cdot q_1 \tag{3.10}$$

or we transform Fig. 3.1*c* to Fig. 3.1*e* with

$$e^- \text{ outgoing} \rightarrow e^+ \text{ incoming}$$
$$\nu_e \text{ incoming} \rightarrow \bar{\nu} \text{ outgoing}$$

which gives

$$|\bar{\mathcal{M}}|^2 = 32G^2 p_1 \cdot q_2' p_1' \cdot q_2 \tag{3.11}$$

In the last equation, we average over initial spins of e^- and e^+, and the primed four-momenta refer to the particles as labeled in Fig. 3.1*e*.

3.2 THE CROSS SECTIONS FOR $l^- \nu_e \rightarrow e^- \nu_l$ AND $e^- \bar{\nu}_e \rightarrow e^- \bar{\nu}_e$

Let us now consider a collision of the general form

$$A(p_A, m_A) + B(p_B, m_B) \rightarrow C(p_C, m_C) + D(p_D, m_D) \tag{3.12}$$

where p and m refer to the four-momentum and rest mass of the respective particle. The differential cross section $d\sigma$ is defined as

$$d\sigma = \frac{dW}{j} \tag{3.13}$$

where dW is the transition probability per unit time [Eq. (2.10)]

$$dW = \frac{(2\pi)^4 V\, \delta^4(p_f - p_i)}{\prod_i (2E_i V)} \left[\prod_f \frac{d^3\mathbf{p}_f}{(2\pi)^3(2E_f)}\right] |\mathcal{M}|^2 \tag{3.14}$$

and j is the flux density. In the laboratory frame, where particle B is at rest, we have

$$j = \frac{v_A}{V} \tag{3.15}$$

where v_A is the velocity of particle A. Therefore,

$$d\sigma = \frac{(2\pi)^4\, \delta^4(p_f - p_i)}{v_A 2m_B 2E_A} \frac{d^3\mathbf{p}_C\, d^3\mathbf{p}_D}{(2\pi)^6(2E_C)(2E_D)} |\mathcal{M}|^2 \tag{3.16}$$

However, we may write the factor $E_A v_A m_B$ in invariant form as follows:

$$E_A v_A m_B = |\mathbf{p}_A| m_B = \sqrt{(p_A p_B)^2 - m_A{}^2 m_B{}^2} \quad ?$$

Thus we obtain the well-known formula

$$d\sigma = \frac{(2\pi)^4\, \delta^4(p_A + p_B - p_C - p_D)}{4\sqrt{(p_A p_B)^2 - m_A{}^2 m_B{}^2}} \frac{d^3\mathbf{p}_C\, d^3\mathbf{p}_D}{(2\pi)^6(2E_C)(2E_D)} |\mathcal{M}|^2 \tag{3.17}$$

which is valid in any frame. For $l\nu$ scattering, the spin-averaged cross section becomes, from (3.9) and (3.17),

$$d\sigma(l\nu) = \frac{4G^2}{(2\pi)^2} \frac{p_2 \cdot q_2}{E_2 \omega_2} d^3\mathbf{p}_2 \, d^3\mathbf{q}_2 \, \delta^4(p_2 + q_2 - p_1 - q_1)$$

Integrating over $\mathbf{p}_2$, with

$$\int d^3\mathbf{p}_2 \, \delta^3(\mathbf{p}_2 + \mathbf{q}_2 - \mathbf{p}_1 - \mathbf{q}_1) = 1$$

we obtain in the center-of-mass (CM) frame

$$d\sigma = \frac{4G^2}{(2\pi)^2} \omega_2^{\,2}\left(1 + \frac{\omega_2}{E_2}\right) d\omega_2 \, \delta(\omega_1 + E_1 - \omega_2 - E_2) \, d\Omega_{\nu 2} \tag{3.18}$$

with $E_1 = \sqrt{\omega_1^{\,2} + m^2}$ and $E_2 = \sqrt{\omega_2^{\,2} + m^2}$

Now recalling that

$$\int f(x) \, \delta[g(x)] = \left. \frac{f(x)}{\partial g(x)/\partial x} \right|_{g(x)=0}$$

?

we obtain

$$\frac{d\sigma}{d\Omega_{\nu 2}} = \frac{4G^2}{(2\pi)^2} \omega_1^{\,2} \tag{3.19a}$$

or

$$\sigma_{\mathrm{CM}}(l\nu) = \frac{4G^2}{\pi} \omega_1^{\,2} \tag{3.19b}$$

for the cross section in the CM frame. It is also useful to have an expression for $\sigma(l\nu)$ in the laboratory frame. For this purpose, we label the neutrino laboratory energy by ω_L. It is related to ω_1 by the following Lorentz transformation:

$$\begin{aligned}
\omega_L &= (\omega_1 + v_l|\mathbf{k}_1|)\gamma \\
&= (\omega_1 + v_l \omega_1)\gamma \\
&= \omega_1\left(1 + \frac{\omega_1}{\sqrt{\omega_1^{\,2} + m^2}}\right) \frac{\sqrt{\omega_1^{\,2} + m^2}}{m} \\
&= \frac{\omega_1}{m}(\omega_1 + \sqrt{\omega_1^{\,2} + m^2})
\end{aligned}$$

or

$$\omega_1 = \frac{\omega_L}{(1 + 2\omega_L/m)^{1/2}} \tag{3.20}$$

Here v_l is the lepton velocity in the CM frame. Therefore we obtain

$$\sigma_{\text{LAB}}(l\nu) = \sigma_0 \frac{2\omega_L{}^2/m^2}{1 + 2\omega_L/m} \tag{3.21}$$

where $$\sigma_0 = \frac{2G^2m^2}{\pi} = 8.3 \times 10^{-45}\ \text{cm}^2 \qquad \text{for } m = m_e$$

To calculate the cross section for $e\bar{\nu}_e$ scattering we have to insert Eq. (3.10) into (3.17), in which case we obtain

$$d\sigma(e\bar{\nu}_e) = \frac{(2\pi)^4\,\delta^4(p_1 + q_1 - p_2 - q_2)}{4p_1q_1}\,\frac{64G^2 p_1\cdot q_2\, p_2\cdot q_1}{(2\pi)^6(2E_2)(2\omega_2)}\,d^3\mathbf{p}_2\,d^3\mathbf{q}_2$$

In the CM frame, in which $\mathbf{p}_1 = -\mathbf{q}_1$, this becomes

$$d\sigma(e\bar{\nu}_e) = \frac{G^2}{\pi^2}\,\frac{(E_1\omega_2 - \mathbf{p}_1\cdot\mathbf{q}_2)(E_2\,\omega_1 - \mathbf{p}_2\cdot\mathbf{q}_1)}{E_2\,\omega_2(E_1\omega_1 + \omega_1{}^2)}\,\delta^3(\mathbf{p}_2 + \mathbf{q}_2)$$
$$\times\,\delta(E_1 + \omega_1 - E_2 - \omega_2)\,d^3\mathbf{p}_2\,d^3\mathbf{q}_2$$

Then writing $\mathbf{p}_1\cdot\mathbf{q}_2 = \omega_1\omega_2\cos\theta$ and $d^3\mathbf{q}_2 = 2\pi\omega_2{}^2\,d\omega_2\sin\theta\,d\theta$, and performing the integrations, we obtain for the differential cross section

$$d\sigma_{\text{CM}}(e\bar{\nu}) = \frac{2G^2}{\pi}\,\frac{\omega_1{}^2}{(E_1 + \omega_1)^2}\sin\theta\,d\theta(E_1 - \omega_1\cos\theta)^2 \tag{3.22}$$

and for the total cross section

$$\sigma_{\text{CM}}(e\bar{\nu}_e) = \frac{4G^2}{\pi}\,\omega_1{}^2\left[1 - \frac{(\frac{2}{3}\omega_1{}^2 + 2E_1\omega_1)}{(\omega_1 + E_1)^2}\right] \tag{3.23}$$

In the laboratory frame, this becomes

$$\sigma_{\text{LAB}}(e\bar{\nu}_e) = \sigma_0\left(\frac{\omega_L}{3m_e}\right)\left[1 - \frac{1}{(1 + 2\omega_L/m_e)^3}\right] \tag{3.24}$$

It should be noted that the neutrino may have a *charge radius* even though its total charge is zero. This could arise via virtual weak couplings to charged particles. In this case there would be an electromagnetic contribution to the νl scattering cross section of order α^2 times the values derived above. Actually, interference between the electromagnetic and weak amplitudes might bring about a contribution to the cross section which is of order α times (3.21) or (3.24) (Ber 63, Lee 64).

3.3 EXPERIMENTAL UPPER LIMITS ON COUPLING STRENGTHS FOR NEUTRINO-LEPTON INTERACTIONS

We noted earlier that $\bar{\nu}_e e^-$ scattering is of special interest on account of the availability of intense $\bar{\nu}_e$ beams from fission reactors. The energy spectrum $N(\omega_L)$ of such a beam has been calculated (Avi 70) and is shown in Fig. 3.2, together with the effective partial cross section $S(E_{2L})$ for $e\bar{\nu}_e$ scattering as a function of electron recoil energy E_{2L} in the laboratory frame:

$$S(E_{2L}) = \int_{\omega_{L,\,\min}}^{\infty} \frac{(d\sigma/dE_{2L})N(\omega_L)\,d\omega_L}{\int_0^\infty N(\omega_L)\,d\omega_L}$$

From these results a spectrum-averaged cross section $\bar{\sigma}_{V-A}(e^-\bar{\nu}_e)$ may be calculated:

$$\bar{\sigma}_{V-A} = (3.3 \pm 0.3) \times 10^{-47}\ \text{cm}^2 \text{ per fission } \bar{\nu}_e$$

In a recent experiment by Reines and Gurr (Rei 70), the result

$$\bar{\sigma}_{\text{exp}}/\bar{\sigma}_{V-A} < 4$$

was obtained. This may be analyzed in terms of more general derivative-free four-fermion couplings as well, obtaining the limits given in Table 3.1 (Che 70).

The process $\nu_e + e^- \rightarrow \nu_e + e^-$ is unfortunately much less accessible to observation. Its existence is confirmed indirectly by the results of a number of astrophysical tests utilizing statistics of white dwarf stars (Sto 70).[1] However, the limit obtained on the coupling strength

$$g^2 = 10^{0\pm 2}G^2$$

Table 3.1 LIMITS ON $\bar{\nu}_e e^-$ COUPLING STRENGTH AS OBTAINED FROM THE REINES–GURR EXPERIMENT (Che 70)

Interaction type	Experimental upper limit
V − A	${g_{V-A}}^2 < 4.0G^2$
V + A	${g_{V+A}}^2 < 0.2G^2$
S	${g_S}^2 < 2.0G^2$
P	${g_P}^2 < 2.0G^2$
T	${g_T}^2 < 0.6G^2$
V	${g_V}^2 < 0.8G^2$
A	${g_A}^2 < 0.7G^2$

[1] See Chap. 14.

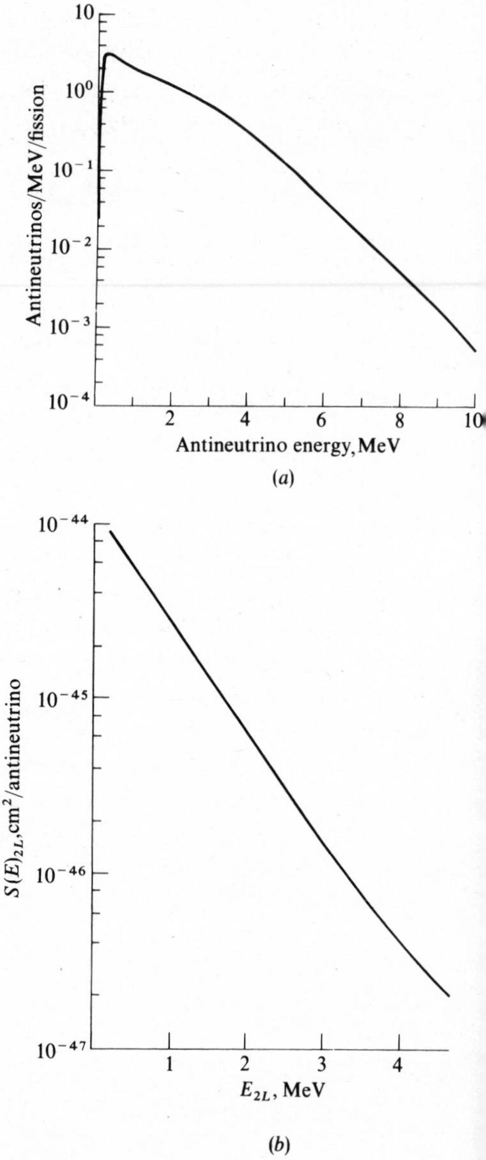

FIGURE 3.2
(*a*) Energy spectrum of antineutrinos from U^{235} fission products in secular equilibrium (Avi 70); (*b*) effective partial cross section $S(E_{2L})$ (Avi 70).

is rather crude. An upper limit is also obtained from analysis of electron production in the CERN neutrino spark chamber, using a ν_e beam obtained from K_{e3} decay. One finds (Ste 70, Cun 70) that

$$g^2 < 40G^2$$

3.4 BEHAVIOR OF NEUTRINO-ELECTRON CROSS SECTIONS AT HIGH ENERGIES; BREAKDOWN OF THE FIRST-ORDER THEORY

Let us recall the differential cross section for $e^- \nu_e$ scattering in the CM frame

$$\frac{d\sigma}{d\Omega_{\nu 2}} = \frac{4G^2}{(2\pi)^2} \omega_1{}^2 \quad (3.19a)$$

In general, we may express such a differential cross section as the absolute square of a scattering amplitude $f(\theta, \phi)$:

$$\frac{d\sigma}{d\Omega_{\nu 2}} = |f(\theta, \phi)|^2$$

And $f(\theta, \phi)$ may be expanded in partial waves:

$$f(\theta, \phi) = \frac{1}{|\mathbf{q}_1|} \sum_{J=0}^{\infty} (J + \tfrac{1}{2}) P_J(\cos\theta) \mathcal{M}_J \qquad (3.25)$$

where $\mathcal{M}_J$ is the amplitude corresponding to the partial wave with angular momentum J.

In the case of $e\nu_e$ scattering as formulated according to the weak lagrangian $\mathcal{L}_w$, only the $J = 0$ partial wave enters for any $|\mathbf{q}_1|$ because we are dealing with a contact interaction of zero range. Thus in our case

$$\frac{d\sigma}{d\Omega_{\nu 2}} = |f(\theta, \phi)|^2 = \frac{1}{4\omega_1{}^2} |\mathcal{M}_0|^2$$

However, unitarity (conservation of probability) requires in the general case that $|\mathcal{M}_J| < 1$ for each J. Therefore we must have

$$\frac{d\sigma(e\nu)}{d\Omega_{\nu 2}} \leq \frac{1}{4\omega_1{}^2} \qquad (3.26)$$

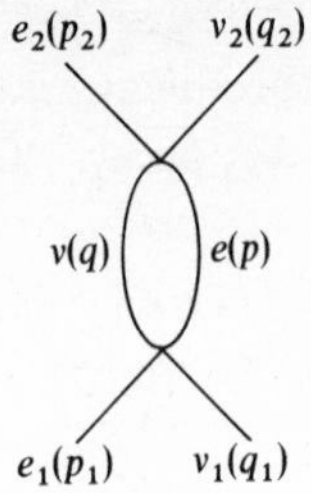

FIGURE 3.3

It is easy to see that (3.26) and (3.19*a*) lead to a contradiction at high energies, i.e., when

$$\frac{4G^2}{(2\pi)^2}\,\omega_1{}^2 \approx \frac{1}{4\omega_1{}^2}$$

or

$$\omega_1{}^4 \approx \frac{\pi^2}{4G^2}$$

$$\omega_1 \approx 300 \text{ GeV}$$

Thus it is clear that our first-order calculation of $\sigma(l\nu)$ fails completely at energies of 300 GeV. We might naturally suppose at first that this difficulty arises from neglect of higher-order corrections, and that if these were included, the cross sections would level out and remain within the bounds imposed by unitarity. We attempt to ascertain if this is the case by considering the contribution to the amplitude which arises from the second-order diagram of Fig. 3.3. It is easy to see that this contribution $\mathcal{M}^{(2)}$ takes the form

$$\mathcal{M}^{(2)} = \frac{G^2}{2}\frac{1}{(2\pi)^4}\int d^4q\left[\bar{u}_{e2}\gamma_\lambda(1+\gamma^5)\frac{1}{\not{q}}\gamma_\mu(1+\gamma^5)u_{e1}\right]$$

$$\times\left[\bar{u}_{\nu 2}\gamma^\lambda(1+\gamma^5)\frac{1}{\not{p}-m_e}\gamma^\mu(1+\gamma^5)u_{\nu 1}\right] \tag{3.27}$$

where $p = p_1 + q_1 - q$ and the factors $1/\not{q}$ and $1/(\not{p} - m_e)$ represent the propagators for intermediate neutrino and electron, respectively, in momentum space. However, the integral of (3.27) diverges as q^2:

$$\mathcal{M}^{(2)} \approx \frac{G^2}{2(2\pi)^4}\int\frac{d^4q}{qp} \approx \frac{G^2}{2(2\pi)^4}\int\frac{q^3\,dq}{q^2} \approx G^2q^2 \tag{3.28}$$

Of course, in quantum electrodynamics one also encounters divergent integrals corresponding to higher-order diagrams (see, for example, Bjo 64, Chap. 8). However, the divergences may be removed to any order by charge and mass renormalization. In the present case, we may attempt to eliminate the divergences in $\mathcal{M}^{(2)}$ by an analogous process, but when we encounter diagrams of third and higher order, the divergences are more and more severe and each requires a new set of renormalization constants. The result is that when all possible diagrams are taken into account, an infinite set of renormalization constants is needed.

Alternatively, we may introduce an arbitrary cutoff Λ^2 in q^2 for each amplitude $\mathcal{M}^{(n)}$. In the absence of all other information this might be chosen as the unitarity limit

$$\Lambda^2 = \frac{\pi}{2G}$$

In this case

$$\mathcal{M}^{(2)} \approx \frac{G^2}{2(2\pi)^4} \frac{\pi}{2G} = \mathcal{O}(G)$$

and indeed each successive order would contribute an amount $\approx G$ to the perturbation series, which show that this procedure is thoroughly unsatisfactory.

3.5 THE INTERMEDIATE VECTOR BOSON W

Another possibility is that the lowest-order weak interaction is itself nonlocal because it is in reality a second-order process mediated by an intermediate vector boson. Such an idea occurs rather naturally by analogy with quantum electrodynamics. In that theory, the interaction between two electrons is represented in lowest order by the Feynman diagrams of Fig. 3.4*a* and *b*. Each diagram contains two vertices coupled by an intermediate, or virtual, photon. Because the photon has zero mass, the coulomb force between the electrons has infinite range ($1/r$ potential). Of course, the photon is a vector particle, being coupled to four-currents.

It seems reasonable to suppose that the weak interaction may also be represented in lowest order by diagrams shown in Fig. 3.5, where W is a hypothetical, so far unobserved, intermediate boson, with the following required properties:

1. W must be charged.
2. W must have very large mass since the weak interaction certainly has very short range. The present limit is $m_W > 2$ GeV (Bie 64).
3. W must be a vector particle since it is coupled to four-currents, but it has no definite parity.

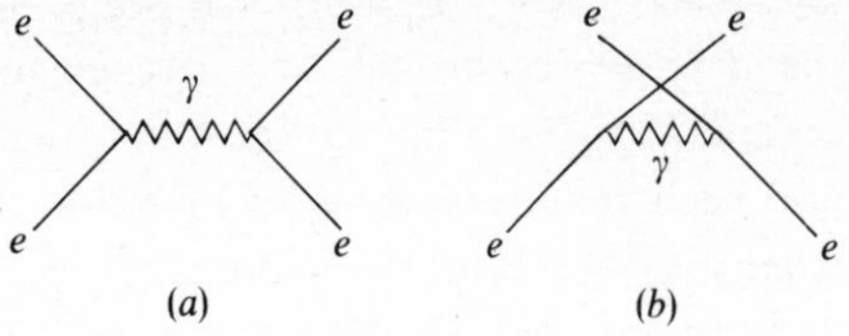

FIGURE 3.4

From these requirements alone we can make some fairly definite statements about the W.

First, we must write a suitable field equation. The problem here is to find a suitable lagrangian. We know that the electromagnetic field, a zero-mass vector field, is described by a lagrangian

$$\mathcal{L}_{\text{EM}} = -\frac{1}{4} F_{\mu\nu} F^{\mu\nu} \qquad \text{where } F_{\mu\nu} = \frac{\partial A_\nu}{\partial x^\mu} - \frac{\partial A_\mu}{\partial x^\nu}$$

We also know that for a free scalar field ϕ associated with particles of mass m, the lagrangian is

$$\mathcal{L}_0 = \frac{m^2 \phi^2}{2} - \frac{1}{2} \frac{\partial \phi}{\partial x^\mu} \frac{\partial \phi}{\partial x_\mu}$$

Plainly the lagrangian for a free vector field ϕ^ν with corresponding particles of mass m_W should be

$$\mathcal{L}_{W0} = -\frac{1}{4} f_{\mu\nu} f^{\mu\nu} + \frac{m_W{}^2}{2} \phi^\nu \phi_\nu \qquad \text{with } f_{\mu\nu} = \frac{\partial \phi_\nu}{\partial x^\mu} - \frac{\partial \phi_\mu}{\partial x^\nu}$$

and if this field is coupled to a current S_ν by means of an interaction $\mathcal{L}_{\text{int}} = -S_\nu \phi^\nu$, we should have altogether

$$\mathcal{L}_W = -\frac{1}{4} f_{\mu\nu} f^{\mu\nu} + \frac{m_W{}^2}{2} \phi_\nu \phi^\nu - S_\nu \phi^\nu \tag{3.29}$$

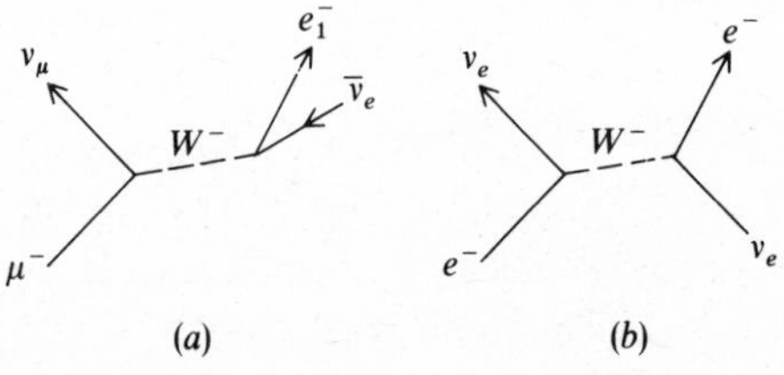

FIGURE 3.5
(a) μ^- decay; (b) $e^-\nu$ scattering.

The field equation is then obtained from the Euler–Lagrange equation:

$$\frac{\partial \mathfrak{L}_W}{\partial \phi^\nu} - \frac{\partial}{\partial x_\mu}\frac{\partial \mathfrak{L}_W}{\partial(\partial \phi^\nu/\partial x_\mu)} = 0 \tag{3.30}$$

Writing

$$-\tfrac{1}{4} f_{\mu\nu} f^{\mu\nu} = -\tfrac{1}{2}\frac{\partial \phi^\nu}{\partial x_\mu}\left(\frac{\partial \phi_\nu}{\partial x^\mu} - \frac{\partial \phi_\mu}{\partial x^\nu}\right)$$

we have

$$\frac{\partial \mathfrak{L}_W}{\partial \phi^\nu} = m_W{}^2 \phi_\nu - S_\nu$$

and

$$-\frac{\partial}{\partial x_\mu}\frac{\partial \mathfrak{L}_W}{\partial(\partial \phi^\nu/\partial x_\mu)} = \Box \phi_\nu - \frac{\partial}{\partial x^\nu}\left(\frac{\partial \phi_\mu}{\partial x_\mu}\right)$$

Thus

$$\Box \phi_\nu - \frac{\partial}{\partial x^\nu}\left(\frac{\partial \phi_\mu}{\partial x_\mu}\right) + m_W{}^2 \phi_\nu = S_\nu \tag{3.31}$$

Now, taking the derivative $\partial/\partial x_\nu$ of both sides, we have

$$\Box\left(\frac{\partial \phi_\nu}{\partial x_\nu}\right) - \Box\left(\frac{\partial \phi_\mu}{\partial x_\mu}\right) + m_W{}^2 \frac{\partial \phi_\nu}{\partial x_\nu} = \frac{\partial S_\nu}{\partial x_\nu}$$

or

$$\frac{\partial \phi_\nu}{\partial x_\nu} = \frac{1}{m_W{}^2}\frac{\partial S_\nu}{\partial x_\nu}$$

Thus we obtain the field equation

$$\Box \phi_\nu + m_W{}^2 \phi_\nu = \left(g_\nu{}^\mu + \frac{1}{m_W{}^2}\frac{\partial}{\partial x^\nu}\frac{\partial}{\partial x_\mu}\right) S_\mu \tag{3.32}$$

Let us now write the Fourier transforms

$$\phi_\nu = \frac{1}{(2\pi)^4}\int F_\nu(k) e^{-ik\cdot x}\, d^4k$$

and

$$S_\mu = \frac{1}{(2\pi)^4}\int s_\mu(k) e^{-ik\cdot x}\, d^4k$$

Thus, from (3.32) we have

$$(\Box + m_W{}^2)\phi_\nu = \frac{1}{(2\pi)^4}\int (m_W{}^2 - k^2) F_\nu(k) e^{-ik\cdot x}\, d^4k$$

$$= \frac{1}{(2\pi)^4}\int \left(g_\nu{}^\mu - \frac{1}{m_W{}^2} k_\nu k^\mu\right) s_\mu(k) e^{-ik\cdot x}\, d^4k$$

Equating the integrands in this last expression, we obtain

$$F_\nu(k) = \frac{g_\nu{}^\mu - \frac{1}{m_W{}^2} k_\nu k^\mu}{m_W{}^2 - k^2} S_\mu(k) \tag{3.33}$$

Thus the momentum-space propagator for W is

$$\frac{g_\nu{}^\mu - \dfrac{1}{m_W{}^2} k_\nu k^\mu}{m_W{}^2 - k^2} \tag{3.34}$$

Of course, in the limit $m_W \to 0$, the lagrangian of (3.29) reduces to the lagrangian of the EM field coupled to a vector current. Thus the W propagator must reduce to the photon propagator $-g_\nu{}^\mu/k^2$ in the $m_W \to 0$ limit. For ordinary weak decays, the momentum transfer k is very small and the propagator reduces to $g_\nu{}^\mu/m_W{}^2$.

Let us consider μ^- decay as described in terms of a W boson. We already know the amplitude according to the first-order, local theory:

$$\mathcal{M} = \frac{G}{\sqrt{2}} \bar{u}_{\nu\mu} \gamma_\alpha (1 + \gamma^5) u_\mu \bar{u}_e \gamma^\alpha (1 + \gamma^5) v_{\bar{\nu}e} \tag{3.35}$$

This may be written as

$$\mathcal{M} = \frac{G}{\sqrt{2}} m_W{}^2 [\bar{u}_{\nu\mu} \gamma_\alpha (1 + \gamma^5) u_\mu] \frac{g_\beta{}^\alpha}{m_W{}^2} [\bar{u}_e \gamma^\beta (1 + \gamma^5) v_{\bar{\nu}e}] \tag{3.36}$$

which is the low-momentum-transfer limit of

$$\mathcal{M} = \frac{G}{\sqrt{2}} m_W{}^2 [\bar{u}_{\nu\mu} \gamma_\alpha (1 + \gamma^5) u_\mu] \left[\frac{g_\beta{}^\alpha - (1/m_W{}^2) k_\beta k^\alpha}{m_W{}^2 - k^2} \right] [\bar{u}_e \gamma^\beta (1 + \gamma^5) v_{\bar{\nu}e}] \tag{3.37}$$

In Eq. (3.37), μ decay is described as a second-order process in which we have a $\mu W \nu_\mu$ vertex with a factor $(G^{1/2}/2^{1/4}) m_W \gamma_\alpha (1 + \gamma^5)$, the W propagator, and the $eW\nu_e$ vertex with a factor $(G^{1/2}/2^{1/4}) m_W \gamma^\beta (1 + \gamma^5)$. When the muon decay rate is calculated with amplitude (3.37), one finds (Lee 57) that the principal change is in ρ:

$$\rho - 0.75 \approx \frac{1}{3} \left(\frac{m_\mu}{m_W} \right)^2$$

The best present value of ρ (see Table 2.1) thus gives a limit on m_W:

$$\left(\frac{m_\mu}{m_W} \right)^2 = 0.0066 \pm 0.0069$$

or

$$m_W = 1.3_{-0.4}{}^{+\infty} \text{GeV}/c^2 \tag{3.38}$$

However, a better lower limit on m_W (>2 GeV) is achieved by methods to be described below.

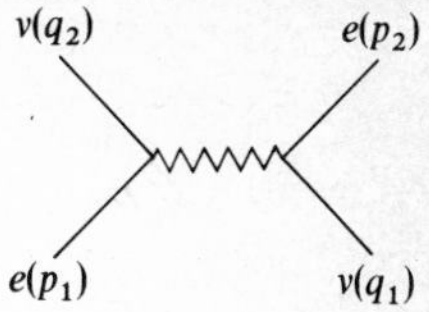

FIGURE 3.6

The existence of the $lW\nu_l$ vertex implies that a $W^\pm$, once produced, can decay spontaneously to $\mu^\pm \nu_\mu(\bar{\nu}_\mu)$ or $e^\pm \nu_e(\bar{\nu}_e)$. It is very easy to show that the rate for such a decay, assuming $m_W \gg m_l$, is

$$R = \frac{G m_W{}^3}{6\pi\sqrt{2}} \tag{3.39}$$

For $m_W \geq 3$ GeV, $R > 10^{19}\ \mathrm{s}^{-1}$, and so W must be very short-lived if it exists at all. The hypothetical W might also be coupled to hadronic weak currents, and it would thus have the decay modes[1]

$$W^\pm \to \pi^\pm, \mathrm{K}^\pm, \mathrm{K} + \pi \qquad \text{and so on}$$

We have seen that the phenomenological lagrangian $\mathfrak{L}_w$ describes a first-order contact interaction which violates unitarity at high energies and for which no satisfactory method is known to calculate higher-order corrections. Does the intermediate vector boson theory ameliorate these difficulties? In order to answer this question, we calculate the cross section for $e\nu_e$ scattering to lowest order (Fig. 3.6), assuming we can neglect m_e in comparison to m_W. The amplitude is

$$\mathcal{M} = \frac{G}{\sqrt{2}} m_W{}^2 [\bar{u}_{\nu 2} \gamma_\alpha (1 + \gamma^5) u_{e1}] \left[\frac{g_\beta{}^\alpha - (1/m_W{}^2) k_\beta k^\alpha}{m_W{}^2 - k^2} \right] [\bar{u}_{e2} \gamma^\beta (1 + \gamma^5) u_{\nu 1}]$$

where

$$q_2 - p_1 = q_1 - p_2 = k$$

Thus we have

$$\mathcal{M} = \frac{G}{\sqrt{2}} \frac{m_W{}^2}{m_W{}^2 - k^2} [\bar{u}_{\nu 2} \gamma_\alpha (1 + \gamma^5) u_{e1}][\bar{u}_{e2} \gamma^\beta (1 + \gamma^5) u_{\nu 1}]$$

$$- \frac{G}{\sqrt{2}} \frac{1}{m_W{}^2 - k^2} [\bar{u}_{\nu 2} (\not{q}_2 - \not{p}_1)(1 + \gamma^5) u_{e1}][\bar{u}_{e2} (\not{q}_1 - \not{p}_2)(1 + \gamma^5) u_{\nu 1}] \tag{3.40}$$

[1] A discussion of these possibilities is given by Y. Yamaguchi in A. Zichichi (ed.), "Strong and Weak Interactions, Present Problems" (1966 International Summer School "Ettore Majorana," Erice, Sicily), Academic, New York, 1966. See also L. M. Lederman and B. G. Pope, *Phys. Rev. Letters*, **27**: 765 (1971).

Since $$\not{p}_1(1+\gamma^5)u_{e1} = (1-\gamma^5)\not{p}_1 u_{e1} = m_e(1-\gamma^5)u_{e1}$$

and $$\bar{u}_{\nu 2}\not{q}_2 = 0$$

and with similar equations for $\bar{u}_{e2}$ and $u_{\nu 1}$, the second term on the right-hand side of (3.40) reduces to

$$-\frac{G}{\sqrt{2}}\frac{m_e^2}{m_W^2 - k^2}[\bar{u}_{\nu 2}(1-\gamma^5)u_{e1}][\bar{u}_{e2}(1+\gamma^5)u_{\nu 1}]$$

which can be neglected.

The remaining term in (3.40) is exactly the same as the amplitude for $\nu_e e$ scattering as calculated earlier by the first-order theory, except for the factor $(m_W^2)/(m_W^2 - k^2)$. Thus, comparing to (3.19*a*), we find for the differential cross section in the present case

$$\frac{d\sigma_W(e_1\nu)}{d\Omega_{\nu 2}} = \frac{G^2}{\pi^2}\frac{m_W^4\omega_1^2}{(m_W^2 - k^2)^2} \tag{3.41}$$

In the limit of zero electron mass, k^2 becomes

$$k^2 = (q_2 - p_1)^2 = q_2^2 + p_1^2 - 2q_2 \cdot p_1 = -2q_2 \cdot p_1$$
$$= -2\omega_1^2(1-\cos\theta)$$

so that (3.41) becomes

$$\frac{d\sigma_W(e_1\nu)}{d\Omega_{\nu 2}} = \frac{G^2}{\pi^2}\frac{m_W^4\omega_1^2}{[m_W^2 + 2\omega_1^2(1-\cos\theta)]^2} \tag{3.42}$$

Integrating over the scattering angle, we find the total cross section

$$\sigma_W(e_1\nu) = \frac{4G^2}{\pi}\frac{\omega_1^2}{1 + 4\omega_1^2/m_W^2} \tag{3.43}$$

Of course, in the limit $m_W \to \infty$, the intermediate boson theory reduces to the ordinary first-order theory [Eq. (3.19*b*)]. For finite m_W, Eq. (3.43) reveals that σ_W approaches a finite limit $(G^2/\pi)\, m_W^2$ as ω_1 increases.

However, like $\sigma(e, \nu)$ for the first-order theory, the cross section $\sigma_W(e, \nu)$ violates unitarity at sufficiently high energy. This may be seen as follows. We again write

$$\frac{d\sigma_W(e_1\nu)}{d\Omega} = |f(\theta)|^2$$

with $$f(\theta) = \frac{G}{\pi}\frac{1}{|\mathbf{q}_1|}\frac{m_W^2\omega_1^2}{m_W^2 + 2\omega_1^2(1-\cos\theta)} \tag{3.44}$$

and utilize the partial wave expansion for $f(\theta)$:

$$f(\theta) = \frac{1}{|\mathbf{q}_1|} \sum_{J=0}^{\infty} (J + \tfrac{1}{2}) P_J(\cos\theta) \mathcal{M}_J$$

As usual, unitarity requires $|\mathcal{M}_J| \leq 1$ for each J, but it is easy to show that Eq. (3.44) leads to

$$\mathcal{M}_0 = \frac{G m_W^2}{2\pi} \log\left(1 + \frac{4\omega_1^2}{m_W^2}\right) \tag{3.45}$$

in violation of the unitarity condition for sufficiently large ω_1. Thus the W boson theory provides no escape from the unitarity problem, and indeed this is a difficulty shared by all perturbation theories. Nor is the W theory capable of renormalization because the term $-(1/m_W^2)k_\mu k^\nu$ in the W propagator [Eq. (3.34)] causes higher-order diagrams to diverge strongly.

Therefore, whether we employ the standard first-order theory, with the phenomenological lagrangian $\mathcal{L}_W$, or opt for the W boson theory, we are in either case confronted by a profound difficulty in the theory of weak interactions. Numerous sophisticated theoretical arguments have been presented in an attempt to explain away this difficulty (see, e.g., Gel 69, Fei 64, Kum 65, Lee 69, and Wei 67, 72), but so far these must be regarded as speculations, albeit ingenious ones, and the basic problem remains unsolved.

3.6 EXPERIMENTAL SEARCHES FOR THE *W*

How would one find the W? It appears that the most practical approach is to utilize the intense high-energy muon-neutrino beams now available at modern accelerators to search for reactions of the form

$$\nu_\mu + Z \to \mu^- + W^+ + Z \tag{3.46}$$

$$\bar{\nu}_\mu + Z \to \mu^+ + W^- + Z \tag{3.47}$$

Here Z stands for a nucleus, which must be present for conservation of momentum (through photon exchange). Feynman diagrams for these reactions are shown in Fig. 3.7. Experimentally, one attempts to observe muons with large transverse momentum arising from W decay.

Calculations of the cross sections for the reactions (3.46) and (3.47) as a function of neutrino energy and m_W were carried out over a restricted range by Lee, Markstein, and Yang (Lee 61) and later extended to cover neutrino energies $\omega_L \leq 500$ GeV and W masses $2 \leq m_W \leq 30$ GeV by Berkov et al. (Ber 69), who present detailed tables. The calculation is somewhat elaborate since it is

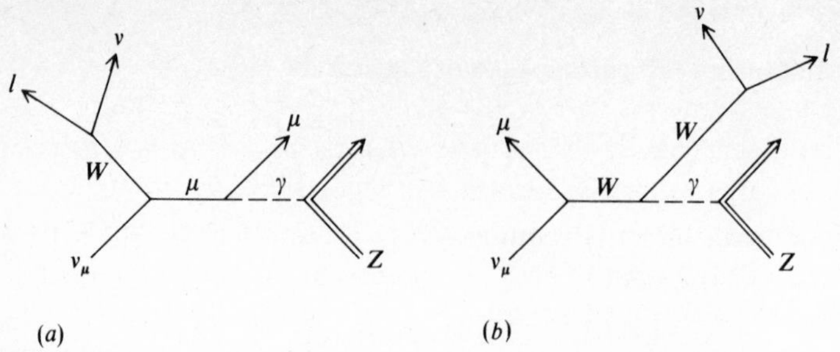

FIGURE 3.7
Feynman diagrams for the reactions (3.46) and (3.47). The leptonic decay of W subsequent to production is also indicated.

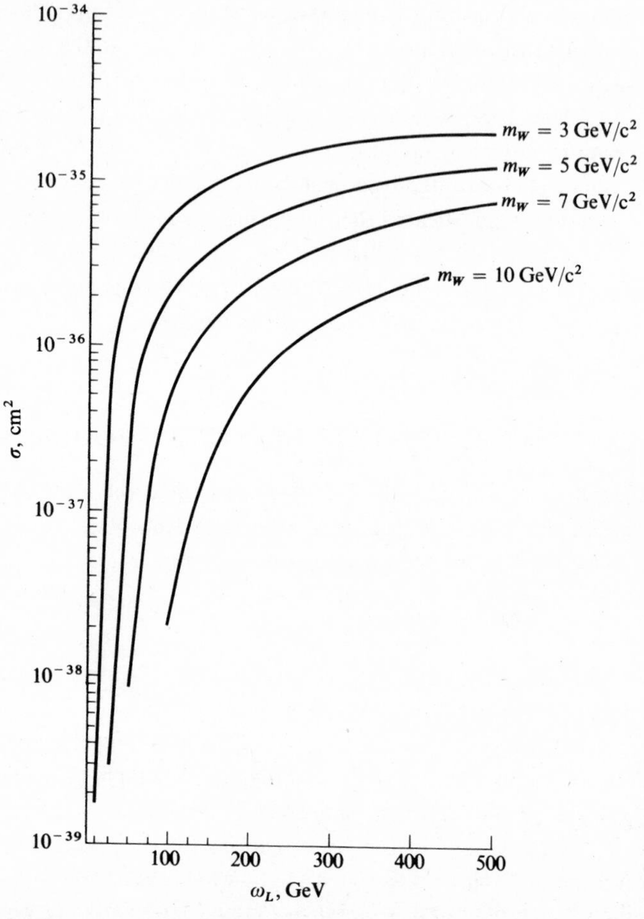

FIGURE 3.8
Cross section for W production as a function of neutrino energy for a proton target (Ber 69).

necessary to take into account the full momentum dependence of the electromagnetic matrix element of the nucleus, and the results depend to some degree on the magnetic moment assumed for W.

In Fig. 3.8 we present for illustration the theoretical cross section for neutrinos on protons as a function of ω_L for four hypothetical boson masses: $m_W = 3$, 5, 7, and 10 GeV. For each of these curves one assumes a "normal" W magnetic moment ($g = 1$). For comparable neutrino energies and boson masses, the cross sections for nuclei (carbon, aluminum, etc.) are somewhat larger. Clearly, for given neutrino energy, the cross section falls with increasing W mass. The W production rate depends, of course, on the neutrino flux as well as the cross section. One expects the following sorts of W production rates at the National Accelerator Laboratory (Batavia, Illinois) for a 300-GeV proton beam (see Chap. 13 for a description of how neutrino beams are produced):

M_W, GeV	W's/ton of detector/3×10^{18} protons on target
4	2×10^3
6	3×10^2
8	≈ 10

As to theoretical estimates of m_W, these can hardly be called more than rough speculations at this time. They vary from 3 to 4 GeV (Moh 68, Gel 69) to 8 GeV (Gla 67) to approximately 30 GeV (Lee 71).

PROBLEMS

1 Suppose the weak interaction were pure V, or pure A. In each case derive the cross section for $e\nu_e$ scattering.

2 Assuming $m_W \gg m_l$, prove that the rate for $W^\pm$ decay to an $l^\pm\nu$ pair would be

$$R = Gm_W{}^3(6\pi\sqrt{2})^{-1}$$

3 Integrate Eq. (3.42) over the solid angle to obtain Eq. (3.43).

4 Show that Eq. (3.44) leads to Eq. (3.45).

REFERENCES

Avi 70 Avignone, F. T.: *Phys. Rev.*, **2D**:2609 (1970).

Ber 63 Bernstein, J., G. Feinberg, and M. Ruderman: *Phys. Rev.*, **132**:1227 (1963).

Ber 69 Berkov, A. V., G. G. Bunatyan, E. D. Zhizhin, and Yu. P. Nikitin: *Sov. J. Nucl. Phys.*, **9**:348 (1969).

Bie 64 Bienlien, H., et al.: *Phys. Letters,* **13**:80(1964).

Bjo 64 Bjorken, J. D., and S. D. Drell: "Relativistic Quantum Mechanics," McGraw-Hill, New York, 1964.

Che 70 Chen, H.: *Phys. Rev. Letters,* **25**:768 (1970).

Cun 70 Cundy, D. C., et al.: *Phys. Letters,* **31B**:478 (1970).

Fei 64 Feinberg, G., and A. Pais: *Phys. Rev.,* **133**:B477 (1964).

Gel 69 Gell–Mann, M., M. Goldberger, N. M. Kroll, and F. E. Low: *Phys. Rev.,* **179**:1518 (1969).

Gla 67 Glashow, S. L., H. J. Schnitzer, and S. Weinberg: *Phys. Rev. Letters,* **19**:205 (1967).

Kum 65 Kummer, W., and G. Segrè: *Nucl. Phys.,* **64**:585 (1965).

Lee 57 Lee, T. D., and C. N. Yang: *Phys. Rev.,* **108**:1611 (1957). See also T. D. Lee and C. N. Yang: *Phys. Rev. Letters,* **4**:307 (1960); *Phys. Rev.,* **119**:1410 (1960).

Lee 61 Lee T. D., P. Markstein, and C. N. Yang: *Phys. Rev. Letters,* **7**:4 (1961).

Lee 64 Lee, T. D., and A. Sirlin: *Rev. Mod. Phys.,* **36**:666 (1964).

Lee 69 Lee, T. D.: *Topical Conf. Weak Interactions, CERN, Geneva,* 427 (1969).

Lee 71 Lee, T. D.: *Phys. Rev. Letters,* **26**:801 (1971).

Moh 68 Mohapatra, R. N., S. J. Rao, and R. E. Marshak: *Phys. Rev. Letters,* **20**:108 (1968).

Rei 70 Reines, F., and H. Gurr: *Phys. Rev. Letters,* **24**:1448 (1970).

Ste 70 Steiner, H.: *Phys. Rev. Letters,* **24**:746 (1970).

Sto 70 Stothers, R. B.: *Phys. Rev. Letters,* **24**:538 (1970).

Wei 67 Weinberg, S.: *Phys. Rev. Letters,* **19**:1264 (1967).

Wei 72 Weinberg, S.: *Phys. Rev.,* **5D**:1412 (1972).

4

GENERAL PROPERTIES OF HADRONIC WEAK CURRENTS

4.1 DECAY OF CHARGED PIONS: $\pi_{\mu 2}$, π_{e2}

A complete basic theory of hadronic interactions does not exist. Thus we are unable to calculate matrix elements of the hadronic weak currents from first principles. However, by exploiting kinematics and symmetry wherever possible, we can do a great deal that is useful. We begin our discussion with the simplest example of a weak interaction involving hadrons, namely, *charged pion decay.* The principal decay mode is $\pi_{\mu 2}$:

$$\pi^{\pm} \to \mu^{\pm} \nu_\mu(\bar{\nu}_\mu) \tag{4.1}$$

According to our phenomenological lagrangian, the amplitude for $\pi^- \to \mu^- \bar{\nu}_\mu$ must be of the form

$$\mathcal{M} = \frac{G}{\sqrt{2}} \bar{u}_\mu \gamma^\alpha (1 + \gamma^5) v_{\nu\mu} \langle 0 | \mathfrak{J}_{0\alpha}{}^\dagger | \pi^- \rangle \tag{4.2}$$

Now, the leptonic portion of $\mathcal{M}$ is a vector plus an axial vector. Therefore the factor $\langle 0 | \mathfrak{J}_{0\alpha}{}^\dagger | \pi^- \rangle$ must be a vector, an axial vector, or a combination of these; and it must be constructed from available kinematic quantities. However, the pion is a pseudoscalar particle. Therefore, in the present case, the only available vector is the momentum transfer q_α:

$$q_\alpha = k_\alpha + p_\alpha \tag{4.3}$$

where k_α and p_α are the neutrino and muon four-momenta, respectively. We conclude that

$$\langle 0 | \mathcal{J}_{0\alpha}{}^\dagger | \pi^- \rangle = i f_\pi q_\alpha \tag{4.4}$$

where f_π is some numerical constant, called the *pion decay constant*. A basic theory of hadronic interactions would provide the means for computing f_π. As it is, we must calculate the π decay rate in terms of f_π as a parameter.[1]

Note that since π^- is pseudoscalar and q_α is a polar vector, only the *axial* vector portion of $\mathcal{J}_{0\alpha}{}^\dagger$ is operative in Eq. (4.4):

$$\langle 0 | \mathcal{A}_{0\alpha}{}^\dagger | \pi^- \rangle = i f_\pi q_\alpha \tag{4.5}$$

Let us now substitute (4.4) into (4.2):

$$\begin{aligned}\mathcal{M} &= \frac{G}{\sqrt{2}} i f_\pi (p_\alpha + k_\alpha) \bar{u}_\mu \gamma^\alpha (1 + \gamma^5) v_{\nu\mu} \\ &= \frac{G}{\sqrt{2}} i f_\pi \bar{u}_\mu (\not{p} + \not{k})(1 + \gamma^5) v_{\nu\mu}\end{aligned}$$

This simplifies as follows:

$$\bar{u}_\mu \not{p}_\mu = m_\mu \bar{u}_\mu$$

and

$$\not{k}(1 + \gamma^5) v_{\nu\mu} = (1 - \gamma^5) \not{k} v_{\nu\mu} = 0$$

since $m_{\nu\mu} = 0$. Therefore $\mathcal{M}$ becomes

$$\mathcal{M} = \frac{G}{\sqrt{2}} i f_\pi m_\mu \bar{u}_\mu (1 + \gamma^5) v_{\nu\mu}$$

and we have

$$\begin{aligned}|\mathcal{M}|^2 &= \frac{G^2}{2} f_\pi{}^2 m_\mu{}^2 \,\mathrm{tr}[(\not{p} + m_\mu)(1 + \gamma^5) \not{k} (1 - \gamma^5)] \\ &= G^2 f_\pi{}^2 m_\mu{}^2 \,\mathrm{tr}[\not{p}\not{k}(1 + \gamma^5)] \\ &= 4 G^2 f_\pi{}^2 m_\mu{}^2 p \cdot k\end{aligned} \tag{4.6}$$

[1] It is possible to relate f_π to the pion-nucleon coupling strength via the Goldberger–Treiman relation. (See Sec. 4.8.)

To calculate the transition probability per unit time in the rest frame of the pion, we write

$$dW = \frac{4G^2 f_\pi^{\ 2} m_\mu^{\ 2} k \cdot p}{2m_\pi} \frac{d^3\mathbf{k}\, d^3\mathbf{p}}{(2\pi)^6 (2E)(2\omega)} (2\pi)^4\, \delta(m_\pi - E - \omega)\, \delta^3(\mathbf{k} + \mathbf{p})$$

where E and ω are, respectively, the muon and neutrino energies. Integrating over $d^3\mathbf{p}$, we have

$$W = \frac{2\pi G^2 f_\pi^{\ 2} m_\mu^{\ 2}}{(2\pi)^2 m_\pi} \int \frac{(E\omega + \omega^2)\omega^2\, d\omega}{E\omega} \delta[m_\pi - \omega - (\omega^2 + m_\mu^{\ 2})^{1/2}]$$

or

$$\boxed{W = \frac{G^2 f_\pi^{\ 2}}{8\pi} m_\mu^{\ 2} m_\pi \left(1 - \frac{m_\mu^{\ 2}}{m_\pi^{\ 2}}\right)^2} \tag{4.7}$$

We may solve this equation for f_π. Inserting the known values of m_μ, m_π, and G (the latter obtained from muon decay) and also the experimental values $W^{-1} = \tau = (2.604 \pm 0.007) \times 10^{-8}$ s, we obtain

$$f_\pi \simeq 0.93 m_\pi \tag{4.8}$$

We may easily obtain the transition rate for the decay $\pi^- \to e^- \bar{\nu}_e$ merely by substituting m_e for m_μ in formula (4.7). Thus we obtain the ratio

$$R_0 = \frac{W(\pi \to e\nu)}{W(\pi \to \mu\nu)} = \left(\frac{m_e}{m_\mu}\right)^2 \left(\frac{m_\pi^{\ 2} - m_e^{\ 2}}{m_\pi^{\ 2} - m_\mu^{\ 2}}\right)^2 \tag{4.9}$$

Actually, it is necessary to correct R_0 for radiative effects before this ratio can be compared to experiment. These radiative corrections have been computed by Berman (Ber 58) and Kinoshita (Kin 59), and they turn out to be surprisingly large. One finds that

$$R_{\text{corr}} = R_0(1 + \delta)(1 + \varepsilon) \tag{4.10}$$

where $\delta = -16.0 \dfrac{\alpha}{\pi}$ and $\varepsilon = -0.92 \dfrac{\alpha}{\pi}$

Thus

$$R_{\text{corr}} = 1.24 \times 10^{-4} \tag{4.11}$$

The best experimental value of R, obtained by De Capua et al. (DeCap 64), is

$$R_{\text{exp}} = (1.247 \pm 0.028)10^{-4} \tag{4.12}$$

The excellent agreement between (4.11) and (4.12) confirms that our hypothesis of electron-muon universality is reasonable.

What is the qualitative explanation for the very small $\pi \to e/\pi \to \mu$ branching ratio? Naively it would appear that, if anything, the ratio should be larger than unity because of electron-muon universality and the fact that more phase space is available for the electron than the muon. [In fact, if the effective hadronic weak current were pseudoscalar (the other a priori possibility) rather than axial vector, we would have $R \approx 5.5$.] However, since angular momentum and linear momentum are conserved and since the π has spin zero, the neutrino and charged lepton must be emitted in opposite directions and with opposite spins and therefore with the same helicities for $\pi_{\mu 2}$ or π_{e2} decay. Since we always have $h(\bar{\nu}) = +1$ in the VA theory, we therefore must have $h(l^-) = +1$ in π_{l2}^- decay. However, in the limit $v \to c$ a negative lepton must have $h = -1$ according to the VA theory. Also, the electron in π_{e2} decay is quite relativistic, whereas the muon in $\pi_{\mu 2}$ decay is not relativistic at all. Therefore the electron decay is strongly inhibited, whereas the muon is more easily "forced" into the wrong helicity state.

4.2 $K_{\mu 2}$ AND K_{e2} DECAYS

The decays

$$K^\pm \to \mu^\pm \nu_\mu(\bar{\nu}_\mu) \tag{4.13}$$

and

$$K^\pm \to e^\pm \nu_e(\bar{\nu}_e) \tag{4.14}$$

are somewhat similar to the decays $\pi_{\mu 2}$ and π_{e2}. Of course, K^- has strangeness $S = -1$, so for K_{l2}^- decay the relevant current is $\mathfrak{J}_{1\alpha}{}^\dagger$ instead of $\mathfrak{J}_{0\alpha}{}^\dagger$. Furthermore, the $K^\pm$, like the $\pi^\pm$, are pseudoscalar particles. Thus once more it is only the axial vector portion of the hadronic current that contributes to the amplitude. We have

$$\langle 0 | \mathcal{A}_{1\alpha}{}^\dagger | K^- \rangle = i f_K q_\alpha \tag{4.15}$$

where f_K is another constant not necessarily equal to f_π. Just as for $\pi_{\mu 2}$ decay, we find

$$W(K^\pm \to l^\pm \nu) = \frac{G^2 f_K{}^2 m_l{}^2 m_K}{8\pi} \left[1 - \frac{m_l{}^2}{m_K{}^2}\right]^2 \tag{4.16}$$

Comparing with π decay we have

$$\frac{W(K \to \mu\nu)}{W(\pi \to \mu\nu)} = \frac{f_K{}^2 m_K}{f_\pi{}^2 m_\pi} \frac{[1 - (m_\mu/m_K)^2]^2}{[1 - (m_\mu/m_K)^2]^2} \tag{4.17}$$

Inserting the known masses and the experimental values for $W(\pi \to \mu\nu)$ and

$$W(K \to \mu\nu) = 5.32 \times 10^7 \text{ s}^{-1}$$

we find

$$\frac{f_K}{f_\pi} = 0.2755 \pm 0.0008 \tag{4.18}$$

This ratio defines the *Cabibbo angle:*

$$\frac{f_K}{f_\pi} = \frac{\sin \Theta_A{}^M}{\cos \Theta_A{}^M} = \tan \Theta_A{}^M \tag{4.19}$$

Here, A stands for *axial* and M for *meson.* It should be noted that the Cabibbo angle may also be determined from the hyperon semileptonic decays, K_{e3} decay, and certain nuclear beta decays. From Eq. (4.16) we also obtain the branching ratio for K_{e2} decay. Including radiative corrections we have

$$R = \left[\frac{W(K^+ \to e^+\nu_e)}{W(K^+ \to \mu^+\nu_\mu)}\right]_{\text{theo}} = 2.09 \times 10^{-5} \tag{4.20}$$

The experimental values are in good agreement (Bot 67, 68; Mac 69), again in confirmation of $e\mu$ universality:

$$R_{\text{exp}} = 1.9^{+0.5}_{-0.4} \times 10^{-5} \tag{4.21}$$

4.3 GENERAL FORMS FOR MORE COMPLEX MATRIX ELEMENTS

We turn next to the problem of constructing matrix elements for the following types of weak decays.

1 Meson → meson + leptons, for example,

$$\pi^+ \to \pi^0 e^+ \nu_e$$

$$K^+ \to \pi^0 \mu^+ \nu_\mu$$

2 Meson → 2 mesons + leptons, for example,

$$K^+ \to \pi^+ \pi^- e^+ \nu_e$$

3 Baryon → baryon + leptons, for example,

$$n \to p e^- \bar{\nu}_e$$

$$\Lambda^0 \to p e^- \bar{\nu}_e$$

Let us consider each of these cases in turn.

4.3.1 Meson $(a) \rightarrow$ Meson $(b)+$ Leptons

The hadronic portion of the matrix element must be formed from the available vectors in the decay, and it must itself be a vector, an axial vector, or a linear combination of the two. In the present case the only four-vectors available are the four-momenta of the initial and final mesons: k_a and k_b. We cannot construct an axial vector from these alone. Thus the matrix element is of the form[1]

$$\langle M_b | \mathfrak{J}_0{}^\alpha \text{ or } \mathfrak{J}_1{}^\alpha | M_a \rangle = (\cos\Theta \text{ or } \sin\Theta)(f_a k_a{}^\alpha + f_b k_b{}^\alpha) \qquad (4.22)$$

where f_a and f_b are form factors, i.e., scalar functions of the various kinematic invariants involved in the transition. In the present case, these are $k_a{}^2$, $k_b{}^2$, and $(k_a - k_b)^2 = q^2 =$ momentum transfer. Actually, however, $k_a{}^2 = m_a{}^2$ and $k_b{}^2 = m_b{}^2$ are merely constants; thus

$$f_a = f_a(q^2) \qquad f_b = f_b(q^2)$$

In all cases of interest here, the initial and final mesons have the same intrinsic parity. Therefore, only the *vector* portion of the weak hadronic current is operative:

$$\langle M_b | \mathcal{V}_0{}^\alpha \text{ or } \mathcal{V}_1{}^\alpha | M_a \rangle = (\cos\Theta \text{ or } \sin\Theta)[f_a(q^2) k_a{}^\alpha + f_b(q^2) k_b{}^\alpha]$$

Frequently it is more convenient to write

$$\langle M_b | \mathcal{V}_0{}^\alpha \text{ or } \mathcal{V}_1{}^\alpha | M_a \rangle = (\cos\Theta \text{ or } \sin\Theta)[f_+(q^2)(k_a{}^\alpha + k_b{}^\alpha) + f_-(q^2)(k_a{}^\alpha - k_b{}^\alpha)] \qquad (4.23)$$

where $f_+(q^2), f_-(q^2)$ are new form factors (linear combinations of f_a and f_b).

4.3.2 Meson $(a) \rightarrow$ Meson $(b)+$ Meson $(c)+$ Leptons

This time there are three vectors k_a, k_b, and k_c. From these we can also form an axial vector:

$$\varepsilon_{\lambda\phi\sigma\delta} k_a{}^\phi k_b{}^\sigma k_c{}^\delta \qquad (4.24)$$

Therefore, the most general matrix elements are

$$\langle M_b M_c | \mathcal{A}_1{}^\alpha | M_a \rangle = \sin\Theta[c_1 k_a{}^\alpha + c_2 k_b{}^\alpha + c_3 k_c{}^\alpha] \qquad (4.25)$$

$$\langle M_b M_c | \mathcal{V}_1{}^\alpha | M_a \rangle = \sin\Theta[c_4 \varepsilon^{\alpha\phi\sigma\delta} k_{a\phi} k_{b\sigma} k_{c\delta}] \qquad (4.26)$$

[1] From now on, we make explicit the dependence of the hadronic amplitudes on Θ.

In Eq. (4.25) only operator $\mathcal{A}_1{}^\alpha$ appears because in all cases of interest mesons a, b, and c all have the same intrinsic parity (odd) and also there is always a strangeness change (that is, $K^+ \to \pi^+\pi^- e^+ \nu_e$). Similarly, only $\mathcal{V}_1{}^\alpha$ appears in (4.26). The quantities c_1, c_2, c_3, and c_4 are form factors.

4.3.3 Baryon (B)→ Baryon (B')+ Leptons

Here we can construct vectors and axial vectors from the momentum transfer $q = p_B - p_{B'}$, the baryon spinors, and the γ matrices. The quantities

$$
\begin{aligned}
&\bar{u}(B')\gamma^\alpha u(B) \\
&\bar{u}(B')\sigma^{\alpha\nu} q_\nu u(B) \qquad (4.27)\\
&\bar{u}(B')u(B)q^\alpha
\end{aligned}
$$

are all four-vectors. In fact, they are the only three independent four-vectors that can be constructed. All others may be shown to reduce to a linear combination of these three by suitable application of the Dirac equation. Similarly, the only three axial vectors we can form are

$$
\begin{aligned}
&\bar{u}(B')\gamma^\alpha\gamma^5 u(B) \\
&\bar{u}(B')\sigma^{\alpha\nu} q_\nu \gamma^5 u(B) \qquad (4.28)\\
&\bar{u}(B')\gamma^5 u(B)q^\alpha
\end{aligned}
$$

The intrinsic parities of p, n, and Λ are chosen as positive by convention, and Σ is found by experiment to have the same parity. Therefore in all baryon semileptonic decays involving these particles in initial or final states, the most general hadron matrix element is $\mathcal{M} = \mathcal{M}_V{}^\alpha + \mathcal{M}_A{}^\alpha$:

$$
\begin{aligned}
\mathcal{M}_V{}^\alpha &= \begin{pmatrix}\cos\Theta\\ \sin\Theta\end{pmatrix} \bar{u}(B')[f_1(q^2)\gamma^\alpha + if_2(q^2)\sigma^{\alpha\nu}q_\nu + f_3(q^2)q^\alpha]u(B) \\
\mathcal{M}_A{}^\alpha &= \begin{pmatrix}\cos\Theta\\ \sin\Theta\end{pmatrix} \bar{u}(B')[g_1(q^2)\gamma^\alpha\gamma^5 - ig_2(q^2)\sigma^{\alpha\nu}\gamma^5 q_\nu + g_3(q^2)\gamma^5 q^\alpha]u(B)
\end{aligned} \qquad (4.29)
$$

In Eqs. (4.29) the vector form factors f_1, f_2, and f_3 and the axial vector form factors g_1, g_2, and g_3 are all functions of q^2. (Of course, each of these form factors also depends on which baryons B, B' participate in the interaction.)

In the remainder of this chapter, we shall see how invariance under time reversal, the *conserved vector current hypothesis*, the *Goldberger–Treiman relation*, and invariance under *G-parity* transformations restrict the possible values of the six functions $f_1, f_2, f_3, g_1, g_2, g_3$. Further restrictions obtained from the Cabibbo hypothesis will be discussed in Chap. 8.

4.4 RESTRICTIONS ON HADRONIC MATRIX ELEMENTS IMPOSED BY TIME-REVERSAL INVARIANCE

Let us consider a reaction of the general form

$$A(\mathbf{p}_i, \mathbf{s}_i) \to B(\mathbf{p}_f, \mathbf{s}_f) \qquad \text{amplitude } \mathcal{M}$$

where A is some initial state consisting of one or more particles and B is the final state, again consisting of one or more particles. This transition is characterized by some definite amplitude $\mathcal{M}$:

$$\mathcal{T}_{fi} = \langle f|\mathcal{T}|i\rangle = \frac{(2\pi)^4\,\delta^4(p_f - p_i)\mathcal{M}}{\sqrt{\prod(2E_i V)\prod_f(2E_f V)}} \tag{4.30}$$

We now reverse all spins and momenta and interchange initial and final states. The amplitude for this *time-reversed* process will be called $\mathcal{M}'$:

$$B(-\mathbf{p}_f, -\mathbf{s}_f) \to A(-\mathbf{p}_i, -\mathbf{s}_i) \qquad \text{amplitude } \mathcal{M}'$$

Time-reversal (T) invariance holds if $|\mathcal{M}'| = |\mathcal{M}|$.

For certain transitions involving strong or electromagnetic interactions, it is feasible to test for T invariance directly by comparing the forward and inverse reaction rates, e.g.,

$$n + p \rightleftharpoons d + \gamma$$

Such detailed balance comparisons are obviously a practical impossibility for weak interactions, however, and in the latter case we are forced to rely on a more indirect argument, which depends on the fact that in a *weak* transition the operator $\mathcal{T}$ appearing in (4.30) is hermitian: $\mathcal{T}^\dagger = \mathcal{T}$.

To illustrate, we consider neutron beta decay with the initial neutron at rest and spin in a definite direction (see Fig. 4.1*a*) and the final particle momenta also specified. In our hypothetical experiment we do not observe the final spins, and these may be disregarded. The amplitude for this decay is $\mathcal{M}$, as in Eq. (4.30). Now we take the complex conjugate of both sides of Eq. (4.30) and, since $\mathcal{T} = \mathcal{T}^\dagger$,

$$\langle f|\mathcal{T}|i\rangle^* = \langle i|\mathcal{T}|f\rangle \tag{4.31}$$

Physically we have interchanged initial and final (plane-wave) states without altering spins and momenta, and the new process is described by amplitude $\mathcal{M}^*$ (see Fig. 4.1*b*). Next, we perform a time-reversal transformation on the system of Fig. 4.1*b*. This has the effect of reversing spins and momenta and

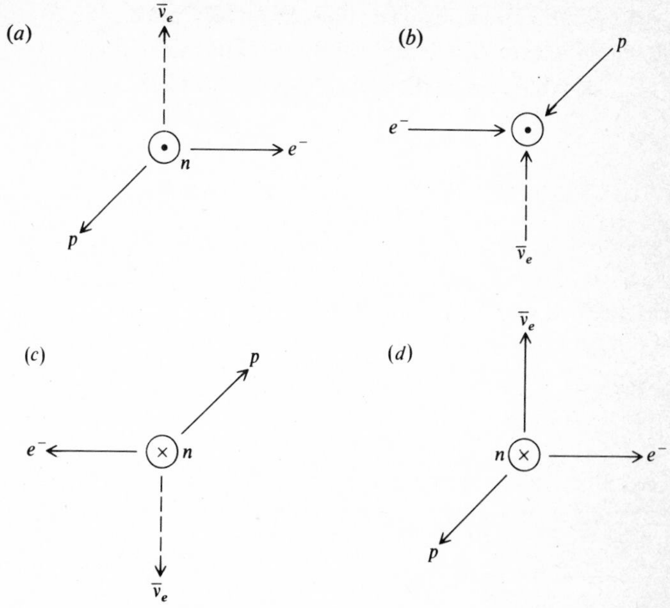

FIGURE 4.1
(*a*) Original state. Amplitude $\mathcal{M}$. Vectors $p, e^-, \bar{\nu}_e$ in plane of page. Neutron spin out of page. (*b*) Initial and final states interchanged because of hermitian conjugation. Amplitude $\mathcal{M}^*$. (*c*) Spins and momenta reversed and initial and final states again interchanged because of time reversal on (*b*). Amplitude $\mathcal{M}'$. (*d*) Rotation of final state in (*c*) by 180° about neutron-spin axis. Amplitude $\mathcal{M}'$. The final state is identical with that of (*a*) except for neutron-spin flip. If time reversal invariance holds, $\mathcal{M}' = \mathcal{M}^*$.

interchanging initial and final states once again. The result (Fig. 4.1*c*) is described by some new amplitude $\mathcal{M}'$; and after a rotation of 180° (Fig. 4.1*d*), it is seen to be equivalent to the configuration of Fig. 4.1*a*, except that the neutron spin is reversed. If $|\mathcal{M}'| \neq |\mathcal{M}^*|$, T invariance is violated and the probabilities for decays with $\boldsymbol{\sigma}_n \cdot \mathbf{p}_e \times \mathbf{k}_\nu > 0$ and $\boldsymbol{\sigma}_n \cdot \mathbf{p}_e \times \mathbf{k}_\nu < 0$ are unequal. The test for T invariance thus amounts to a search for the existence of a *triple* correlation such as $\boldsymbol{\sigma}_n \cdot \mathbf{p}_e \times \mathbf{p}_\nu$ in neutron beta decay or, by an analogous argument, $\boldsymbol{\sigma}_\mu \cdot \mathbf{p}_\pi \times \mathbf{p}_\mu$ in $K_{\mu 3}$ decay, etc.

It is important to point out that the simple argument given above is valid only in the absence of final-state interactions, such as the coulomb scattering of electron and proton after neutron decay. It can be seen that these are not accountable for by the simple plane-wave description, and their inclusion introduces phase shifts which can masquerade as violation of T invariance.

We now seek to determine the conditions on the form factors $f_1, f_2, f_3, g_1, g_2, g_3$ which yield $\mathcal{M}' = \mathcal{M}^*$ for neutron beta decay. As shown in Appendix 3, under Time Reversal, a Dirac spinor u is transformed to $i\gamma^1\gamma^3 u^*$:

$$u(p, s) \rightarrow i\gamma^1\gamma^3 u^*(p, s)$$

Thus the bilinear form $\bar{u}_b F u_a$, where F is some 4×4 matrix, transforms as follows:

$$\bar{u}_a(p, s)Fu_b(p, s) \rightarrow \bar{u}_b(p, s)\gamma^0\gamma^1\gamma^3\tilde{F}\gamma^3\gamma^1\gamma^0 u_a(p, s)$$

We thus find that under T

$$\mathcal{M} = \frac{G \cos \Theta}{\sqrt{2}} \bar{u}(p)[f_1\gamma^\alpha + if_2\sigma^{\alpha\nu}q_\nu + f_3 q^\alpha + g_1\gamma^\alpha\gamma^5$$
$$- ig_2\sigma^{\alpha\nu}q_\nu\gamma^5 + g_3 q^\alpha\gamma^5]u(n)\bar{u}_e\gamma_\alpha(1 + \gamma^5)v_{\bar{\nu}e}$$

transforms to

$$\mathcal{M}' = \frac{G \cos \Theta}{\sqrt{2}} \bar{u}(n)[f_1\gamma^\alpha - if_2\sigma^{\alpha\nu}q_\nu + f_3 q^\alpha + g_1\gamma^\alpha\gamma^5$$
$$- ig_2\sigma^{\alpha\nu}q_\nu\gamma^5 - g_3 q^\alpha\gamma^5]u(p)\bar{v}_{\bar{\nu}e}\gamma_\alpha(1 + \gamma^5)u_e$$

However, the hermitian conjugate of $\mathcal{M}$ is

$$\mathcal{M}^* = \frac{G \cos \Theta}{\sqrt{2}} \bar{u}(n)[f_1^*\gamma^\alpha - if_2^*\sigma^{\alpha\nu}q_\nu + f_3^* q^\alpha + g_1^*\gamma^\alpha\gamma^5$$
$$- ig_2^*\sigma^{\alpha\nu}q_\nu\gamma^5 - g_3^* q^\alpha\gamma^5]u(p)\bar{v}_{\bar{\nu}e}\gamma_\alpha(1 + \gamma^5)u_e$$

Clearly, time reversal and hermitian conjugation are completely equivalent if the form factors $f_{1,2,3}$, $g_{1,2,3}$ are all *real*. A similar statement applies to meson weak decays.

4.5 THE CONSERVED VECTOR CURRENT HYPOTHESIS

Let us digress momentarily to consider the interaction of a baryon (e.g., proton) with an electromagnetic field. The hamiltonian density is

$$\mathcal{H}(x) = + ej_{\mathrm{EM}}{}^\alpha(x)A^\alpha(x) \qquad (4.32)$$

where $j_{\mathrm{EM}}{}^\alpha(x)$ is the electromagnetic current operator, a four-vector, and $A_\alpha(x)$ is the electromagnetic field four-vector potential. We shall need to develop an expression for the matrix element of $j_{\mathrm{EM}}{}^\alpha(x)$ between initial and final proton states $|p, s\rangle$ and $|p', s'\rangle$. If the proton had no strong interactions and could be regarded as a point particle like the electron or muon, then this matrix element would simply be

$$\langle p's' | j_{\text{EM}}{}^{\alpha}(x) | p, s\rangle = e^{i(p'-p)\cdot x}\bar{u}(p', s')\gamma^{\alpha}u(p, s) \tag{4.33}$$

However, because of strong interactions the matrix element is not given by (4.33) but instead includes momentum-transfer-dependent terms. Since $j_{EM}{}^{\alpha}$ is a polar vector, the most general matrix element is

$$\langle p', s' | j_{\text{EM}}{}^{\alpha} | p, s\rangle = e^{i(p'-p)\cdot x}\bar{u}(p', s')[C_p(q^2)\gamma^{\alpha} + iM_p(q^2)\sigma^{\alpha\nu}q_\nu + F_3(q^2)q^{\alpha}]u(p, s) \tag{4.34}$$

where C_p, M_p, and F_3 are form factors to be determined.

The electromagnetic current is conserved, however:

$$\frac{\partial j_{\text{EM}}{}^{\alpha}(x)}{\partial x^{\alpha}} = 0 \tag{4.35}$$

and this has important implications for the matrix element (4.34). We take the four-divergence of both sides of (4.34) and obtain

$$0 = \bar{u}(p', s')[C_p(q^2)(\not{p}' - \not{p}) - iM_p(q^2)\sigma^{\alpha\nu}q_\nu q_\alpha - F_3(q^2)q^2]u(p, s) \tag{4.36}$$

However,

$$\bar{u}(p', s')(\not{p}' - \not{p})u(p, s) = (m_p - m_p)\bar{u}(p's')u(p, s) = 0$$

and

$$\sigma^{\alpha\nu}q_\nu q_\alpha = 0$$

since $\sigma^{\alpha\nu}$ is an antisymmetric tensor. Therefore,

$$q^2F_3(q^2) = 0$$

or

$$F_3(q^2) = 0 \tag{4.37}$$

We have thus shown that conservation of the EM current leads to $F_3 = 0$, and our matrix element is now reduced to

$$\mathcal{M}_p = \langle p's' | j_{\text{EM}}{}^{\alpha} | ps\rangle = e^{-iq\cdot x}\bar{u}_p(p', s')[C_p(q^2)\gamma^{\alpha} + iM_p(q^2)\sigma^{\alpha\nu}q_\nu]u_p(p, s) \tag{4.38}$$

where, for time-reversal invariance, $C_p(q^2)$ and $M_p(q^2)$ are real. In order that expression (4.38) will correctly describe the interaction of a proton with a static external field, we must require that

$$\begin{aligned} C_p(0) &= 1 \\ eM_p(0) &= \frac{g-2}{2}\mu_0 = 1.79\frac{e\hbar}{2m_pc} = \frac{1.79e}{2} \end{aligned} \tag{4.39}$$

Then, the term $C_p(0)\bar{u}'\gamma^{\alpha}u$ contains the interaction of the proton charge with the external static electric field plus the interaction of the normal (Dirac) portion of the spin magnetic moment with the external magnetic field, while the term in $M_p(0)$ accounts for the anomalous portion of the magnetic moment, as indicated in (4.39).

In exactly the same fashion we can construct a matrix element for the electromagnetic current between two neutron states:

$$\mathcal{M}_n = e^{-iq\cdot x}\bar{u}_n(p', s')[C_n(q^2)\gamma^\alpha + iM_n(q^2)\sigma^{\alpha\nu}q_\nu]u_n(p, s) \qquad (4.40)$$

Of course, the neutron has no charge and all of its magnetic moment is anomalous; therefore

$$\begin{aligned} C_n(0) &= 0 \\ M_n(0) &= -\frac{1.91}{2} \end{aligned} \qquad (4.41)$$

We may go one step further and unite the two matrix elements $\mathcal{M}_p$, $\mathcal{M}_n$ with the aid of the notion of isotopic spin. We regard the neutron and proton as $T_3 = -\frac{1}{2}$ and $+\frac{1}{2}$ components, respectively, of the nucleon isodoublet. The nucleon is then described by a spinor u (with no suffix p or n) which contains a discrete charge variable as well as the usual momentum and spin variables.

Introducing the 2 × 2 Pauli matrices for isospin

$$\begin{aligned} \tau_+ &= \tfrac{1}{2}(\tau_x + i\tau_y) \\ \tau_- &= \tfrac{1}{2}(\tau_x - i\tau_y) \\ \tau_3 &= \tau_z \end{aligned}$$

with commutation relations

$$\begin{aligned} [\tau_+, \tau_-] &= \tau^3 \\ \left[\tau_+, \frac{\tau_3}{2}\right] &= -\tau_+ \\ \left[\tau_-, \frac{\tau_3}{2}\right] &= +\tau_- \end{aligned} \qquad (4.42)$$

we recover the spinors u_p and u_n from u with the aid of the projection operators $(1 \pm \tau^3)/2$:

$$\begin{aligned} \tfrac{1}{2}(1 + \tau_3)u &= u_p \\ \tfrac{1}{2}(1 - \tau_3)u &= u_n \end{aligned} \qquad (4.43)$$

Then, defining

$$\begin{aligned} F_S(q^2) &= \tfrac{1}{2}[C_p(q^2) + C_n(q^2)] \\ F_V(q^2) &= \tfrac{1}{2}[C_p(q^2) - C_n(q^2)] \\ M_S(q^2) &= \tfrac{1}{2}[M_p(q^2) + M_n(q^2)] \\ M_V(q^2) &= \tfrac{1}{2}[M_p(q^2) - M_n(q^2)] \end{aligned} \qquad (4.44)$$

it is easy to see that

$$\mathcal{M}_{EM} = e^{-iq\cdot x}\bar{u}'\{[F_S(q^2)\gamma^\alpha + iM_S(q^2)\sigma^{\alpha\nu}q_\nu] + [F_V(q^2)\gamma^\alpha + iM_V(q^2)\sigma^{\alpha\nu}q_\nu]\tau_3\}u \tag{4.45}$$

may be used in place of $\mathcal{M}_p$ (4.38) *or* $\mathcal{M}_n$ (4.40). The symbols S and V here refer to *isoscalar* and *isovector*, respectively.

Let us now return to weak interactions. We know the matrix element of the strangeness-conserving vector hadronic weak current $\mathcal{V}_0^{\alpha\dagger}$ taken between neutron and proton states:

$$\begin{aligned}\langle p|\mathcal{V}_0^{\alpha\dagger}|n\rangle &= \cos\Theta\langle p|j_1^\alpha + ij_2^\alpha|n\rangle \\ &= \cos\Theta e^{-iq\cdot x}\bar{u}_p[f_1(q^2)\gamma^\alpha + if_2(q^2)\sigma^{\alpha\nu}q_\nu + f_3(q^2)q^\alpha]u_n\end{aligned} \tag{4.46}$$

or, in terms of the spinors u, $\bar{u}'$,

$$\langle p|j_1^\alpha + ij_2^\alpha|n\rangle = e^{-iq\cdot x}\bar{u}'[f_1(q^2)\gamma^\alpha + if_2(q^2)\sigma^{\alpha\nu}q_\nu + f_3(q^2)q^\alpha]\tau_+ u \tag{4.47}$$

We take the limit of the isovector portion of j_{EM} and of $\langle p|j_1^\alpha + ij_2^\alpha|n\rangle$ as $q^2 \to 0$:

$$\begin{aligned}\lim_{q^2\to 0} j_{EM,\,\text{isov}} &= \bar{u}'F_V(0)\gamma^\alpha\tau_3 u \\ &= \tfrac{1}{2}\bar{u}'\gamma^\alpha[C_p(0) - C_n(0)]\tau_3 u \\ &= C_p(0)\left[\bar{u}'\gamma^\alpha\left(\frac{\tau_3}{2}\right)u\right]\end{aligned} \tag{4.48}$$

$$\lim_{q^2\to 0}\langle p|j_1^\alpha + ij_2^\alpha|n\rangle = \bar{u}'\gamma^\alpha f_1(0)\tau^+u = C_V^\beta[\bar{u}'\gamma^\alpha\tau^+u] \tag{4.49}$$

Here, the quantity $f_1(0)$ is called the *vector coupling constant* C_V^β in neutron (or nuclear) beta decay.

Now, it is a remarkable experimental fact that $C_V^\beta \approx 1$. This is the same value, essentially, as the vector coupling constant in muon decay, where there is no strong interaction. The weak coupling constant C_V^β is thus unmodified (*unrenormalized*) by strong interactions. Another quantity that is similarly unrenormalized is the proton electric charge, which has precisely the same value as the charge of μ^+ or e^+. This is expressed by the condition $C_p(0) = 1$. Thus we have a striking analogy between the matrix elements of $j_{EM,\,\text{isov}}$ and $j_1 + ij_2$, both taken between nucleon states, at zero momentum transfer:

$$\lim_{q^2\to 0}\langle p's'|j_{EM,\,\text{isov}}^\alpha|ps\rangle = \bar{u}'\gamma^\alpha\frac{\tau_3}{2}u \tag{4.50}$$

$$\lim_{q^2\to 0}\langle p's'|j_1^\alpha + ij_2^\alpha|ps\rangle = \bar{u}'\gamma^\alpha\tau^+u \tag{4.51}$$

Equations (4.50) and (4.51) suggest an important extrapolation—*that* $j_1^\alpha - ij_2^\alpha$, $j_1^\alpha + ij_2^\alpha$, and $j_{EM,\mathrm{isov}}^\alpha$ *are members of a single isotriplet vector of current operators.* This is called the *conserved vector current hypothesis*, or CVC (although a better name might be the *isotriplet vector hypothesis*), because with this assumption

$$\frac{\partial j_{\mathrm{EM}}^\alpha}{\partial x^\alpha} = 0$$

implies $\partial \mathcal{V}_0^\alpha / \partial x^\alpha$ also. In turn, this implies that $f_3(q^2)$ in Eq. (4.46) is zero. Furthermore, the CVC hypothesis implies that

$$\begin{aligned} f_1(q^2) &= F_V(q^2) \\ f_2(q^2) &= M_V(q^2) \end{aligned} \tag{4.52}$$

for all q^2 and in particular that

$$f_2(0) = M_V(0) = \tfrac{1}{2}[\mu_{\mathrm{proton}} - \mu_{\mathrm{neutron}} - \mu_0] \tag{4.53}$$

where μ_0 is the nuclear Bohr magneton.

We have thus obtained a most remarkable connection between the weak vector amplitude in neutron decay and the magnetic moments of the initial and final baryons. This is discussed in further detail in Chap. 5. For now, however, we turn to another example which corroborates the CVC hypothesis: *pion beta decay*.

4.6 PION BETA DECAY

Let us consider the matrix element between two $\pi^\pm$ states of the EM current $j_{\mathrm{EM}}^\alpha(x)$. We know that such a matrix element must take the general form

$$\langle \pi^\pm(p')| j_{\mathrm{EM}}^\alpha(x) | \pi^\pm(p)\rangle = \pm [F_1(q^2)(p' + p)^\alpha + F_2(q^2)(p' - p)^\alpha] e^{-iq\cdot x} \tag{4.54}$$

where $q = p - p'$ and F_1 and F_2 are form factors. Now

$$\frac{\partial j_{\mathrm{EM}}^\alpha(x)}{\partial x^\alpha} = 0$$

Therefore, taking the four-divergence of both sides of (4.54), we obtain

$$0 = F_1(q^2)(p'^2 - p^2) + F_2(q^2)(p' - p)^2$$

However, since $p'^2 = p^2 = m_\pi^2$, we have $F_2(q^2) = 0$. Therefore,

$$\mathcal{M}_{\mathrm{EM}}^\alpha = \langle \pi^\pm(p')| j_{\mathrm{EM}}^\alpha(x) | \pi^\pm(p)\rangle = \pm F_1(q^2)(p'^\alpha + p^\alpha) e^{-iq\cdot x} \tag{4.55}$$

This can be expressed in terms of isospin. We write a three-component isospinor ϕ for the initial pion and another ϕ' for the final pion. Then

$$\mathcal{M}_{EM}{}^{\alpha} = F_1(q^2)(p'^{\alpha} + p^{\alpha})(\phi'^{\dagger} I_3 \phi)e^{-iq\cdot x} \tag{4.56}$$

where
$$I_3 = \begin{pmatrix} 1 & 0 & 0 \\ 0 & 0 & 0 \\ 0 & 0 & -1 \end{pmatrix}$$

According to the CVC hypothesis, we can then write for the weak interaction matrix element

$$\langle \pi(p') | \mathcal{V}_0{}^{\mu} | \pi(p) \rangle = \cos\Theta F_1(q^2)(p'^{\mu} + p^{\mu})\phi'^{\dagger} I_- \phi e^{-iq\cdot x} \tag{4.57}$$

where $I_{\pm} = I_x \pm iI_y$. In particular, for $\pi(p') = \pi^0$, $\pi(p) = \pi^+$, we have

$$\langle \pi^0 | \mathcal{V}_0{}^{\mu} | \pi^+ \rangle = \sqrt{2}\cos\Theta F_1(q^2)(p'^{\mu} + p^{\mu})e^{-iq\cdot x} \tag{4.58}$$

[the factor $\sqrt{2}$ comes from $I_- \begin{pmatrix} 1 \\ 0 \\ 0 \end{pmatrix} = \sqrt{2}\begin{pmatrix} 0 \\ 1 \\ 0 \end{pmatrix}$]. Now, we are really interested in expression (4.58) in the limit where $q^2 \to 0$ since the momentum transfer in pion beta decay is very small: $m_{\pi^+} - m_{\pi^0} = 4.60$ MeV/c^2 = 9.015 m_e. Thus, we want to evaluate $F_1(0)$. But, it is easy to show, by comparing (4.56) to the ordinary expression for the interaction of a charged pion with a static electric field, that $F_1(0) = 1$.

Therefore, $\lim_{q^2\to 0} \langle \pi^0 | \mathcal{V}_0{}^{\alpha} | \pi^+ \rangle = \sqrt{2}(p'^{\mu} + p^{\mu})\cos\Theta$, and we have for the entire weak amplitude

$$\mathcal{M}(\pi^+ \to \pi^0 e^+ \nu_e) = \frac{G\cos\Theta}{\sqrt{2}}[\bar{u}_{\nu_e}\gamma_{\alpha}(1+\gamma^5)v_e]\sqrt{2}(p'^{\mu} + p^{\mu}) \tag{4.59}$$

Because $\Delta = m_{\pi^+} - m_{\pi^0} = 4.60$ MeV/$c^2 \ll m_{\pi^0} = 135$ MeV/c^2, the kinetic energy of the emitted π^0 is very small and may be neglected. Thus, to a good approximation we can write, for the π^+ rest frame,

$$(p_{\pi^+} + p_{\pi^0})^0 = 2m_{\pi}$$
$$(p_{\pi^+} + p_{\pi^0})^i = 0 \qquad i = 1, 2, 3$$

Thus $\mathcal{M}$ becomes

$$\mathcal{M} = 2G\cos\Theta\bar{u}_{\nu}\gamma^0(1+\gamma^5)v_e \cdot m_{\pi}$$

Summing over e^+ spins, we have

$$\begin{aligned}\overline{|\mathcal{M}|^2} &= 4G^2\cos^2\Theta m_{\pi}{}^2 \operatorname{tr}[\not{p}_{\nu}\gamma^0(1+\gamma^5)(\not{p}_e - m_e)\gamma^0(1+\gamma^5)] \\ &= 8G^2\cos^2\Theta m_{\pi}{}^2 p_{\nu\alpha}p_{e\beta}\operatorname{tr}[\gamma^{\alpha}\gamma^0\gamma^{\beta}\gamma^0(1+\gamma^5)]\end{aligned}$$

Now

$$\mathrm{tr}[\gamma^\alpha\gamma^0\gamma^\beta\gamma^0(1+\gamma^5)] = 4\chi^{\alpha\beta 00}$$
$$= 4[2g^{\alpha 0}g^{\beta 0} - g^{\alpha\beta}g^{00}]$$

Therefore

$$|\overline{\mathcal{M}}|^2 = 32G^2\cos^2\Theta m_\pi{}^2[2p_\nu{}^0 p_e{}^0 - p_\nu p_e]$$
$$= 32G^2\cos^2\Theta m_\pi{}^2[E_e E_\nu + \mathbf{p}_e\cdot\mathbf{p}_\nu]$$
$$= 32G^2\cos^2\Theta m_\pi{}^2 E_e E_\nu[1 + \mathbf{v}_e\cdot\hat{\mathbf{p}}_\nu] \tag{4.60}$$

where $\mathbf{v}_e$ is the electron velocity and $\hat{\mathbf{p}}_\nu$ is a unit vector in the direction of neutrino momentum. To compute the total rate for π^+ beta decay we need $|\overline{\mathcal{M}}|^2$ integrated over all angles, and so the term in $\mathbf{v}_e\cdot\hat{\mathbf{p}}_\upsilon$ contributes nothing. Therefore, the total rate is

$$W = \frac{32G^2\cos^2\Theta}{(2\pi)^5}\frac{m_\pi{}^2}{2m_\pi}\int\frac{d^3\mathbf{p}_e\,d^3\mathbf{p}_\nu\,d^3\mathbf{p}_0\,\delta^4(p_+ - p_0 - p_e - p_\nu)}{2E_e 2E_\nu 2E_0}E_e E_\nu \tag{4.61}$$

where the subscript 0 refer to π_0.

According to our previous approximation $E_0 = m_\pi$. Thus

$$W = \frac{2G^2\cos^2\Theta}{(2\pi)^5}\int d^3\mathbf{p}_e\,d^3\mathbf{p}_\nu\,d^3\mathbf{p}_0\,\delta^3(\mathbf{p}_0 + \mathbf{p}_e + \mathbf{p}_\nu)\,\delta(\Delta - E_e - E_\nu)$$

However,

$$\int d^3\mathbf{p}_0\,\delta^3(\mathbf{p}_0 + \mathbf{p}_e + \mathbf{p}_\nu) = 1$$

and

$$\int d^3\mathbf{p}_\nu\,\delta(\Delta - E_e - E_\nu) = 4\pi(\Delta - E_e)^2$$

Thus

$$W = \frac{8\pi G^2\cos^2\Theta}{(2\pi)^5}\int_0^{p_{e,\max}} 4\pi p_e{}^2\,dp_e(\Delta - E_e)^2 \tag{4.62}$$

Neglecting the electron mass ($\Delta = 9.015\ m_e$), we find

$$W \approx \frac{32\pi^2 G^2\cos^2\Theta}{(2\pi)^5}\int_0^\Delta E_e{}^2(\Delta - E_e)^2\,dE_e = \frac{G^2\Delta^5}{30\pi^3}\cos^2\Theta \tag{4.63}$$

If we had not neglected the electron mass, we would have obtained

$$W \approx \frac{G^2\cos^2\Theta\,\Delta^5}{30\pi^3}\left(1 - \frac{3}{2}\frac{\Delta}{m_{\pi^+}} + \frac{5m_e{}^2}{\Delta^2} + \cdots\right) \tag{4.64}$$

Inserting numerical values, we obtain

$$W_{\text{theo}} = 0.393^{-1} \tag{4.65}$$

Experimentally the following result is obtained (DePom 68):

$$\frac{R(\pi^+ \to \pi^0 e^+ \nu)}{R(\mu^+ \to \mu^+ \nu)} = 1.00\,^{+0.08}_{-0.10} \times 10^{-8}$$

which implies

$$W_{\text{exp}}(\pi^+ \to \pi^0 e^+ \nu) = 0.38^{+0.03}_{-0.40}\ \text{s}^{-1} \tag{4.66}$$

The excellent agreement between (4.65) and (4.66) confirms the CVC hypothesis.

4.7 SECOND-CLASS CURRENTS AND G PARITY; CHARGE SYMMETRY

A *G-parity transformation* is compounded of charge conjugation and a rotation of 180° in isospin space about the I_2 axis. The strong interaction is separately invariant under charge conjugation and arbitrary isospin rotations, and it is therefore G invariant.

What implications does G invariance of the strong interaction have for the weak interaction? Let us suppose that it makes sense to think of "bare" hadrons which have weak interactions but no strong interactions. The latter interaction is supposed to be *turned off* for the moment. In this case the $n \to p$ vector and axial vector matrix elements would take the following form:

$$(\mathcal{M}_V)_0 = \bar{u}_p \gamma^\alpha u_n \tag{4.67}$$

$$(\mathcal{M}_A)_0 = \bar{u}_p \gamma^\alpha \gamma^5 u_n \tag{4.68}$$

i.e., the neutron and proton would enter into the weak interaction exactly as do their neutral and charged leptonic counterparts. If the strong interaction is now *turned on*, the matrix elements of (4.67) and (4.68) become

$$(\mathcal{M}_V)_0 \to \mathcal{M}_V = \bar{u}_p(f_1\gamma^\alpha + if_2\,\sigma^{\alpha\beta}q_\beta + f_3\,q^\alpha)u_n \tag{4.69}$$

$$(\mathcal{M}_A)_0 \to \mathcal{M}_A = \bar{u}_p(g_1\gamma^\alpha\gamma^5 - ig_2\,\sigma^{\alpha\beta}q_\beta\gamma^5 + g_3\,q^\alpha\gamma^5)u_n \tag{4.70}$$

Now, since the strong interaction is G invariant, the G-transformation properties of $\mathcal{M}_{V_0}(\mathcal{M}_{A_0})$ should be the same as the G-transformation properties of $\mathcal{M}_V(\mathcal{M}_A)$. In other words, we should require that the G symmetry of the terms in f_2 and f_3 be the same as that of the term in f_1. Similarly, the terms in g_2 and g_3 should have the same G symmetry as the term in g_1. Under an isospin rotation of 180° about the I_2 axis, all six terms transform in the same way—as isovectors.

Thus the question of G symmetry reduces to a question of charge-conjugation symmetry. Under C,[1]

$$\bar{u}_p u_n \rightarrow -\bar{u}_n u_p$$
$$\bar{u}_p \gamma^5 u_n \rightarrow -\bar{u}_n \gamma^5 u_p$$
$$\bar{u}_p \gamma^\mu \gamma^5 u_n \rightarrow -\bar{u}_n \gamma^\mu \gamma^5 u_p$$

However

$$\bar{u}_p \gamma^\mu u_n \rightarrow \bar{u}_n \gamma^\mu u_p$$
$$\bar{u}_p \sigma^{\mu\nu} u_p \rightarrow \bar{u}_n \sigma^{\mu\nu} u_p$$
$$\bar{u}_p \sigma^{\mu\nu} \gamma^5 u_n \rightarrow \bar{u}_n \sigma^{\mu\nu} \gamma^5 u_p$$

Therefore, the term in f_2 does, in fact, have the same G symmetry as the term in f_1, but the term in f_3 has opposite G symmetry. Similarly, the terms in g_1 and g_3 transform the same way, but the term in g_2 transforms oppositely. Therefore we require $f_3 = g_2 = 0$. Of course, we already have $f_3 = 0$ by the CVC hypothesis. The new requirement is that $g_2 = 0$. Vector terms which transform like $(\mathcal{M}_V)_0$ and axial terms which transform like $(\mathcal{M}_A)_0$ are called *first-class terms;* the other terms are called *second-class terms.* The selection rule we have obtained (by somewhat questionable reasoning) says: *no second-class terms* (Wei 58).

A related question concerns so-called *charge symmetry.* The currents $\mathfrak{J}_0{}^\lambda(x)$ and $\mathfrak{J}_0{}^{\lambda\dagger}(x)$ may be written

$$\mathfrak{J}_0{}^\lambda(x) = \tfrac{1}{2}[\mathfrak{J}_0{}^\lambda(x) + e^{-i\pi I_2}\mathfrak{J}_0{}^{\lambda\dagger}(x)e^{i\pi I_2}] + \tfrac{1}{2}[\mathfrak{J}_0{}^\lambda(x) - e^{-i\pi I_2}\mathfrak{J}_0{}^{\lambda\dagger}(x)e^{i\pi I_2}]$$

and

$$\mathfrak{J}_0{}^{\lambda\dagger}(x) = \tfrac{1}{2}[\mathfrak{J}_0{}^{\lambda\dagger}(x) + e^{-i\pi I_2}\mathfrak{J}_0{}^\lambda(x)e^{i\pi I_2}] + \tfrac{1}{2}[\mathfrak{J}_0{}^{\lambda\dagger}(x) - e^{-i\pi I_2}\mathfrak{J}_0{}^\lambda(x)e^{i\pi I_2}]$$

If

$$e^{-i\pi I_2}\mathfrak{J}_0{}^\lambda(x)e^{i\pi I_2} = \pm\mathfrak{J}_0{}^{\lambda\dagger}(x)$$

we have charge symmetry by definition. It is easy to show, by arguments very similar to those already employed, that charge symmetry implies

$$f_1(q^2), f_2(q^2), g_1(q^2), g_3(q^2) \qquad \text{real}$$
$$g_2(q^2), f_3(q^2) \qquad \text{imaginary}$$

Thus time-reversal invariance *and* charge symmetry imply that $g_2(q^2) = f_3(q^2) = 0$.

[1] See Appendix 3.

4.8 THE GOLDBERGER–TREIMAN RELATION

If we assume $g_2 = 0$, the axial vector contribution to the matrix element for $n \to p$ decay is

$$\langle p | \mathcal{A}_\mu^{0\dagger}(x) | n \rangle = \cos \Theta e^{i(p_p - p_n)\cdot x} \bar{u}(p)[g_1(q^2)\gamma_\mu \gamma^5 + g_3(q^2) q_\mu \gamma^5] u(n) \tag{4.71}$$

We may analyze the various contributions to (4.71) if we think first of beta decay of a "bare" neutron (strong interactions turned off) and then add various corrections due to strong interactions. Bare neutron decay corresponds to Fig. 4.2 and contributes a term $\bar{u}(p)\gamma_\mu \gamma^5 u(n)$ to the factor in square brackets in (4.71). To this we have to add *pion vertex* corrections, such as those shown in Fig. 4.3. Of course, no one knows how to calculate the total contribution of Fig. 4.3 and higher-order meson (vertex, etc.) corrections to $\langle p | \mathcal{A}_0^\dagger | n \rangle$, but we do know that all diagrams of this form (Figs. 4.2, 4.3, etc.) make their contribution to the form factor $g_1(q^2)$.

On the other hand, there are the *one-pion exchange* diagrams (Fig. 4.4) and in addition the various vertex corrections to Fig. 4.4. According to the usual Feynman rules, Fig. 4.4 corresponds to

Amplitude =

$$(n \to p\pi^- \text{ vertex}) \times (\pi^- \text{ propagator}) \times (\pi^- \to e^- \bar{\nu}_e \text{ vertex}) \tag{4.72}$$

Now we can write

$$(n \to p\pi^- \text{ vertex}) = g_0 \sqrt{2} \bar{u}(p) \gamma^5 u(n)$$

where g_0 is the physical pion-nucleon coupling constant.[1] Also we have

$$(\pi^- \text{ propagator}) = \frac{+i}{q^2 - m_\pi^2}$$

and $$(\pi^- \to e^- \bar{\nu}_e \text{ vertex}) = \frac{G}{\sqrt{2}} i f_\pi q_\mu [\bar{u}_e \gamma^\mu (1 + \gamma^5) v_\nu] = \frac{G}{\sqrt{2}} i f_\pi q_\mu L^\mu$$

[1] See, for example, Bjo 64, p. 222, Eq. (10.27).

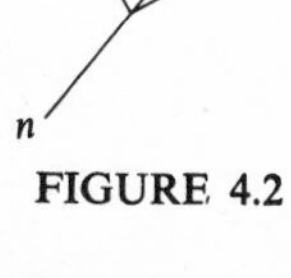

FIGURE 4.2

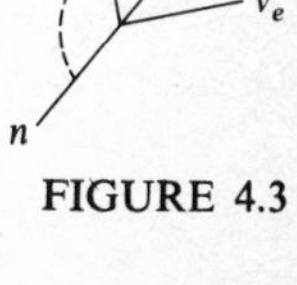

FIGURE 4.3

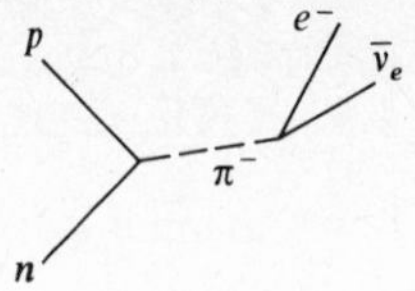

FIGURE 4.4

Putting all this into Eq. (4.72), we have

$$\text{Amplitude} = -\frac{G}{\sqrt{2}} f_\pi g_0 \sqrt{2} \frac{1}{q^2 - m_\pi^2} \bar{u}(p) q_\mu \gamma^5 u(n) L^\mu \tag{4.73}$$

The right-hand side of (4.73) contains the factor $\bar{u}(p)q_\mu\gamma^5 u(n)$. Thus, Fig. 4.4 contributes to the form factor $g_3(q^2)$, and if this figure represented the *only* contribution to $g_3(q^2)$ we would have

$$g_3(q^2) = \frac{f_\pi g_0 \sqrt{2}}{m_\pi^2 - q^2}$$

Actually, in order to take into account vertex and other corrections, we must write

$$g_3(q^2) = \frac{f_\pi g_0 \sqrt{2}}{m_\pi^2 - q^2} F(q^2) \tag{4.74}$$

where $F(q^2)$ is a smooth function of q^2, with

$$F(q^2 = m_\pi^2) = 1$$

since g_0 is the physical pion-nucleon coupling constant. Higher-order (multi-pion) exchange diagrams would give another contribution $H(q^2)$ to $g_3(q^2)$:

$$g_3(q^2) = H(q^2) + \frac{f_\pi g_0 \sqrt{2}}{m_\pi^2 - q^2} F(q^2) \tag{4.75}$$

but we neglect such diagrams and assume $H = 0$. Thus we have

$$\langle p | \mathcal{A}_\mu^{0\dagger}(x) | n \rangle = \cos \Theta e^{i(p_p - p_n)\cdot x} \bar{u}(p) \left[g_1(q^2) \gamma_\mu \gamma^5 - \frac{f_\pi g_0 \sqrt{2}}{q^2 - m_\pi^2} F(q^2) q_\mu \gamma^5 \right] u(n) \tag{4.76}$$

Now, we know that

$$\frac{\partial \mathcal{V}_\mu^0(x)}{\partial x^\mu} = 0$$

by CVC. Is it also true that

$$\frac{\partial \mathcal{A}_\mu{}^0(x)}{\partial x^\mu} = 0$$

If the latter were the case, we would have for π^- decay

$$\langle 0 | \mathcal{A}_\mu{}^{0\dagger}(x) | \pi^- \rangle = i f_\pi q_\mu e^{-iq\cdot x}$$
$$\left\langle 0 \left| \frac{\partial \mathcal{A}_\mu{}^{0\dagger}(x)}{\partial x^\mu} \right| \pi^- \right\rangle = f_\pi q^2 e^{-iq\cdot x}$$
$$= f_\pi m_\pi{}^2 e^{-iq\cdot x} \qquad (4.77)$$

However, $m_\pi \neq 0$ and $f_\pi \neq 0$ since the pion decays and the axial current is not conserved. Nevertheless, it may be described as conserved in an approximate sense, namely, in the limit in which $m_\pi \to 0$:

$$\lim_{m_\pi \to 0} \frac{\partial \mathcal{A}_\mu{}^{0\dagger}(x)}{\partial x^\mu} = 0 \qquad (4.78)$$

Now, using (4.78), let us consider the four-divergence of (4.76). We have

$$\lim_{m_\pi \to 0} \cos \Theta e^{i(p_p - p_n)\cdot x} \bar{u}(p) \left[i g_1(q^2)(\not{p}_p - \not{p}_n)\gamma^5 - i \frac{f_\pi g_0 \sqrt{2}}{q^2 - m_\pi{}^2} F(q^2) q^2 \gamma^5 \right] u(n)$$

$$= \lim_{m_\pi \to 0} \cos \Theta i e^{i(p_p - p_n)\cdot x} \bar{u}(p) \left[g_1(q^2)(m_p + m_n) - \frac{f_\pi g_0 \sqrt{2}}{q^2 - m_\pi{}^2} F(q^2) q^2 \right] \gamma^5 u(n)$$

or

$$g_1(q^2) = + \frac{f_\pi g_0 \sqrt{2}}{m_p + m_n} F(q^2) \qquad (4.79)$$

Also we know that $F(q^2 = m_\pi{}^2) = 1$. If we assume further that $F(q^2)$ is a slowly varying function, we may make the approximation $F(q^2 = 0) = 1$. Then,

$$g_1(0) \approx \frac{f_\pi g_0 \sqrt{2}}{m_p + m_n} \qquad (4.80)$$

The quantity $-g_1(0)$ is just the axial vector coupling constant C_A in nuclear beta decay:

$$g_1(0) = -C_A = +1.23$$

Since g_0 is also known, we may use Eq. (4.80) [known as the Goldberger–Treiman relation (Gol 58)] to estimate f_π. This important result agrees, within about 10 percent, with the value of f_π determined from the half-life for $\pi^\pm$ decay.

Let us recall the key assumptions on which (4.80) is based:

1 Single-pion exchange dominates the induced pseudoscalar form factor $g_3(q^2)$.

2 The axial current is conserved in the limit $m_\pi \to 0$. This is called the *partially conserved axial current* hypothesis (PCAC).

3 $F(q^2)$ is a slowly varying function.

From Eqs. (4.75) and (4.79) we may finally obtain a useful formula for $g_3(q^2)$:

$$g_3(q^2) \approx \frac{2g_1(q^2)}{m_\pi{}^2 - q^2} \tag{4.81}$$

The induced pseudoscalar coupling g_3 is not large enough to manifest itself in beta decay, where $q^2 \approx 0$, but it does produce observable effects in muon capture where q^2 is considerable.

REFERENCES

Ber 58 Berman, S. M.: *Phys. Rev. Letters*, **1**:468 (1958).

Bjo 64 Bjorken, J. D., and S. D. Drell: "Relativistic Quantum Mechanics," McGraw-Hill, New York, 1964.

Bot 67 Botterill, D. R., R. M. Brown, I. F. Corbett, G. Culligan, J. McL. Emmerson, R. C. Field, J. Garvey, P. B. Jones, N. Middlemas, D. Newton, T. W. Quirk, G. L. Salmon, P. Steinberg, and W. S. C. Williams: *Phys. Rev. Letters*, **19**:982 (1967).

Bot 68 Botterill, D. R., R. M. Brown, I. F. Corbett, G. Culligan, J. McL. Emmerson, R. C. Field, J. Garvey, P. B. Jones, N. Middlemas, D. Newton, T. W. Quirk, G. L. Salmon, P. Steinberg, and W. S. C. Williams: *Phys. Rev.*, **171**:1402 (1968).

DeCap 64 De Capua, E., R. Garland, L. Pondrom, and A. Strelzoff: *Phys. Rev.*, **133B**:1333 (1964).

DePom 68 DePommier, P., Du Clos, Heintze, and Kleinknecht: *Nucl. Phys.*, **B4**:189 (1968).

Gol 58 Goldberger, M. L., and S. B. Treiman: *Phys. Rev.*, **110**:1178 (1958); *ibid.*, **111**:354 (1958).

Kin 59 Kinoshita, T.: *Phys. Rev. Letters*, **2**:477 (1959).

Mac 69 Macek, R., A. K. Mann, W. K. McFarlane, J. B. Roberts, K. W. Rothe, C. H. West, and L. B. Auerbach: *Phys. Rev. Letters*, **22**:32 (1969).

Wei 58 Weinberg, S.: *Phys. Rev.*, **112**:1375 (1958).

5
NUCLEAR BETA DECAY

5.1 INTRODUCTION

In this chapter our goal is to survey those aspects of the immense subject of nuclear beta decay which are of fundamental significance in the broader context of weak interactions. Because of the limitations of space, it will be necessary to give only passing reference to many important topics and omit others entirely. Readers who are interested in more detailed treatments are urged to consult recent comprehensive monographs, for example, the works by Konopinski (Kon 66), Wu and Moszkowski (Wu 66), and Schopper (Sch 66).

5.2 GENERAL FORMULA FOR BETA DECAY TRANSITION PROBABILITY

If each of the particles participating in neutron beta decay

$$n \to pe^- \bar{\nu}_e$$

were in plane-wave states of definite momentum and spin, then, as in Chap. 4, the transition matrix element could be written

$$\mathcal{T}_{fi}(n \to pe^-\bar{\nu}_e) = \frac{G_\beta}{\sqrt{2}} \int \langle pe^-\bar{\nu}_e | \mathfrak{J}_0^{\dagger}(x) j_e(x) | n \rangle \, d^4x$$

$$= \frac{G_\beta}{\sqrt{2}} \int \frac{1}{(2E_n V 2E_p V 2EV 2\omega V)^{1/2}}$$

$$\times \, d^4x [\bar{u}_p e^{ip_p \cdot x} \Gamma^\mu u_n e^{-ip_n \cdot x}][\bar{u}_e e^{ip_e \cdot x} \gamma_\mu (1 + \gamma^5) v_\nu e^{ip_{\bar{\nu}} \cdot x}] \tag{5.1}$$

or
$$\mathcal{T}_{fi} = \frac{(2\pi)^4 \, \delta^4(p_p + p_e + p_{\bar{\nu}} - p_n) \mathcal{M}}{\sqrt{2E_n V 2E_p V 2EV 2\omega V}} \tag{5.2}$$

where $G_\beta = G \cos \Theta$

and
$$\mathcal{M} = \frac{G_\beta}{\sqrt{2}} \bar{u}_p \Gamma^\mu \bar{u}_n \bar{u}_e \gamma_\mu (1 + \gamma^5) v_\nu \tag{5.3}$$

and
$$\Gamma^\mu = f_1(q^2)\gamma^\mu + if_2(q^2)\sigma^{\mu\alpha}(p_e + p_{\bar{\nu}})_\alpha$$
$$+ g_1(q^2)\gamma^\mu\gamma^5 - ig_2(q^2)\sigma^{\mu\alpha}(p_e + p_{\bar{\nu}})_\alpha \gamma^5 + g_3(q^2)(p_e + p_{\bar{\nu}})^\mu \gamma^5 \tag{5.4}$$

[In (5.4) we have included a possible second-class term in g_2 but have omitted the *induced scalar* form factor f_3, which should be zero according to the CVC hypothesis.]

In actual fact, we cannot describe the particles participating in neutron beta decay (or, more generally, in nuclear beta decay) as plane waves because the nuclei are well localized and require spatial wave functions for their description. Thus, in the integrand of (5.1), we must make the following replacements:

$$\begin{aligned}
\frac{1}{\sqrt{2E_n}} u(n) e^{-iE_n t} e^{i\mathbf{p}_n \cdot \mathbf{x}} &\to \psi_n(\mathbf{x}) e^{-iE_n t} \\
\frac{1}{\sqrt{2E_p}} \bar{u}(p) e^{iE_p t} e^{-i\mathbf{p}_p \cdot \mathbf{x}} &\to \bar{\psi}_p(\mathbf{x}) e^{iE_p t} \\
\frac{1}{\sqrt{2\omega}} v(\bar{\nu}) e^{i\omega t} e^{-i\mathbf{p}_{\bar{\nu}} \cdot \mathbf{x}} &\to \psi_{\bar{\nu}}(\mathbf{x}) e^{i\omega t} \\
\frac{1}{\sqrt{2E}} \bar{u}(e) e^{iEt} e^{i\mathbf{p}_e \cdot \mathbf{x}} &\to \bar{\psi}_e(\mathbf{x}) e^{iEt}
\end{aligned} \tag{5.5}$$

where the ψ's are four-component spatial-wave functions and the particle energies E_n, E_p, ω, and E are assumed to be still well defined. In formulas (5.5) and henceforth, we set $V = 1$.

Furthermore, the quantities p_e and $p_{\bar{\nu}}$ appearing in Eq. (5.4) for Γ_λ must now be regarded as operators on the lepton wave functions. Thus, writing

$$j_\mu(\mathbf{x}) = \bar{\psi}_{\bar{\nu}}(\mathbf{x})\gamma_\mu(1+\gamma^5)\psi(\mathbf{x})$$

we obtain as a result

$$\mathcal{T}_{fi} = \frac{G_\beta}{\sqrt{2}} \int dt \int d^3x \bar{\psi}_p(\mathbf{x}) e^{iE_p t} \Big\{ f_1(q^2)\gamma^\mu j_\mu(\mathbf{x}) e^{i(E+\omega)t}$$

$$+ f_2(q^2)\sigma^{\mu\alpha} \frac{\partial}{\partial x^\alpha}[j_\mu(\mathbf{x})e^{i(E+\omega)t}] + g_1(q^2)\gamma^\mu\gamma^5 j_\mu(\mathbf{x})e^{i(E+\omega)t} - g_2(q^2)\sigma^{\mu\alpha}\gamma^5$$

$$\times \frac{\partial}{\partial x^\alpha}[j_\mu(\mathbf{x})e^{i(E+\omega)t}] - ig_3(q^2)\gamma^5 \frac{\partial}{\partial x_\mu}[j_\mu(\mathbf{x})e^{i(E+\omega)t}]\Big\} \psi_n(\mathbf{x})e^{-iE_n t} \qquad (5.6)$$

Fortunately, this formula may be simplified considerably because in nuclear beta decay the momentum imparted to the leptons is generally very small: Typically, the mass difference between initial and final nuclear states is of order 1 MeV/c^2, and so lepton three-momenta are generally of order $m_0 c$, where m_0 is the electron rest mass. The final nucleus recoil momentum must also be of this order of magnitude, and therefore its kinetic energy in the rest frame of the initial nucleus is of order $({m_0}^2c^2)/(2Am_p)$, where A is the mass number. Even for small A, this energy is only in the kilovolt range. The kinetic energy of nucleons inside nuclei is larger (approximately 5 MeV) but still extremely small compared to the nucleon rest energy. Therefore we are justified in making a "large-small" component reduction of the four-component nucleon wave functions. Thus we write for the neutron and proton, respectively,

$$\psi_n = \begin{pmatrix} \chi_n \\ \dfrac{\boldsymbol{\sigma}\cdot\mathbf{P}}{2}\chi_n \end{pmatrix}$$

$$\psi_p = \begin{pmatrix} \chi_p \\ \dfrac{\boldsymbol{\sigma}\cdot\mathbf{P}}{2}\chi_p \end{pmatrix}$$

where χ_n, χ_p are two-component Pauli wave functions. Here $\mathbf{P}$ is the operator $-i\boldsymbol{\nabla}$, and $E_n \approx E_p \approx m_n \approx m_p = 1$ since the nucleons are always very non-relativistic in nuclear beta decay.

Now writing $E_n - E_p = \Delta$, we find that (5.6) becomes

$$\mathcal{T}_{fi} = \int dt\, e^{i(E+\omega-\Delta)t}\mathfrak{A}(n-p)$$

$$= 2\pi\delta(E+\omega-\Delta)\mathfrak{A}(n-p) \qquad (5.7)$$

where
$$\begin{aligned}\mathfrak{A}(n \to p) &= \frac{G_\beta}{\sqrt{2}} \int d^3x \chi_p{}^\dagger \mathbf{O} \chi_n \\ &= \frac{G_\beta}{\sqrt{2}} \int d^3x \Big\{ f_1(0) \chi_p{}^\dagger j_0 \chi_n \\ &\quad + \frac{i}{2} f_1(0) \chi_p{}^\dagger (\boldsymbol{\sigma} \cdot \nabla \boldsymbol{\sigma} \cdot \mathbf{j} \chi_n + \boldsymbol{\sigma} \cdot \mathbf{j} \boldsymbol{\sigma} \cdot \nabla \chi_n) \\ &\quad - f_2(0) \chi_p{}^\dagger \boldsymbol{\sigma} \cdot (\nabla \times \mathbf{j}) \chi_n + g_1(0) \chi_p{}^\dagger \boldsymbol{\sigma} \cdot \mathbf{j} \chi_n \\ &\quad + \frac{i}{2} g_1(0) \chi_p{}^\dagger (\boldsymbol{\sigma} \cdot \nabla j_0 \chi_n + j_0 \boldsymbol{\sigma} \cdot \nabla \chi_n) \\ &\quad + i g_2(0) \chi_p{}^\dagger (\boldsymbol{\sigma} \cdot \nabla j_0) \chi_n - g_2(0)(E + \omega) \chi_p{}^\dagger \boldsymbol{\sigma} \cdot \mathbf{j} \chi_n \\ &\quad - \frac{i}{2} g_3(0)(E + \omega) \chi_p{}^\dagger (\boldsymbol{\sigma} \cdot \nabla j_0 \chi_n - j_0 \boldsymbol{\sigma} \cdot \nabla \chi_n) \\ &\quad - \frac{i}{2} g_3(0) \chi_p{}^\dagger (\boldsymbol{\sigma} \cdot \nabla \nabla \cdot j \chi_n - \nabla \cdot j \boldsymbol{\sigma} \cdot \nabla \chi_n) \Big\} \end{aligned} \tag{5.8}$$

In (5.8), only the lowest-order energy-momentum transfer terms have been retained for $f_2(0)$ and $g_2(0)$.

Equations (5.7) and (5.8) are easily generalized to nuclear beta decay $(Z, N) \to (Z + 1, N - 1)e^- \bar{\nu}_e$ if we allow Δ to denote the energy difference between initial and final *nuclei* and write

$$\mathfrak{A}(Z, N \to Z + 1, N - 1) = \frac{G_\beta}{\sqrt{2}} \sum_{i=1}^{A} \int d^3x \chi'^\dagger \mathbf{O}_i \tau_i{}^+ \chi \tag{5.9}$$

In (5.9), χ stands for the initial nuclear wave function, χ' is the final nuclear wave function, $\tau_i{}^+$ is the isospin raising operator for the ith nucleon, and $\mathbf{O}_i$ is the same operator that appears in (5.8), except that $\boldsymbol{\sigma}$ is replaced by $\boldsymbol{\sigma}_i$, the spin operator for the ith nucleon. Neglecting the final nucleus recoil kinetic energy in evaluation of the phase-space factor, we obtain the transition probability per unit time for nuclear beta decay by a standard calculation from (5.7) and (5.9):

$$dW = (2\pi)^{-5} \, d^3\mathbf{p}_e \, d^3p_{\bar{\nu}} \, \delta(E + \omega - \Delta) |\mathfrak{A}(Z, N \to Z + 1, N - 1)|^2 \tag{5.10}$$

5.3 ALLOWED DECAYS

The de Broglie wavelength $\lambda\!\!\!^{-}$ of either lepton in beta decay is typically much larger than the nuclear radius R. We have, with $p \approx m_0 c$,

$$\lambda\!\!\!^{-} \approx \frac{\hbar}{m_0 c} \approx 4 \times 10^{-11} \text{ cm}$$

while

$$R \approx 1.3 A^{1/3} \times 10^{-13} \text{ cm}$$

Thus, even for the largest nuclei, $R/\lambda\!\!\!\bar{} \approx \frac{1}{40}$. Consequently the lepton wave functions vary only very slightly over the nuclear volume. Furthermore, the nucleon momenta are very small, as already noted. It is therefore a very good first approximation to neglect nuclear momentum and variation of the lepton wave functions over the nuclear volume, and this is equivalent to discarding all terms in the integrand of (5.8) [or (5.9)] which contain the operator $\boldsymbol{\nabla}$ or are proportional to $E + \omega$. This is called the *allowed* approximation. We see that in this case (5.8) simplifies to

$$\mathfrak{A} = \frac{G_\beta}{\sqrt{2}} \left[\sum_{i=1}^{A} f_1(0) \int \chi'^\dagger(\mathbf{x}) \tau_i^{\,+} \chi(\mathbf{x})\, d^3x\, j_0(0) + \sum_{i=1}^{A} g_1(0) \int \chi'^\dagger(\mathbf{x}) \boldsymbol{\sigma}_i \tau_i^{\,+} \chi(\mathbf{x})\, d^3x \mathbf{j}(0) \right] \tag{5.11}$$

Here the quantities j_0 and $\mathbf{j}$ are evaluated at the origin of nuclear coordinates. For short, we frequently write

$$C_V = f_1(0)$$

$$-C_A = g_1(0)$$

$$\langle 1 \rangle = \sum_{i=1}^{A} \int \chi'^\dagger(\mathbf{x}) \tau_i^{\,+} \chi(\mathbf{x})\, d^3x \tag{5.12}$$

and

$$\langle \boldsymbol{\sigma} \rangle = \sum_{i=1}^{A} \int \chi'^\dagger(\mathbf{x}) \boldsymbol{\sigma}_i \tau_i^{\,+} \chi(\mathbf{x})\, d^3x \tag{5.13}$$

Thus, in the allowed approximation,

$$\mathfrak{A} = \frac{G \cos \Theta}{\sqrt{2}} [C_V \langle 1 \rangle j_0(0) - C_A \langle \boldsymbol{\sigma} \rangle \cdot \mathbf{j}(0)] \tag{5.14}$$

It is easy to see that neglect of the variation of j_0, $\mathbf{j}$ across the nucleus is equivalent to neglect of all total lepton orbital angular momenta $L > 0$. Thus in the allowed approximation the lepton orbital angular momentum is zero and the total angular momentum carried off by the leptons is just their total spin: $S = 1$ or 0 since each lepton has spin $\frac{1}{2}$. If $S = 0$, we have a *Fermi* transition; if $S = 1$, a *Gamow–Teller* transition. Conservation of angular momentum dictates the following selection rules for the initial and final nuclear spins J_i, J_f:

Fermi: $$J_i = J_f \tag{5.15a}$$

Gamow–Teller: $$J_i = J_f,\ J_f \pm 1 \qquad J_i = 0 \nleftrightarrow J_f = 0 \tag{5.15b}$$

Also, since in either case the leptons carry off no orbital angular momentum, the initial and final nuclear parities must be the same:

$$\pi_i \pi_f = +1 \tag{5.16}$$

Nature provides examples of pure Fermi transitions, e.g.,

$$O^{14}(0^+) \to N^{14*}(0^+)$$

pure Gamow–Teller transitions, e.g.,

$$He^6(0^+) \to Li^6(1^+)$$

and "mixed" Fermi, Gamow–Teller transitions, e.g.,

$$n(\tfrac{1}{2}^+) \to p(\tfrac{1}{2}^+)$$

Obviously, the Gamow–Teller amplitude must be identified with the nuclear matrix element $\langle \boldsymbol{\sigma} \rangle$ since $\boldsymbol{\sigma}$ is a first-rank tensor whose matrix elements are capable of satisfying the selection rules (5.15b). The Fermi matrix element is clearly $\langle 1 \rangle$. Writing $T^+ = \sum_{i=1}^{A} \tau_i^+$, we have for the Fermi matrix element

$$\langle 1 \rangle = \int \chi'^{\dagger} T^+ \chi \, d^3x$$

Since T^+ has no nonvanishing matrix elements between states of different total isospin, we conclude that $\Delta T = 0$ for all Fermi transitions. No such selection rule applies in the case of Gamow–Teller transitions.

In evaluating the lepton factors $j_0(0)$ and $\mathbf{j}(0)$ in (5.14) we must be careful to take account of the distortion of the electron wave function by the nuclear coulomb field. We thus define

$$|\mathfrak{A}|^2 = F(Z, E)|\mathfrak{A}_0|^2 \frac{1}{2E2\omega} \tag{5.17}$$

where $F(Z, E)$ is a coulomb correction factor and

$$\mathfrak{A}_0 = \frac{G_\beta}{\sqrt{2}} [C_V \langle 1 \rangle \bar{u}_e \gamma^0 (1 + \gamma^5) v_{\bar{\nu}} - C_A \langle \boldsymbol{\sigma} \rangle \cdot \bar{u}_e \boldsymbol{\gamma} (1 + \gamma^5) v_{\bar{\nu}}] \tag{5.18}$$

Here the $\bar{u}_e$ and $v_{\bar{\nu}}$ are lepton four-spinors normalized conventionally. The coulomb correction factor $F(Z, E)$ is just the ratio

$$\left| \frac{\psi_{e,c}(0)}{\psi_{e,0}(0)} \right|^2$$

where $\psi_{e,c}(0)$ is the electron wave function with coulomb field on, and $\psi_{e,0}(0)$ is the electron wave function without the coulomb field. For a point nucleus with charge Ze, we find that[1]

$$\left|\frac{\psi_{e,c}(r)}{\psi_{e,0}(r)}\right|^2 = 2(1+\gamma_0)(2p_e r)^{-2(1-\gamma_0)}e^{\pi\nu}\frac{|\Gamma(\gamma_0+i\nu)|^2}{|\Gamma(2\gamma_0+1)|^2} \tag{5.19}$$

where $\nu = \pm Ze^2/v_e$ for $e^\pm$ emission, $\gamma_0 = [1-(\alpha Z)^2]^{1/2}$, and p_e, v_e are the electron three-momentum and velocity, respectively. The right-hand side of (5.19) diverges at $r = 0$. However, if the nucleus is taken to be of finite extent, the divergence disappears, the electron density is essentially flat inside the nuclear radius R, and $F(Z, E)$ is given accurately by (5.19) for $r = R$ for all but the heaviest nuclei.

It is frequently useful to express $\langle\boldsymbol{\sigma}\rangle$ in terms of a reduced matrix element $\langle\sigma\rangle$ by means of the Wigner–Eckhart theorem. We write

$$\boldsymbol{\sigma} = \frac{\sigma_x - i\sigma_y}{\sqrt{2}}\frac{\hat{\imath}+i\hat{\jmath}}{\sqrt{2}} + \frac{\sigma_x + i\sigma_y}{\sqrt{2}}\frac{\hat{\imath}-i\hat{\jmath}}{\sqrt{2}} + \sigma_z\hat{k} \tag{5.20}$$

Then we have

$$-\frac{1}{\sqrt{2}}\langle f|\sigma_x + i\sigma_y|i\rangle = \langle\sigma\rangle\langle j_i, m_i, 1, 1|j_f, m_i+1\rangle$$

$$\langle f|\sigma_z|i\rangle = \langle\sigma\rangle\langle j_i, m_i, 1, 0|j_f, m_i\rangle$$

$$\frac{1}{\sqrt{2}}\langle f|\sigma_x - i\sigma_y|i\rangle = \langle\sigma\rangle\langle j_i, m_i, 1, -1|j_f, m_i-1\rangle$$

where the $\langle j_i, m_i, 1, M|j_f, m_i+M\rangle$ are Clebsch–Gordan coefficients.

Forbidden beta decay transitions do occur in violation of the selection rules (5.15a), (5.15b), and (5.16), although they are generally orders of magnitude slower than allowed transitions. These forbidden decays must be accounted for by terms so far neglected in $\mathfrak{A}$ [i.e., those in (5.9) in which the $\boldsymbol{\nabla}$ operator appears].[2] Even for allowed decays, it is necessary to include *forbidden corrections* in order to account for certain fine details. At the end of this chapter we shall briefly discuss two examples: the effect of *weak magnetism* and the possible effects of a nonzero induced tensor coupling g_2. With these exceptions, however, almost all beta decay effects of basic interest for the weak interactions occur in the "allowed" approximation.

[1] See, for example, Kon 66, p. 91.
[2] See, for example, Kon 66, chap. 7.

5.4 DETAILED CALCULATIONS OF THE TRANSITION PROBABILITY FOR ALLOWED DECAYS

We now consider in detail the differential transition probability per unit time for allowed decay:

$$dW = (2\pi)^{-5}\frac{\delta(E+\omega-\Delta)}{2E2\omega}F(Z, E)|\mathfrak{A}_0|^2\, d^3\mathbf{p}_e\, d^3\mathbf{p}_\nu \tag{5.21}$$

where
$$\mathfrak{A}_0 = \frac{G_\beta}{\sqrt{2}}[C_V\langle 1\rangle \bar{u}_e\gamma^0(1+\gamma^5)v_{\bar{\nu}} - C_A\langle\boldsymbol{\sigma}\rangle\cdot\bar{u}_e\boldsymbol{\gamma}(1+\gamma^5)v_{\bar{\nu}}] \tag{5.22}$$

It is useful to distinguish here between two broad categories of experiments: (1) those in which the longitudinal polarization of electrons is observed in decay from unpolarized samples of initial nuclei and (2) those in which the initial nucleus may be polarized and we sum over all final spins but otherwise examine the decay in full detail.

In the case of longitudinal electron polarization, all the information we seek is contained in the factors $\bar{u}_e\gamma^0(1+\gamma^5)v_{\bar{\nu}}$, $\bar{u}_e\boldsymbol{\gamma}(1+\gamma^5)v_{\bar{\nu}}$. In each case, since $[(1+\gamma^5)/2]$ is a projection operator, the electron four-spinor appears in the form

$$u_{eL} = \tfrac{1}{2}(1+\gamma^5)\begin{pmatrix}\chi_e \\ \dfrac{\boldsymbol{\sigma}\cdot\mathbf{p}_e}{E+m_e}\chi_e\end{pmatrix}$$

Let us compute the relative probabilities that the electron is found to have spin along the $\pm z$ directions if its momentum is $\mathbf{p} = +p\hat{k}$. If the probability for the spin to be up (down) is denoted by $P_+(P_-)$, we have

$$\begin{aligned}
P_\pm &= u_{eL}{}^\dagger(\pm)u_{eL}(\pm) \\
&= \left(\chi_e{}^\dagger, \pm\frac{p}{E+m_e}\chi_e{}^\dagger\right)\tfrac{1}{2}(1+\gamma^5)\begin{pmatrix}\chi_e \\ \pm\dfrac{p}{E+m_e}\end{pmatrix} \\
&= \tfrac{1}{2}\left[1 \mp \frac{2p}{E+m_e} + \frac{p^2}{(E+m_e)^2}\right] \\
&= \left[\frac{E\mp p}{E+m_e}\right]
\end{aligned}$$

where we have set $\chi_e{}^\dagger\chi_e = 1$. The polarization is

$$\mathfrak{P} = \frac{P_+ - P_-}{P_+ + P_-} = \frac{(E-p)-(E+p)}{(E-p)+(E+p)} = -\frac{p}{E} = -v \tag{5.23}$$

Thus we obtain the important result that the longitudinal polarization of electrons emitted by unpolarized nuclei must be $\mathcal{P} = -v/c$ regardless of whether the decay is pure Fermi, pure Gamow–Teller, or mixed. Of course, $\mathcal{P} = +v/c$ for positron emission. A wide variety of experiments confirms this conclusion. For example, a recent and particularly precise measurement of the velocity dependence of e^- helicity from Co^{60} decay (a pure Gamow–Teller transition) has been reported by Lazarus and Greenberg (Laz 70). They find

$$h(e^-) = -\frac{v}{c}(1.014 \pm 0.018) \tag{5.24}$$

over an extended range of electron energies ($0.4 < v/c < 0.7$). Detailed summaries of earlier helicity measurements are presented by many authors, e.g., by Frauenfelder and Steffen (Fra 65).

As for the second category of experiments, we seek to compare the theory with observation by calculating dW of (5.21), with $|\mathfrak{A}_0|^2$ summed over final spins but for a given initial nuclear polarization. From (5.22), we have

$$\begin{aligned}\overline{|\mathfrak{A}_0|^2} = &\frac{G_\beta{}^2}{2}|C_V|^2|\langle 1\rangle|^2 \operatorname{tr}[\Lambda_e\gamma^0(1+\gamma^5)\Lambda_{\bar\nu}\gamma^0(1+\gamma^5)] \\ &+ \frac{G_\beta{}^2}{2}|C_A|^2[\langle\sigma_i\rangle^*\langle\sigma_j\rangle]\operatorname{tr}[\Lambda_e\gamma_i(1+\gamma^5)\Lambda_{\bar\nu}\gamma_j(1+\gamma^5)] \\ &- \frac{G_\beta{}^2}{2}C_V^* C_A\langle 1\rangle^*\langle\sigma_i\rangle\operatorname{tr}[\Lambda_e\gamma_i(1+\gamma^5)\Lambda_{\bar\nu}\gamma^0(1+\gamma^5)] \\ &- \frac{G_\beta{}^2}{2}C_A C_V^*\langle 1\rangle\langle\sigma_i\rangle^* \operatorname{tr}[\Lambda_e\gamma^0(1+\gamma^5)\Lambda_{\bar\nu}\gamma_i(1+\gamma^5)]\end{aligned} \tag{5.25}$$

Evaluating the traces by standard methods, we find that the first term in (5.25) is

$$4G_\beta{}^2|C_V|^2|\langle 1\rangle|^2 E\omega(1 + \mathbf{v}_e\cdot\hat{\mathbf{p}}_{\bar\nu}) \tag{5.26}$$

If the initial and final nuclear spins are 0, the Gamow–Teller matrix element vanishes and (5.26) represents the only contribution to $\overline{|\mathfrak{A}_0|^2}$. In this case the transition probability becomes

$$dW(0^+ \to 0^+) = \frac{G_\beta{}^2}{(2\pi)^5}\delta(E + \omega - \Delta)F(Z, E)|C_V|^2|\langle 1\rangle|^2(1 + \mathbf{v}_e\cdot\hat{\mathbf{p}}_{\bar\nu})\, d^3\mathbf{p}_e\, d^3\mathbf{p}_{\bar\nu} \tag{5.27}$$

Suppose instead that $J_i{}^P = \frac{1}{2}{}^+ = J_f{}^P$. Then we must include contributions to $\overline{|\mathfrak{A}_0|^2}$ arising from the Gamow–Teller matrix element as well. The second term in (5.25) is then found to be

$$4G_\beta{}^2|\langle\sigma\rangle|^2|C_A|^2E\omega(1 - \tfrac{1}{3}\mathbf{v}_e\cdot\hat{\mathbf{p}}_{\bar\nu} + \tfrac{2}{3}\hat{\mathbf{J}}\cdot\hat{\mathbf{p}}_{\bar\nu} - \tfrac{2}{3}\hat{\mathbf{J}}\cdot\mathbf{v}_e) \tag{5.28}$$

After a simple manipulation, the third and fourth terms of (5.25) become

$$-\frac{4G_\beta{}^2}{\sqrt{3}}(C_V^* C_A + C_A^* C_V)\langle 1\rangle\langle\sigma\rangle(\hat{\mathbf{J}}\cdot\mathbf{v}_e + \hat{\mathbf{J}}\cdot\hat{\mathbf{p}}_{\bar{\nu}})$$

$$+\frac{4iG_\beta{}^2}{\sqrt{3}}(C_V^* C_A - C_A^* C_V)\langle 1\rangle\langle\sigma\rangle\hat{\mathbf{J}}\cdot\mathbf{v}\times\hat{\mathbf{p}}_{\bar{\nu}} \tag{5.29}$$

Collecting all of these terms together, we finally obtain, for $J_i = J_f = \frac{1}{2}$, e^- emission

$$dW(\tfrac{1}{2}\to\tfrac{1}{2}) = \frac{G_\beta{}^2}{(2\pi)^5}\,\delta(E+\omega-\Delta)\,d^3\mathbf{p}_e\,d^3\mathbf{p}_\nu\,F(Z,E)\xi \times\{1 + a\mathbf{v}\cdot\hat{\mathbf{p}}_{\bar{\nu}} + A\hat{\mathbf{J}}\cdot\mathbf{v} + B\hat{\mathbf{J}}\cdot\hat{\mathbf{p}}_{\bar{\nu}} + D\hat{\mathbf{J}}\cdot\mathbf{v}\times\hat{\mathbf{p}}_{\bar{\nu}}\} \tag{5.30}$$

where

$$\xi = |C_V|^2|\langle 1\rangle|^2 + |C_A|^2|\langle\sigma\rangle|^2 \tag{5.31}$$

$$\xi a = |C_V|^2|\langle 1\rangle|^2 + \tfrac{1}{3}|C_A|^2|\langle\sigma\rangle|^2 \tag{5.32}$$

$$\xi A = -\tfrac{2}{3}|C_A|^2|\langle\sigma\rangle|^2 - \frac{1}{\sqrt{3}}(C_V^*C_A + C_A^*C_V)\langle 1\rangle\langle\sigma\rangle \tag{5.33}$$

$$\xi B = +\tfrac{2}{3}|C_A|^2|\langle\sigma\rangle|^2 - \frac{1}{\sqrt{3}}(C_V^*C_A + C_A^*C_V)\langle 1\rangle\langle\sigma\rangle \tag{5.34}$$

and

$$\xi D = \frac{i}{\sqrt{3}}(C_A^* C_V - C_V^* C_A)\langle 1\rangle\langle\sigma\rangle \tag{5.35}$$

One may actually assume $\langle 1\rangle$ and $\langle\sigma\rangle$ to be real without loss of generality. However, C_A/C_V is real only if time-reversal invariance holds. In this case $D = 0$, neglecting final-state effects.

Formulas (5.30) to (5.35) apply to e^- emission with $J_i = J_f = \frac{1}{2}$, but these formulas may be generalized quite easily to $e^\pm$ emission and to arbitrary initial and final spins (with $J_i = J_f$ or $J_f \pm 1$, of course). In this case we find that (5.30) is unchanged except for an additional *tensor* correlation term within the braces, which exists for $J_f, J_i \geq 1$:

$$\frac{|C_A|^2|\langle\sigma\rangle|^2}{\xi}\bar{\Lambda}[\tfrac{1}{3}\hat{\mathbf{p}}_\nu - (\hat{\mathbf{p}}_\nu\cdot\hat{\mathbf{J}})\hat{\mathbf{J}}]\cdot\mathbf{v}$$

with

$$\bar{\Lambda} = \Lambda\left[\frac{3\langle M_{J_i}{}^2\rangle - J_i(J_i+1)}{J_i(2J_i-1)}\right]$$

and	$\Lambda = 1$	$-\frac{(2J_i-1)}{J_i+1}$	$\frac{J_i(2J_i-1)}{(J_i+1)(2J_i+3)}$
when	$J_f = J_i - 1$	J_i	$J_i + 1$

Coefficients ξ and a remain unchanged in the general case. However, for A, B, and D we now have

$$\xi A = \pm\kappa |C_A|^2 |\langle\sigma\rangle|^2 - (C_V C_A^* + C_A C_V^*)\langle 1\rangle\langle\sigma\rangle\sqrt{\frac{J_i}{J_i+1}} \tag{5.36}$$

$$\xi B = \mp\kappa |C_A|^2 |\langle\sigma\rangle|^2 - (C_V C_A^* + C_A C_V^*)\langle 1\rangle\langle\sigma\rangle\sqrt{\frac{J_i}{J_i+1}} \tag{5.37}$$

$$\xi D = i(C_V C_A^* - C_A C_V^*)\langle 1\rangle\langle\sigma\rangle\sqrt{\frac{J_i}{J_i+1}} \tag{5.38}$$

where $\kappa = 1$ for $J_f = J_i - 1$

$\kappa = (J_1 + 1)^{-1}$ for $J_f = J_i$

and $\kappa = -\dfrac{J_i}{J_i + 1}$ for $J_f = J_i + 1$

where $\langle 1\rangle = 0$ for $J_i \neq J_f$ and we use $\pm$ for $e^{\pm}$. We are now in a position to discuss most of the important physical effects in allowed decay.

5.5 TOTAL DECAY RATE; *ft* VALUES

We first integrate dW [Eq. (5.27) or (5.30)] over electron and neutrino directions and average over nuclear spin. Then we obtain

$$\begin{aligned} dW &= \frac{G_\beta^{\,2}}{(2\pi)^5}\,\xi F(Z, E)p_e^{\,2}\,dp_e \int \delta(\Delta - E - \omega)\omega^2\,d\omega \int d\Omega_e \int d\Omega_\nu \\ &= \frac{G_\beta^{\,2}}{(2\pi)^5}\,\xi F(Z, E)p_e^{\,2}\,dp_e(\Delta - E)^2(4\pi)^2 \end{aligned} \tag{5.39}$$

Note that (5.39) applies to all allowed decays regardless of initial and final nuclear spin. This is because the definition of ξ in (5.31) is broad enough to cover pure Fermi, pure Gamow–Teller, and mixed decays.

Since $p_e^{\,2}\,dp_e = p_e E\,dE$, the electron energy spectrum for allowed decay is

$$N(E) = \frac{dW}{dE} = \frac{G_\beta^{\,2}}{2\pi^3}\,\xi F(Z, E)p_e E(\Delta - E)^2 \tag{5.40}$$

One usually exhibits the experimental electron energy spectrum by making a *Kurie plot* in which

$$\left|\frac{N(E)_{\text{exp}}}{F(Z, E)p_e E}\right|^{1/2}$$

is plotted against E. Of course, this should be a straight line with negative slope, intersecting the E axis at $E = \Delta$, provided that the neutrino mass is really zero. In the case $H^3 \to He^3 e^- \bar{\nu}_e$, Δ is known with high precision independently and a careful measurement of the electron energy spectrum yields an upper limit on the (anti-) neutrino mass (Ber 69, Dar 69):

$$m_{\bar{\nu}e} < 60 \text{ eV}/c^2$$

(The rest mass of ν_e is not known so well. From the positron spectrum of Na^{22} and an independent determination of the masses of parent and daughter nuclei, we conclude that $m_{\nu e} < 4.1$ keV. See Bee 68.)

Let us calculate the total decay rate for an allowed transition. First we integrate (5.40) over electron energy:

$$W_0 = \frac{G_\beta^{\ 2}}{2\pi^3} \xi \int_{E=m_e}^{\Delta} F(Z, E) p_e E(\Delta - E)^2 \, dE \tag{5.41}$$

The Fermi integral of (5.41)

$$f_0(Z, \Delta) = \int_{E=m_e}^{\Delta} F(Z, E) p_e E(\Delta - E)^2 \, dE \tag{5.42}$$

has been tabulated for a wide range of Δ and Z (see, for example, NBS 52). It is a rapidly varying function of Δ, as illustrated in Fig. 5.1. Next, we must correct f_0 for radiative effects (Käl 68) and also apply forbidden corrections (Bli 69). The total correction to f_0 may amount to several percent. Then, the mean lifetime for the transition is

$$\tau = \frac{2\pi^3}{G_\beta^{\ 2} \xi f_{0,\,\text{corr}}} \tag{5.43}$$

It is customary to define the *comparative* half-life $f_{0,\,\text{corr}}\, t_{1/2}$ (or simply ft). Obviously,

$$ft = \frac{2\pi^3 \ln 2}{G_\beta^{\ 2} \xi} \qquad \text{(natural units)} \tag{5.44}$$

or, in cgs units,

$$ft = (1.23018) \times 10^{-94} \frac{1}{G_\beta^{\ 2} \xi} \tag{5.45}$$

For certain allowed transitions, namely, the $0^+ \to 0^+$ decays

$C^{10} \to B^{10}$	$O^{14} \to N^{14}$	$Al^{26m} \to Mg^{26}$	$Cl^{34} \to S^{34}$
$Sc^{42} \to Ca^{42}$	$V^{46} \to Ti^{46}$	$Mn^{50} \to Cr^{50}$	$Co^{54} \to Fe^{54}$

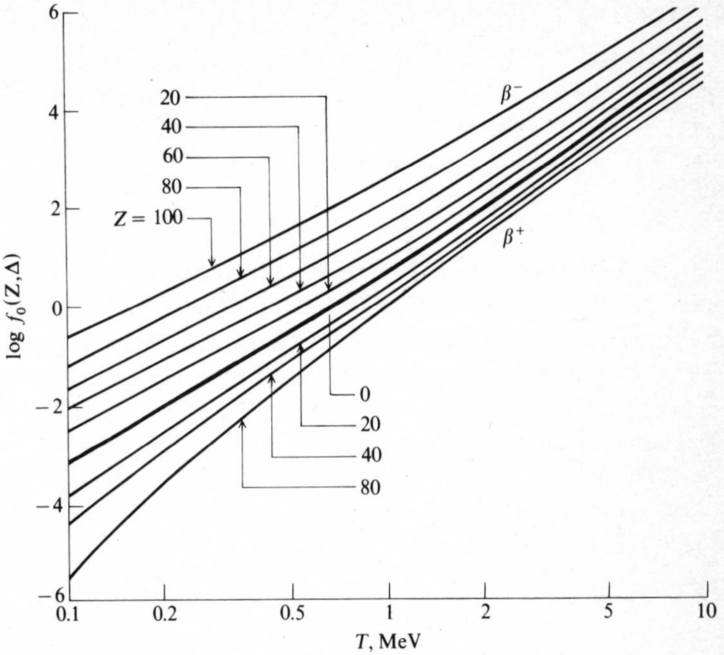

FIGURE 5.1
The Fermi integral f_0 as a function of the maximum kinetic energy $T = \Delta - m_e$ as calculated for a point charge nucleus.

and the mirror nucleus decays

$$\begin{array}{ll} n \to p & (J_i^p = J_f^p = \tfrac{1}{2}^+) \\ Ne^{19} \to F^{19} & (J_i^p = J_f^p = \tfrac{1}{2}^+) \\ Ar^{35} \to Cl^{35} & (J_i^p = J_f^p = \tfrac{3}{2}^+) \end{array} \tag{5.46}$$

measurements of ft are of special importance since in these cases ξ can be determined accurately. Thus evaluation of ft leads immediately to a precise determination of G_β.

For the above $0^+ \to 0^+$ transitions, in each case the initial and final nuclei belong to the same $T = 1$ isomultiplet. Thus the Fermi matrix element is

$$\langle 1 \rangle = \sqrt{2}$$

within about 1 percent, charge-dependent effects being approximately calculable and small.[1] Therefore, assuming $C_V = 1$, we have $\xi(0^+ \to 0^+) = 2$. For the

[1] See, for example, Bli 69.

decays of mirror nuclei n, Ne^{19}, and Ar^{35}, the Gamow–Teller amplitudes are nonzero. However, we can write

$$\xi = |C_V|^2 |\langle 1 \rangle|^2 (1 + |\rho|^2)$$

where
$$\rho = \frac{C_A \langle \sigma \rangle}{C_V \langle 1 \rangle} \tag{5.47}$$

The quantity ρ has been determined from measurements of the electron asymmetry coefficient A [cf. Eq. (5.30)] in the cases of neutron decay (Chr 69), Ne^{19} decay (Cala 69), and Ar^{35} decay (Cala 67). Since in each of these cases, the initial and final nuclei belong to the same *isodoublet*, we now have $\langle 1 \rangle = 1$ to within about 1 percent, charge-dependent effects again being very small. Therefore, assuming $C_V = 1$ as before, we have $\xi(\text{mirror}) = 1 + |\rho|^2$.

The results for G_β are shown in Fig. 5.2.[1] We see that there is a large (and so far unexplained) discrepancy between the Ar^{35} value and the other values. Neglecting the Ar^{35} result, the Al^{26m} value differs slightly from the rest. The

[1] See Fre 68, 69a, and 69b; Cala 69; Wic 69; DeWit 69. These values include radiative corrections.

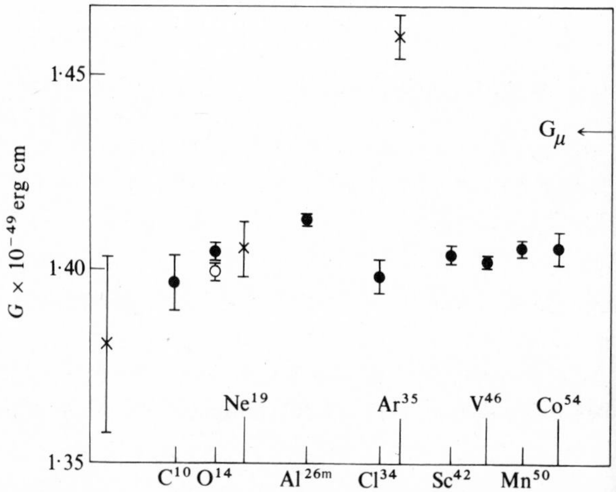

FIGURE 5.2
Values of G_β from $0^+ \to 0^+$ (pure Fermi) decays and mirror transitions for which $|\rho|^2 = |(C_A\langle\sigma\rangle)/(C_V\langle 1\rangle)|^2$ has been determined independently. [*From J. M. Freeman et al., Phys. Letters,* **30B:** *240 (1969).*]

remaining values are consistent with a value of G less than that obtained from muon decay. Taking the mean of the $0^+ \to 0^+ G_\beta$'s, we find

$$\frac{G_\mu - \bar{G}_\beta}{G_\mu} = +(2.2 \pm 0.5)\% \qquad (5.48)$$

Or, writing $G_\mu = G$, $G_\beta = G \cos \Theta$, we have

$$\Theta_V = 0.188 \pm 0.007 \qquad (5.49)$$

In the case of neutron decay, $\langle 1 \rangle = 1$ and $\langle \sigma \rangle = +\sqrt{3}$ exactly. Thus, by determining ft for the neutron and using G_β from the $0^+ \to 0^+$ decays, we may obtain the ratio $|C_A/C_V|$:

$$\left| \frac{C_A}{C_V} \right| = 1.23 \pm 0.01 \qquad (5.50)$$

The end-point energy in neutron decay is accurately known: $\Delta = M_n - M_p$, and the main experimental problem is determination of the neutron half-life. For this purpose, the decay in flight of a well-collimated beam of thermal neutrons of known intensity is carefully measured with calibrated counters. The result is (Chr 67)

$$t_{1/2} = 10.80 \pm 0.16 \text{ min} \qquad (5.51)$$

5.6 ELECTRON-NEUTRINO ANGULAR CORRELATION

We can average over initial nuclear spins but not integrate over the angle between electron and neutrino, in which case the differential transition probability is

$$dW = \frac{G_\beta^{\ 2}}{2\pi^3} F(Z, E)pE(\Delta - E)^2 \xi(1 + a\mathbf{v} \cdot \hat{\mathbf{p}}_\nu) \qquad (5.52)$$

The electron-neutrino angular correlation term $a\mathbf{v} \cdot \hat{\mathbf{p}}_\nu$ does not violate parity since $\mathbf{v}$ and $\hat{\mathbf{p}}_\nu$ are both polar vectors. We have already seen that for $e^\pm$ decay

$$a = \frac{|C_V|^2 |\langle 1 \rangle|^2 - \frac{1}{3}|C_A|^2 |\langle \sigma \rangle|^2}{|C_V|^2 |\langle 1 \rangle|^2 + |C_A|^2 |\langle \sigma \rangle|^2} = \frac{1 - \frac{1}{3}|\rho|^2}{1 + |\rho|^2}$$

Defining the Fermi fraction $x = 1/(1 + \rho^2)$, $(0 < x < 1)$, we have

$$a_{V,A} = \tfrac{4}{3}x - \tfrac{1}{3}$$

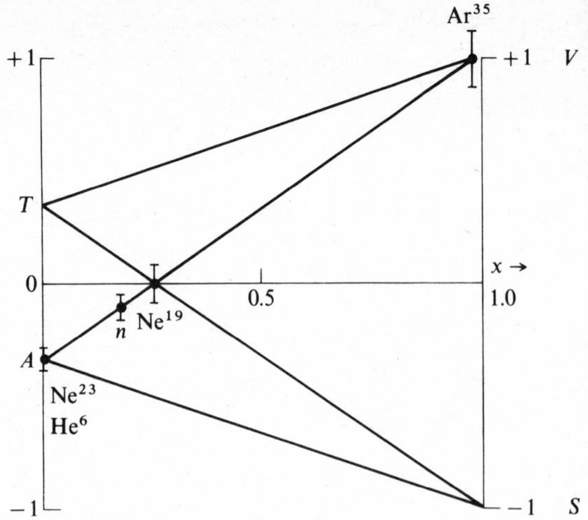

FIGURE 5.3
The $e\nu$ correlation coefficient a versus $x = 1/(1 + \rho^2)$

In Fig. 5.3 we plot $a_{V,A}$ versus x and also include the experimental a values for the following decays:

$$\begin{array}{lll} n \to pe^-\bar{\nu}_e{}^{\ddagger} & J_i = J_f = \frac{1}{2} & a = -0.091 \pm 0.039 \\ \mathrm{Ne}^{19} \to \mathrm{F}^{19}e^+\nu_e{}^{\S} & J_i = J_f = \frac{1}{2} & a = 0.00 \pm 0.08 \\ \mathrm{Ar}^{35} \to \mathrm{Cl}^{35}e^+\nu_e{}^{\P} & J_i = J_f = \frac{3}{2} & a = 0.97 \pm 0.14 \\ \mathrm{He}^6 \to \mathrm{i}^6e^-\bar{\nu}_e{}^{\ddagger\ddagger} & J_i = 0,\ J_f = 1 & a = -0.3343 \pm 0.0030 \\ \mathrm{Ne}^{23} \to \mathrm{Na}^{23}e^-\bar{\nu}_e{}^{\ddagger\ddagger} & J_i = \frac{5}{2},\ J_f = \frac{3}{2} & a = -0.33 \pm 0.03 \end{array} \tag{5.53}$$

The figure also shows the angular correlations to be expected if the decay interaction were S and/or T rather than V and A.[1] As Fig. 5.3 clearly shows, the data are consistent only with VA couplings.

‡ Gri 68.
§ All 59.
¶ All 59.
‡‡ Joh 63.
[1] Once the possibility of ST couplings is admitted, we must generalize the definition of x:

$$x = \frac{\xi_F{}^2}{\xi_F{}^2 + \xi_{GT}{}^2}$$

where $\xi_F{}^2$, $\xi_{GT}{}^2$ are defined by Eqs. (5.61) and (5.62) in Sec. 5.8.

Qualitatively it is easy to see why a takes the values shown in Fig. 5.3 for pure S, V, T, or A coupling. Consider, for example, Ar^{35} decay. The Gamow–Teller matrix element happens to be very small here, and so the transition is practically pure vector. Thus the leptons are essentially in a singlet spin state. Now, we know that e^+ has helicity $+v/c$ in beta decay, and since $v \approx c$, the e^+ is essentially completely right-handed. Also, the neutrino is left-handed. Therefore, the neutrino and e^+ must come off in the same direction since their total angular momentum is zero (see Fig. 5.4). Incidentally, this implies that the recoil momentum of the final nucleus will be rather large. (In fact, one actually determines the $e\nu$ correlation coefficient a by measuring the spectrum of recoil momenta.) On the other hand, suppose the positron and neutrino were both right-handed. Then, because the total angular momentum is zero, the e^+ and ν_e would come off in opposite directions, giving $a = -1$. In this case the average ion recoil momentum would be quite low. Similar considerations apply in the Gamow–Teller cases. In summary, since electron and positron helicities in beta decay are already known to be $-v/c$ and $+v/c$, respectively, from electron polarization measurements, the measurements of electron-neutrino correlations may be regarded as determinations of neutrino helicity.

Another very elegant method for determining the neutrino helicity is the experiment of Goldhaber, Grodzins, and Sunyar (Gol 58). The experiment concerns Eu^{152} which decays to Sm^{152*} by electron capture:

$$e^- + Eu^{152}(J = 0) \rightarrow Sm^{152*}(J = 1) + \nu_e$$
$$\qquad\qquad\qquad\qquad \llcorner\!\!\longrightarrow Sm^{152}(J = 0) + \gamma$$

Since the initial angular momentum is zero, ν_e and Sm^{152*} must have opposite projections of angular momentum on the recoil axis and the projection of Sm^{152*} must be transferred to the γ ray emitted in the transition to the ground state. It was found that the circular polarization of the γ rays is left-handed; hence the helicity of the neutrinos is -1. The helicity of antineutrinos has been shown to be $+1$ in a somewhat similar experiment involving the emission of a γ ray following the beta decay of Hg^{203} (Pal 69).

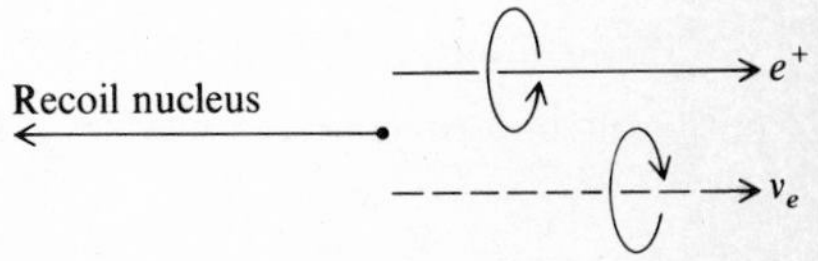

FIGURE 5.4

5.7 CORRELATIONS DEPENDING ON INITIAL NUCLEAR POLARIZATION

We turn next to the remaining terms in dW [Eq. (5.30)]. These are of the form

$$\hat{\mathbf{J}} \cdot (A\mathbf{v} + B\hat{\mathbf{p}}_\nu + D\mathbf{v} \times \hat{\mathbf{p}}_\nu)$$

and require polarized nuclei for their detection.

Observation of the beta decay asymmetry coefficient A in the decay of polarized Co^{60} constituted the first experimental demonstration of parity violation in 1957. Since then the beta decay asymmetry has become a useful experimental tool for studies of nuclear polarization in solids and gases, nuclear magnetic resonance, etc. Since detection of neutrinos emitted in beta decay is virtually impossible, observation of the *neutrino* asymmetry B is much more difficult than observation of A. The electron-recoil ion coincidences must be detected and correlated with nuclear polarization. The neutrino correlation is then inferred from conservation of linear momentum.

Observations of A and B are of fundamental importance in the case of neutron decay. We have already noted that measurement of A allows the determination of G_β from the ft values for each of the three decays: $n \to p$, $Ne^{19} \to F^{19}$, $Ar^{35} \to Cl^{35}$. In the case of free neutron decay, A and B are given by the formulas (already derived)

$$\xi A = -\tfrac{2}{3}|C_A|^2|\langle\sigma\rangle|^2 - \frac{1}{\sqrt{3}}(C_V C_A^* + C_A C_V^*)\langle 1\rangle\langle\sigma\rangle \qquad (5.54)$$

$$\xi B = +\tfrac{2}{3}|C_A|^2|\langle\sigma\rangle|^2 - \frac{1}{\sqrt{3}}(C_V^* C_A + C_A^* C_V)\langle 1\rangle\langle\sigma\rangle \qquad (5.55)$$

Inserting the results[1] $A(n) = -0.115 \pm 0.008$ and $B(n) = 1.01 \pm 0.05$ and the values $\langle 1\rangle = +1$, $\langle\sigma\rangle = +\sqrt{3}$ into (5.54), (5.55) and assuming C_A/C_V is real, we obtain

$$\frac{C_A}{C_V} = -1.26 \pm 0.02 \qquad (5.56)$$

in agreement with the result obtained from the neutron ft value, using $G_\beta(O^{14})$ [Eq. (5.50)].

Determinations of coefficient D in neutron and Ne^{19} decays are important as tests of time-reversal invariance. It is convenient for this purpose to express

[1] Chr 69.

D by the following formula [equivalent to Eq. (5.38)], valid for neutron or Ne^{19} decay:

$$D = \frac{2}{\sqrt{3}} \frac{|\rho| \sin(\theta + \phi)}{1 + |\rho|^2} \tag{5.57}$$

where θ is the relative phase of $\langle\sigma\rangle$ and $\langle 1\rangle$, ϕ is the relative phase of C_A and C_V, and $\rho = (C_A\langle\sigma\rangle)/(C_V\langle 1\rangle)$. If time-reversal invariance holds and we neglect electromagnetic final-state effects, ρ must be real and $\sin(\theta + \phi) = 0$. [In practice, it is assumed $\theta(n) = 0$, $\theta(Ne^{19}) = \pi$, in accord with established conventions and the nuclear wave function of Ne^{19}, and that a T violation manifests itself by a departure of ϕ from 0 or π.]

Now, it can be shown that the coulomb corrections to D are extremely small.[1] The first-order correction, proportional to $Z\alpha/p$, vanishes (fortuitously) for a pure VA beta interaction. If this correction is calculated for a general beta interaction with S and T couplings given their maximum possible values by limiting experiments, then

$$|D_{\text{coul}}^{(1)}(Ne^{19})| \leq 0.0017$$

A second-order coulomb correction should be of order 10^{-3} even if there is no fortuitous VA cancellation. Finally a correction which takes into account the

[1] Jac 57.

Table 5.1 EXPERIMENTAL VALUES OF CORRELATION COEFFICIENTS AND OTHER DECAY PARAMETERS FOR THE $\frac{1}{2}^+ \rightarrow \frac{1}{2}^+$ ALLOWED "MIRROR" TRANSITIONS $n \rightarrow pe^-\bar{\nu}_e$, $Ne^{19} \rightarrow F^{19}e^+\nu_e$

	$n \rightarrow pe^-\bar{\nu}_e$	$Ne^{19} \rightarrow F^{19}e^+\nu_e$
$\tau_{1/2}$	10.80 ± 0.16 min‡	17.36 ± 0.06 s#
Max electron KE	0.78 MeV	2.206 ± 0.001 MeV#
$\|\rho\| = \left\|\frac{C_A\langle\sigma\rangle}{C_V\langle 1\rangle}\right\|$	2.13	1.603 ± 0.006#
A	0.115 ± 0.008§	-0.039 ± 0.002††
B	1.01 ± 0.04§	-0.90 ± 0.13#
D	0.01 ± 0.01‖	0.001 ± 0.003††

‡ Chr 67.
§ Chr 69.
‖ Ero 70.
Cala 67.
†† D. Girvin, thesis, Univ. of California Physics Dept., Berkeley, Calif., 1972 (unpublished). Also, F. P. Calaprice, E. D. Commins, and D. Girvin, to be published.

interference between *weak magnetism* and the coulomb interaction in the final state yields (Call 67)

$$D_{\text{weak mag}}^{(1)}(\text{Ne}^{19}) = \frac{0.00026\, p_e}{p_{e,\,\max}}$$

Thus, altogether,

$$|D_{\text{coul}}(\text{Ne}^{19})| < 0.002 \qquad (5.58)$$

The coulomb correction in neutron decay should be even smaller.

The quantity D, like B, is measured by correlating electron-recoil ion coincidences with nuclear spin. Recent measurements of $D(n)$ and $D(\text{Ne}^{19})$ establish to rather high precision that $D = 0$ consistent with T invariance. See Table 5.1.[1]

5.8 BETA DECAY COUPLING CONSTANTS

We have now established that, neglecting momentum transfer-dependent terms, the amplitude $\mathfrak{A}$ takes the form

$$\mathfrak{A} = \frac{G \cos \Theta}{\sqrt{2}} \left[\sum_{i=1}^{A} \int d^3x\, \bar{\psi}'(\mathbf{x})\gamma_\mu(1 + \lambda\gamma^5)\tau_i^{\,+}\psi(\mathbf{x}) \cdot \bar{\psi}_e(\mathbf{x})\gamma^\mu(1 + \gamma^5)\psi_{\bar{\nu}}(\mathbf{x}) \right] \qquad (5.59)$$

where

$$\lambda = -\frac{C_A}{C_V} = +1.23 \pm 0.01$$

and

$$\Theta \approx 12^0$$

If λ were unity instead of $+1.23$, each nucleon bilinear form in (5.59) would contain the same factor $\gamma_\mu(1 + \gamma^5)$ as appears in the lepton current. The factor 1.23 occurs because the axial vector current, unlike the vector current, is renormalized by strong interaction. As we shall see in Chap. 8, the method of current commutators provides a means of calculating λ.

The amplitude (5.59) accounts successfully for all known results in beta decay where the momentum transfer dependence of the hadronic portion of the matrix element can be neglected. This includes not only the experimental results we have already mentioned, but others involving $\beta - \gamma$ circular polarizations, etc. However, we may ask whether (5.59) is necessary as well as sufficient.

[1] It should be noted that even if T violation occurs, the *charge symmetry* condition would require $D = 0$. See Sec. 4.7.

Could there be another formulation which describes the experimental results equally well?

Let us write down the most general derivative-free, four-fermion beta decay amplitude, consistent with proper Lorentz invariance and lepton conservation. What we shall do here is very much in the spirit of Sec. 2.4, devoted to muon decay. The most general amplitude is then

$$\mathfrak{A}' = \frac{G_\beta}{\sqrt{2}} \sum_i \int d^3x [\bar{\psi}_p O_i \psi_n][\bar{\psi}_e O_i (C_i + C_i' \gamma^5)\psi_\nu] \tag{5.60}$$

where the coefficients C_i, C_i' are to be determined by experiment and may be complex if we do not assume T invariance a priori. (The C_i, C_i' used here are *not* the same as in Sec. 2.4.) Since $\mathfrak{A}'$ may contain an arbitrary overall phase, we have 19 real parameters to be determined by experiment. We shall now see how the coefficients C_i, C_i' are limited.

1 Elimination of pseudoscalar coupling in the nonrelativistic limit. If we deal only with allowed decays, the pseudoscalar couplings (terms in C_p, C_p') give no contribution since $\bar{\psi}_p \gamma^5 \psi_n$ is zero in the nonrelativistic limit.

2 Restrictions due to spectrum shapes.[1] Retaining the other couplings a priori, we are led to the expectation of an allowed spectrum which departs from the usual statistical shape by an extra factor

$$1 \pm \gamma_0 b E^{-1} \qquad \text{for } e^\pm$$

where

$$\gamma_0 = [1 - (\alpha Z)^2]^{1/2}$$

and coefficient b, the *Fierz interference coefficient*, is given by

$$b = (\xi_F b_F + \xi_{GT} b_{GT})\xi^{-1}$$

where

$$\xi = \xi_F + \xi_{GT}$$

$$\xi_F = (\overset{*}{C_S} C_S + \overset{*}{C'_S} C'_S + \overset{*}{C_V} C_V + \overset{*}{C'_V} C'_V)|\langle 1 \rangle|^2 \tag{5.61}$$

$$\xi_{GT} = (\overset{*}{C_A} C_A + \overset{*}{C'_A} C'_A + \overset{*}{C_T} C_T + \overset{*}{C'_T} C'_T)|\langle \sigma \rangle|^2 \tag{5.62}$$

and

$$\xi_F b_F = \tfrac{1}{2}[\overset{*}{C_S} C_V + \overset{*}{C'_S} C'_V + cc]|\langle 1 \rangle|^2$$

$$\xi_{GT} b_{GT} = \tfrac{1}{2}[\overset{*}{C_T} C_A + \overset{*}{C'_A} C'_A + cc]|\langle \sigma \rangle|^2$$

Measured allowed spectrum shapes restrict the coefficients b_F and b_{GT} to

$$\begin{aligned} |b_F| &< 0.1 \\ b_{GT} &= -0.01 \pm 0.02 \end{aligned} \tag{5.63}$$

[1] For a detailed treatment, see Dan 68.

Thus, from spectrum measurements alone, we may conclude that the combinations $C_S^* C_V$, $C_S'^* C_V'$, $C_T^* C_A$, and $C_T'^* C_A'$ must be very small in magnitude.

3 Restrictions due to electron-neutrino angular correlations. In general, the $e\nu$ correlation coefficient a is given by

$$a = (\xi_F a_F + \xi_{GT} a_{GT})\xi^{-1}$$

where

$$\xi_F a_F = \tfrac{1}{2}[|C_V|^2 + |C_V'|^2 - |C_S|^2 - |C_S'|^2]|\langle 1 \rangle|^2$$

$$\xi_{GT} a_{GT} = -\tfrac{1}{6}[|C_A|^2 + |C_A'|^2 - |C_T|^2 - |C_T'|^2]|\langle \sigma \rangle|^2$$

The measurements [Eqs. (5.53)] on the pure GT decays of He^6 and Ne^{23} are sufficiently accurate to exclude all tensor couplings:

$$C_T = C_T' = 0 \tag{5.64}$$

The Ar^{35} result gives

$$|C_S|^2 + |C_S'|^2 < 0.1(|C_V|^2 + |C_V'|^2) \tag{5.65}$$

4 Restrictions due to observed electron helicities. Next we return to measurements of longitudinal electron polarization. First we write the form

$$\bar{\psi}_e O_i(C_i + C_i'\gamma^5)\psi_\nu \tag{5.66}$$

in a slightly different way, as follows:

$$(C_i + C_i'\gamma^5) = (C_i + C_i')\left(\frac{1+\gamma^5}{2}\right) + (C_i - C_i')\left(\frac{1-\gamma^5}{2}\right)$$

Further, we note that

$$O_i\left(\frac{1 \pm \gamma^5}{2}\right) = \left(\frac{1 \pm \gamma^5}{2}\right)O_i \qquad i = S, T, P$$

and

$$O_i\left(\frac{1 \pm \gamma^5}{2}\right) = \left(\frac{1 \mp \gamma^5}{2}\right)O_i \qquad i = V, A$$

Then, (5.66) becomes

$$\bar{\psi}_{eR}(C_i + C_i')O_i\psi_\nu + \bar{\psi}_{eL}(C_i - C_i')O_i\psi_\nu \qquad i = S, T, P \tag{5.67}$$

and

$$\bar{\psi}_{eL}(C_i + C_i')O_i\psi_\nu + \bar{\psi}_{eR}(C_i - C_i')O_i\psi_\nu \qquad i = V, A \tag{5.68}$$

where

$$\psi_{eL} = \left(\frac{1+\gamma^5}{2}\right)\psi_e$$

$$\psi_{eR} = \left(\frac{1-\gamma^5}{2}\right)\psi_e$$

As we have noted, numerous experiments in which the longitudinal $e^\pm$ polarizations are measured in pure F, pure GT, and mixed transitions by a variety of methods, including Møller–Bhabha scattering and Mott scattering, reveal that the longitudinal electron polarization $\mathcal{I}$ is always $\pm v/c$ (for $e^{\pm}$). This result, taken together with (5.67) and (5.68), implies that

$$C_V = C'_V \qquad C_A = C'_A$$

and

$$C_S = -C'_S \qquad C_T = -C'_T$$

Since we already know that $C_T = C'_T = 0$ and C_S, C'_S are extremely small, we are led to the final conclusion that

$$C_V = C'_V \qquad C_A = C'_A \qquad C_S = C'_S = C_T = C'_T = 0$$

It should be noted, however, that the pseudoscalar coupling constants C_P, C'_P could be quite large (C_P, $C'_P \approx C_V$, C_A) and still escape observation. Any appreciable pseudoscalar coupling would certainly reveal itself in $\pi^{\pm}$ decay and therefore can be ruled out in that case. However, universality must be assumed in order to conclude that there is also no pseudoscalar coupling in beta decay as well.

5.9 WEAK MAGNETISM

An important consequence of the conserved vector current hypothesis is the prediction that $f_2(0)$ is related to the magnetic moments

$$\mu_p = 2.79 \frac{e\hbar}{2mc} = \frac{2.79e}{2}$$

and

$$\mu_n = \frac{-1.91e\hbar}{2m_n c} = \frac{-1.91e}{2}$$

of the proton and neutron, respectively, as follows:

$$ef_2(0) = \frac{\mu_p - \mu_n - 1}{2} \tag{5.69}$$

This prediction can be tested in nuclear beta decay by observing the electron energy spectra in the decays $B^{12} \to C^{12}e^-\bar{\nu}_e$, $N^{12} \to C^{12}e^+\nu_e$ (see Fig. 5.5). According to the theory, which we shall sketch below, the usual allowed statistical shape for these decays (pure axial vector, $1^+ \to 0^+$) is modified by a correction factor

$$C_{\pm}(E) \approx 1 \pm \frac{8}{3}aE \qquad \text{for } e^{\pm} \tag{5.70}$$

where

$$a = \frac{\mu_p - \mu_n}{2m_p}\left|\frac{C_V}{C_A}\right|$$

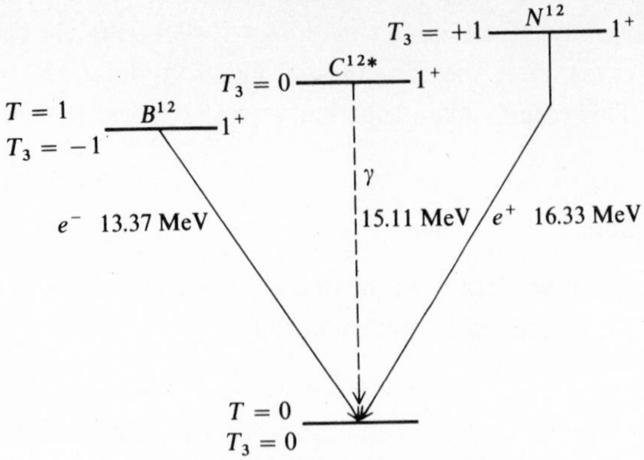

FIGURE 5.5
Decay schemes for $B^{12} \to C^{12}$, $N^{12} \to C^{12}$.

The correction factor arises mainly from interference between forbidden corrections to the vector amplitude (mostly the f_2 term) and the allowed axial vector amplitude. The spectral shapes have been measured accurately for B^{12} and N^{12} decays by Lee, Mo, and Wu (Lee 63). The results are in good agreement with Eq. (5.70) (see Fig. 5.6).

Our starting point for discussion of the theory, first worked out by Gell–Mann (Gel 58), is Eq. (5.9) for amplitude $\mathfrak{A}$. Generalizing $\mathfrak{A}$ to describe positron as well as negatron emissions, we have

$$\begin{aligned}
\mathfrak{A}(e^\pm) = \frac{G_\beta}{\sqrt{2}} \sum_{i=1}^{A} \int d^3r \Big[& C_V \chi'^\dagger j_0 \tau_i^{\pm} \chi \\
& + \frac{i}{2} C_V \chi'^\dagger (\boldsymbol{\sigma}_i \cdot \nabla \boldsymbol{\sigma}_i \cdot \mathbf{j} \tau_i^{\pm} \chi + \boldsymbol{\sigma}_i \cdot \mathbf{j}_i \boldsymbol{\sigma}_i \cdot \nabla \tau_i^{\pm} \chi) \\
& - f_2(0) \chi'^\dagger \boldsymbol{\sigma}_i \cdot (\nabla \times \mathbf{j}) \tau_i^{\pm} \chi - C_A \chi'^\dagger \boldsymbol{\sigma}_i \cdot \mathbf{j} \tau_i^{\pm} \chi \\
& + \frac{i}{2} C_A \chi'^\dagger (\boldsymbol{\sigma}_i \cdot \nabla j_0 \tau_i^{\pm} \chi + j_0 \boldsymbol{\sigma}_i \cdot \nabla \tau_i^{\pm} \chi) \\
& \mp i g_2(0) \chi'^\dagger \boldsymbol{\sigma}_i \cdot \nabla j_0 \tau_i^{\pm} \chi \pm g_2(0)(E + \omega) \chi'^\dagger \boldsymbol{\sigma}_i \cdot \mathbf{j} \tau_i^{\pm} \chi \\
& - \frac{i}{2} g_3(0)(E + \omega) \chi'^\dagger (\boldsymbol{\sigma}_i \cdot \nabla j_0 \tau_i^{\pm} \chi - j_0 \boldsymbol{\sigma}_i \cdot \nabla \tau_i^{\pm} \chi) \\
& - \frac{i}{2} g_3(0) \chi'^\dagger (\boldsymbol{\sigma}_i \cdot \nabla \nabla \cdot \mathbf{j} \tau_i^{\pm} \chi - \nabla \cdot \mathbf{j} \boldsymbol{\sigma}_i \cdot \nabla \tau_i^{\pm} \chi \Big]
\end{aligned} \tag{5.71}$$

It is important to note that in (5.71) the signs of the terms in $g_2(0)$ depend on whether we are considering electron or positron emission. This comes about

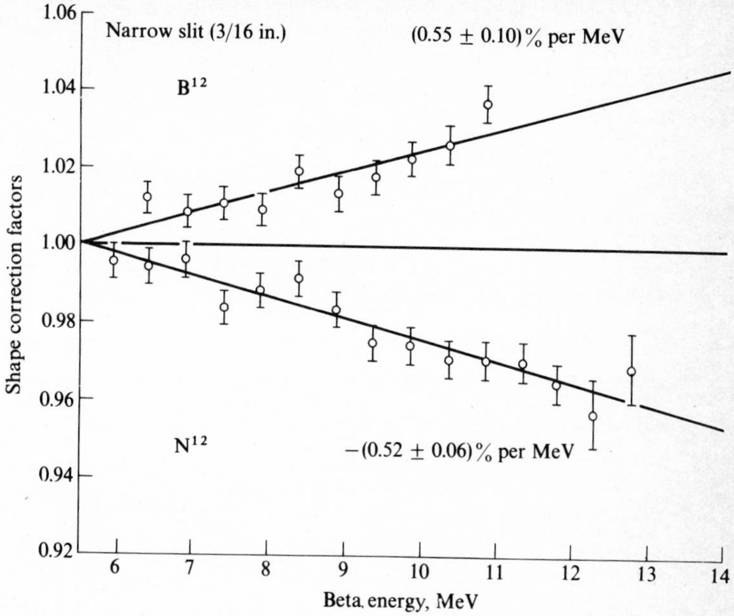

FIGURE 5.6
Shape correction factors for the decays $B^{12} \to C^{12}$, $N^{12} \to C^{12}$ versus beta energy. Solid curves are given by theory; the points are from experimental results (Lee 63).

because under a G transformation the g_2 terms have opposite symmetry from all of the others. Of all the terms in (5.71), only three contribute significantly to the weak magnetism effect. These are the term in $i/2\ C_V$, the term in $-f_2(0)$, and the allowed term in C_A. The allowed term in C_V contributes nothing because we are dealing with $1^+ \to 0^+$ transitions. Also, the contributions to the differential effect on spectral shape from the questionable g_2 terms, the forbidden corrections in $-i/2\ C_A$, and induced pseudoscalar terms in $g_3(0)$ have negligible effect. Thus,

$$\mathfrak{A}_{\text{eff}}(e^\pm) \approx \frac{G_\beta}{\sqrt{2}} \sum_{i=1}^{A} \int d^3\mathbf{r} \Big[-C_A \chi'^\dagger \boldsymbol{\sigma}_i \cdot \mathbf{j} \tau_i^\pm \chi$$

$$+ \frac{i}{2} C_V \chi'^\dagger (\boldsymbol{\sigma}_i \cdot \nabla \boldsymbol{\sigma}_i \cdot \mathbf{j} \tau_i^\pm \chi + \boldsymbol{\sigma}_i \cdot \mathbf{j} \boldsymbol{\sigma}_i \cdot \nabla \tau_i^\pm \chi)$$

$$- f_2(0) \chi'^\dagger \boldsymbol{\sigma}_i \cdot (\nabla \times \mathbf{j}) \tau_i^\pm \chi \Big] \tag{5.72}$$

These terms are reduced as follows. The first is just

$$-\frac{G_\beta}{\sqrt{2}} C_A \langle \boldsymbol{\sigma} \rangle \cdot \mathbf{j}$$

The second is

$$\frac{i}{2} \frac{G_\beta C_V}{\sqrt{2}} \sum_{i=1}^{A} \int d^3\mathbf{r} \cdot \chi'^\dagger (\boldsymbol{\sigma}_i \cdot \nabla \boldsymbol{\sigma}_i \cdot \mathbf{j} \tau_i^{\pm} \chi + \boldsymbol{\sigma}_i \cdot \mathbf{j} \boldsymbol{\sigma}_i \cdot \nabla \tau_i^{\pm} \chi)$$

$$= \frac{i}{2} \frac{G_\beta C_V}{\sqrt{2}} \sum_{i=1}^{A} \int d^3\mathbf{r}\, \chi'^\dagger [\nabla \cdot \mathbf{j} \tau_i^{\pm} \chi + \mathbf{j} \cdot \nabla \tau_i^{\pm} \chi + i \boldsymbol{\sigma}_i \cdot (\nabla \times \mathbf{j} \tau_i^{\pm} \chi) - i \boldsymbol{\sigma}_i \cdot (\nabla \tau_i^{\pm} \chi) \times \mathbf{j}]$$

$$= \frac{i}{2} \frac{G_\beta C_V}{\sqrt{2}} \sum_{i=1}^{A} \int d^3\mathbf{r}\, \chi'^\dagger [2\mathbf{j} \cdot \nabla \tau_i^{\pm} \chi + i \boldsymbol{\sigma}_i \cdot (\nabla \times \mathbf{j}) \tau_i^{\pm} \chi + (\nabla \cdot \mathbf{j}) \tau_i^{\pm} \chi] \tag{5.73}$$

Now we can write $\mathbf{j} = \mathbf{j}(0) e^{-i\mathbf{k} \cdot \mathbf{r}}$ where $\mathbf{k} = \mathbf{p}_e + \mathbf{p}_{\bar{\nu}}$. Then

$$2\mathbf{j} \cdot \nabla \tau_i^{\pm} \approx 2\mathbf{j}(0) \cdot (1 - i\mathbf{k} \cdot \mathbf{r}) \nabla \tau_i^{\pm} \chi$$

$$i \boldsymbol{\sigma}_i \cdot (\nabla \times \mathbf{j}) \tau_i^{\pm} \chi \approx \boldsymbol{\sigma}_i \cdot [\mathbf{k} \times \mathbf{j}(0)] \tau_i^{\pm} \chi$$

Therefore the right-hand side of (5.72) becomes

$$\frac{G_\beta C_V}{\sqrt{2}} \sum_{i=1}^{A} \int d^3\mathbf{r} \chi'^\dagger \left[(\mathbf{k} \cdot \mathbf{r}) \mathbf{j}(0) \cdot \nabla \tau_i^{\pm} \chi + \frac{i}{2} \boldsymbol{\sigma}_i \cdot (\mathbf{k} \times \mathbf{j}(0)) \tau_i^{\pm} \chi \right] \tag{5.74}$$

since the other contributions to (5.73) vanish for $1^+ \to 0^+$ transitions. Now

$$\mathbf{k} \cdot \mathbf{r} \mathbf{j}(0) \cdot \nabla = \tfrac{1}{3} \mathbf{k} \cdot \mathbf{j}(0) \mathbf{r} \cdot \nabla + \tfrac{1}{2} \mathbf{k} \times \mathbf{j}(0) \cdot \mathbf{r} \times \nabla$$
$$+ \tfrac{1}{4} [k_i j_j(0) + k_j j_i(0) - \tfrac{2}{3} \delta_{ij} \mathbf{k} \cdot \mathbf{j}(0)] (r_i \nabla_j + r_j \nabla_i - \tfrac{2}{3} \delta_{ij} \mathbf{r} \cdot \nabla]$$

Of these terms, only the middle one contributes to (5.74) for $1^+ \to 0^+$ transitions. Thus (5.74) becomes

$$\frac{G_\beta}{\sqrt{2}} C_V \mathbf{k} \times \mathbf{j}(0) \cdot \langle \mathbf{L} + \boldsymbol{\sigma} \rangle$$

where
$$\langle \mathbf{L} \rangle = -i \int \sum_{i=1}^{A} \chi'^\dagger \mathbf{r} \times \nabla \tau_i^{\pm} \chi \, d^3 r$$

In fact, this matrix element $\langle \mathbf{L} \rangle$ makes only a negligible contribution to the spectrum shape corrections. Therefore, since $C_V = 1$, and taking into account (5.69) and the $f_2(0)$ term in (5.71), we have finally

$$\mathfrak{A}_{\text{eff}}(e^{\pm}) \approx \frac{G_\beta}{\sqrt{2}} \left[-C_A \langle \boldsymbol{\sigma} \rangle_{\pm} + i \left(\frac{\mu_p - \mu_n}{2} \right) \langle \boldsymbol{\sigma} \rangle_{\pm} \times \mathbf{k} \right] \cdot \mathbf{j}_{\pm} \tag{5.75}$$

Now
$$\mathbf{j}_- = \bar{u}_e \gamma (1 + \gamma^5) v_\nu \tag{5.76}$$

while
$$\mathbf{j}_+ = \bar{u}_\nu \gamma (1 + \gamma^5) v_e \tag{5.77}$$

The matrix elements $\langle\sigma\rangle_-$ (for B^{12} decay) and $\langle\sigma\rangle_+$ (for N^{12} decay) may be assumed equal, to a good approximation, since B^{12} and N^{12} are both members of the same isotriplet. With this assumption, it is straightforward to calculate the transition rate as a function of electron energy, from (5.75), using (5.76) or (5.77). This interference between the allowed term in C_A and the term in $i[(\mu_p - \mu_n)/2]$ of (5.75) accounts for the correction factors of (5.70). The sign difference in the shape correction factors for $e^{\pm}$ emission arises from the $(1 \pm \gamma^5)$ factors in (5.76) and (5.77).

The name *weak magnetism* is appropriate since, in the description of $1^+ \to 0^+$ (magnetic dipole) gamma transitions, a strictly analogous formulation applies in which j is replaced by the vector potential $\mathbf{A}(\mathbf{B} = \nabla \times \mathbf{A})$ and $G_\beta/\sqrt{2}$ is replaced by the electric charge e.

5.10 DO SECOND-CLASS CURRENTS EXIST?

We now consider somewhat further the possible existence of the second-class term

$$-i\bar{u}_p[g_2(q^2)\sigma^{\mu\nu}q_\nu\gamma^5]u_n \qquad (5.78)$$

in the axial vector beta decay amplitude. It will be recalled that if T invariance holds, g_2 is real, while if charge symmetry holds, g_2 is imaginary. (Of course, G invariance would require $g_2 = 0$.) Concerning the existence of nonzero g_2, real or imaginary, unfortunately there is no definite experimental evidence.

If $\operatorname{Im} g_2 \neq 0$ and charge symmetry is valid, we must have a T violation. In fact, Cabibbo (Cab 64), Maiani (Mai 68), and Kim and Primakoff (Kim 69) have speculated that this is where T violation might be found in beta decay. However, present results of the most sensitive tests for T in beta decay ($n \to pe^-\bar{\nu}_e$, $Ne^{19} \to F^{19}e^+\nu_e$) could not reveal the existence of $\operatorname{Im} g_2$ even if it were comparable in magnitude to C_A because of the very small momentum transfer in these decays:

$$\frac{|q|}{m_p} \approx 10^{-3}$$

It has been suggested that $\operatorname{Re} g_2$ might manifest itself in an asymmetry in the ft values of *mirror pairs* of decays (for example, for $A = 8, 9, 12, 13, 17, 18, 20, 24, 28$, and 30). For example, the ft value of N^{12} decay is about 10 percent greater than $ft\ (B^{12} \to C^{12})$. One might expect a difference in ft for these decays simply because of the difference in shape correction factors due to weak magnetism, which we discussed in the last section. In addition, other effects such as isospin mixing, electromagnetic corrections, and binding energy corrections also contribute to the fractional difference in ft values, δ:

$$\delta = \frac{(ft)^+}{(ft)^-} - 1 \qquad (5.79)$$

On the other hand, the above-mentioned effects seem inadequate to account for the large $\delta(B^{12}, N^{12})$, especially since it can be shown that binding energy corrections approximately cancel the other corrections to the ft values mentioned above.

However, basing calculations on the simplest possible nuclear models (independent nucleons) and neglecting the electromagnetic, weak magnetism, binding energy and isospin mixing corrections, one can show that an induced tensor term will make the following contribution to δ (see, e.g., Huf 63):

$$\delta \approx \frac{4}{3}\left|\frac{C_V}{C_A}\right| |g_2(0)| (E_0{}^+ + E_0{}^-)$$

where $E_0{}^+$ and $E_0{}^-$ are the maximum positron (electron) kinetic energies in megaelectron volts. On the other hand, meson exchange effects, which are not taken into account in an independent particle model, could radically alter the dependence of δ on $g_2(0)$ and $E_0{}^+$, $E_0{}^-$ (see, e.g., Del 71).

Although preliminary experimental data with δ for the mirror pairs $A = 8, 9, 12, 13, 17, 18, 20, 24, 28$, and 30 seemed to suggest a linear dependence of δ on $E_0{}^+ + E_0{}^-$ with a slope giving $|g_2(0)| \approx 1$ (see, e.g., Wil 70a,b,c, Alb 70), more recent observations of the $A = 8$ system only confuse the issue. Here, Li^8 and B^8 decay to a very broad excited state of Be^8, which, in turn, decays to two α particles with a spectrum of energies. We can plot δ versus the sum of the beta decay energies, as revealed by the α spectrum, and, again, we expect a linear dependence on $|g_2(0)|$ in the simplest approximation. However, experiment shows that in this case δ is independent of $E_0{}^+ + E_0{}^-$.

In conclusion, there seems to be no definite experimental evidence for $g_2 \neq 0$ but no utterly compelling theoretical argument or experimental demonstration that $g_2 = 0$.

5.11 DOUBLE BETA DECAY

There are two possible varieties of double beta decay: neutrinoless (or $\beta\beta_0$)

$$(Z, N) \rightarrow (Z + 2, N - 2) + 2e^- \qquad (5.80)$$

and two-neutrino (or $\beta\beta_2$)

$$(Z, N) \rightarrow (Z + 2, N - 2) + 2e^- + 2\bar{\nu} \qquad (5.81)$$

If $\beta\beta_0$ decay exists, lepton conservation must be violated, as noted in Chap. 1. Conversely, if the amplitude for $\beta\beta_0$ is strictly zero, there is always a way to define electron lepton number in such a way that it is conserved.

How can we distinguish $\beta\beta_0$ from $\beta\beta_2$ experimentally? If lepton conservation were completely violated so that the matrix elements for $\beta\beta_0$ and $\beta\beta_2$ were about the same order of magnitude, then $\beta\beta_0$ decay would proceed about 10^5 times faster than $\beta\beta_2$ decay for a given nucleus since, in the $\beta\beta_0$ case, much more phase space would be available. Thus, for example, we would expect the following $\beta\beta_0$ lifetime for Ca^{48} decay:

$$\tau_{\frac{1}{2}}(\mathrm{Ca}^{48}, \beta\beta_0) \approx 10^{16} \text{ years} \qquad (5.82)$$

while

$$\tau_{\frac{1}{2}}(\mathrm{Ca}^{48}, \beta\beta_2) \approx 10^{21} \text{ years} \qquad (5.83)$$

Moreover, in $\beta\beta_0$ decay, the electrons would share the full available energy (4.24 MeV in Ca^{48}), whereas in $\beta\beta_2$ decay the sum of the electron energies would vary continuously from 0 to $E_{\max}$. (Detailed theoretical predictions of double-beta-decay rates are extremely complex and difficult, however. See, e.g., Ros 65.)

Recently an attempt to find $\beta\beta_0$ decay of Ca^{48} was carried out by Bardin et al. (Bar 67). A search was made for coincidence electron tracks in a magnetic field in a two-gap discharge counter containing a thin Ca^{48} source as center plate. The sum of the energies of suspected events was required to be >3 MeV. From the small number of genuine events it was concluded that

$$\tau_{\frac{1}{2}}(\mathrm{Ca}^{48}, \beta\beta_0) > 1.6 \times 10^{21} \text{ years} \qquad (5.84)$$

Comparing (5.84) with (5.82), we conclude that any lepton-nonconserving amplitude present in double beta decay must be less than 10^{-3} of the ordinary lepton-conserving amplitude responsible for $\beta\beta_2$.

Actually, unambiguous positive evidence for $\beta\beta_2$ decays has come from an entirely different direction, namely, mass spectrographic analysis of certain ores to determine isotopic excesses of occluded rare gases (over and above those isotopic ratios found in atmospheric gas). For example, in work reported by Kirsten et al. (Kir 68), native tellurium ore from a mine in Colorado, with a rather reliable potassium–argon dated age of $(1.31 \pm 0.14) \times 10^9$ years, was analyzed for relative abundances of the xenon isotopes occluded within. A very large Xe^{130} anomaly was found, unaccompanied by excesses of other Xe isotopes. (See Fig. 5.7.) A careful analysis shows that the Xe^{130} must have been generated by the $\beta\beta_2$ decay of Te^{130} and that the half-life is

$$\tau_{1/2}(\mathrm{Te}^{130}) = 10^{21.34 \pm 0.12} \text{ years} \qquad (5.85)$$

in good agreement with a theoretical estimate of the $\beta\beta_2$ half-life (Ros 65):

$$\tau_{1/2}(\mathrm{Te}^{130}, \beta\beta_2, \text{theo}) = 10^{22.5 \pm 2.0} \text{ years} \qquad (5.86)$$

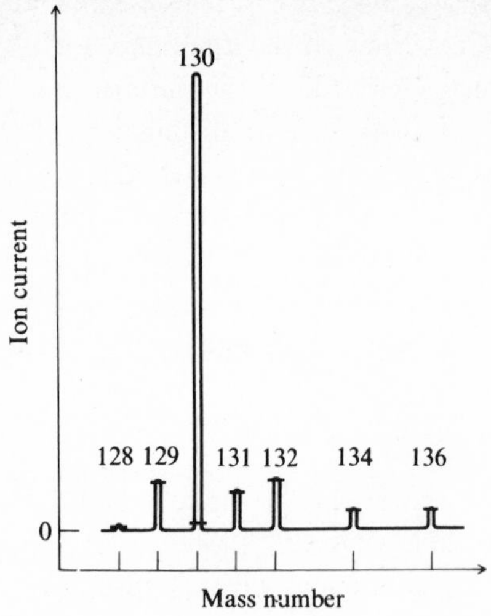

FIGURE 5.7
Isotopic composition of xenon extracted from native tellurium ore. The horizontal lines indicate the maximum contribution of atmospheric xenon. [*From T. Kirsten et. al., Phys. Rev. Letters,* **20**:*1300* (*1968*).]

Other workers [E. C. Alexander et al. (Ale 69)] have confirmed the results of Kirsten et al., so that Eq. (5.85) may be regarded as being very well established.

Good evidence has been found recently for the $\beta\beta_2$ decay—$Se^{82} \to Kr^{82}$—from mass spectroscopic analyses of krypton occluded in six different selenium ores (Kir 69). The preliminary half-life of Se^{82} is 1.4×10^{20} years, with an error of about 20 percent, in reasonable agreement with the theoretical prediction (Ros 65):

$$\tau_{1/2}(\mathrm{Se}^{82}, \beta\beta_2, \mathrm{theo}) = 10^{22.15 \pm 2} \text{ years}$$

PROBLEMS

1 Consider a pure Fermi allowed beta transition

$$Z, N \to Z+1, N-1 + e^- + \bar{\nu}_e$$

where the maximum electron energy is E_0. Calculate the differential transition probability per unit time, to first order in α, for a beta decay accompanied by *inner bremsstrahlung* (emission of a photon in addition to an $e\bar{\nu}_e$ pair) as in the figure.

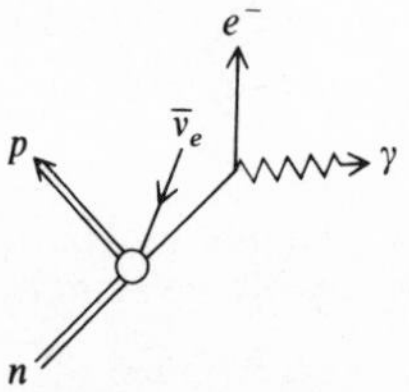

The photon has energy $\hbar\omega$ in range $\hbar\, d\omega$, the electron has energy E in range dE, and the angle between photon and electron is θ in range of solid angle $d\Omega$. Compare the result with the transition probability for ordinary beta decay with no photon emission. Assume the initial nuclei are unpolarized, and neglect all coulomb corrections to the ordinary beta decay.

2 Consider a $\beta\gamma$ cascade as shown in the figure.

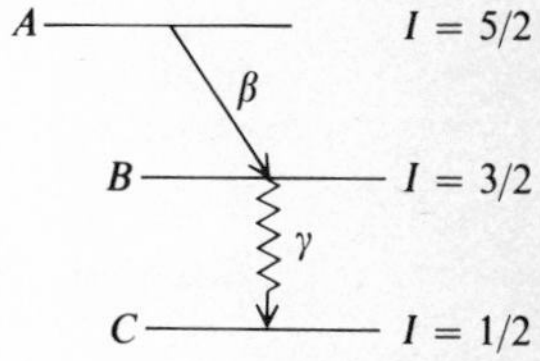

The nuclear spins of levels A, B, C are $\frac{5}{2}$, $\frac{3}{2}$, and $\frac{1}{2}$, respectively. The β transition is allowed, and the γ transition is pure electric dipole. Assume that the helicity of a beta decay electron is $-v/c$ for decay of unpolarized nuclei. Suppose the population distribution of the nuclei in state A, among the various $m_I(A)$ values, is known:

m_I	Relative population
$\frac{5}{2}$	a
$\frac{3}{2}$	b
$\frac{1}{2}$	c
$-\frac{1}{2}$	d
$-\frac{3}{2}$	e
$-\frac{5}{2}$	$f = 1 - a - b - c - d - e$

Use only elementary principles of symmetry in quantum mechanics (Clebsch–Gordan coefficients, rotation matrices, etc.) to show that

(*a*) When averaged over all directions of e^- emission, the polarization of the nuclei after beta decay (in level B) is $P(B) = P(A)$.

(*b*) The angular distribution of decay electrons is $I_e(\theta_e) \approx 1 - v/c\, P(A) \cos\theta_e$ where θ_e is the angle between the electron momentum and the z axis.

(*c*) If a decay electron is observed along the $+z$ axis, the polarization of the corresponding nucleus in level B is

$$P(B) = \frac{P(A)\, v/c(a + b/5 - c/5 - d/5 + e/5 + f)}{1 - v/c\, P(A)}$$

(*d*) If $P(A) = +1$, the total intensity of γ radiation without regard to circular polarization is proportional to $1 + \cos^2\theta_\gamma$ where θ_γ is the angle between the z axis and direction of photon momentum.

3 Consider the shape of the electron spectrum for allowed decay near the *end point.* How is this shape affected if the neutrino is assumed to possess a finite rest mass?

4 Derive an expression for the energy spectrum of the recoil proton in neutron beta decay. (Note that in the usual analysis of beta decay, the final nucleus recoil is neglected.)

REFERENCES

Alb 70 Alburger, D. E., and D. H. Wilkinson: *Phys. Letters,* **32B**:190 (1970).

Ale 69 Alexander, E. C., B. Srinivasan, and O. K. Maund: *Earth & Planetary Sci. Letters,* **5**:478 (1969).

All 59 Allen, J. S., R. L. Burman, W. B. Hermannsfeldt, P. Stähelin, and T. H. Braid: *Phys. Rev.,* **116**:134 (1959).

Bar 67 Bardin, R. K., P. J. Gollon, J. D. Ullman, and C. S. Wu: *Phys. Letters,* **26B**:111 (1967).

Bee 68 Beek, E.: *Z. Physik.,* **216**:229 (1968).

Ber 69 Bergkvist, K. E.: *Proc. Topical Conf. Weak Interactions, CERN Geneva,* p. 91 (1969).

Bli 69 Blin-Stoyle, R. J.: *Phys. Letters,* **29B**:12 (1969).

Cab 64 Cabibbo, N.: *Phys. Letters,* **12**:137 (1964).

Cala 67 Calaprice, F. P.: doctorial dissertation, University of California, Berkeley (University of California Lawrence Radiation Laboratory Report 17551), June 1967. [See also: F. P. Calaprice, E. D. Commins, and D. A. Dobson: *Phys. Rev.,* **137**:B1453 (1965).]

Cala 69 Calaprice, F. P., E. D. Commins, H. M. Gibbs, G. L. Wick, and D. A. Dobson: *Phys. Rev.,* **184**:1117 (1969).

Call 67 Callan, C. G., Jr., and S. B. Treiman: *Phys. Rev.,* **162**:1494 (1967).

Chr 67 Christenson, C. J., et al.: *Phys. Letters,* **26B**:11 (1967).

Chr 69 Christenson, C. J., V. E. Krohn, and G. R. Ringo: *Phys. Letters,* **28B**:411 (1969).

Dan 68 Daniel, H.: *Rev. Mod. Phys.,* **40**:659 (1968).

Dar 69 Daris, R., and C. St. Pierre: *Nucl. Phys.,* **A138**:545 (1969).

Del 71 Delorme, J., and M. Rho: *Phys. Letters,* **34B**:238 (1971).

DeWit 69 DeWit, P., and C. Van der Leun: *Phys. Letters,* **30B**:639 (1969).

Ero 70 Erozolimski, B. G., L. A. Bondarenko, Yu. A. Mostovoi, B. A. Obinyakov, V. P. Zakharova, and V. A. Titov: *Sov. J. Nucl. Phys.,* **11**:583 (1970).

Fra 65 Frauenfelder, H., and R. M. Steffen: The Helicity of Beta Particles, in K. Siegbahn (ed.), "Alpha, Beta, and Gamma Ray Spectroscopy," vol. 2, p. 431, North-Holland, Amsterdam, 1965.

Fre 68 Freeman, J. M., J. G. Jenkin, D. C. Robinson, G. Murray, and W. E. Burcham: *Phys. Letters,* **27B**:156 (1968).

Fre 69a Freeman, J. M., D. C. Robinson, and G. L. Wick: *Phys. Letters,* **30B**:240 (1969).

Fre 69b Freeman, J. M., J. G. Jenkin, and G. Murray: *Nucl. Phys.,* **A124**:393 (1969).

Gel 58 Gell–Mann, M.: *Phys. Rev.,* **111**:362 (1958).

Gol 58 Goldhaber, M., L. Grodzins, and A. W. Sunyar: *Phys. Rev.,* **109**:1015 (1958).

Gri 68 Grigoryev, V. K., A. P. Grishin, V. V. Vladimirsky, E. S. Nikolaevsky, and D. P. Zharkov: *Sov. J. Nucl. Phys.,* **6**:239 (1968).

Huf 63 Huffaker, J. N., and E. Grenling: *Phys. Rev.,* **132**:738 (1963).

Jac 57 Jackson, J. D., S. B. Treiman, and H. W. Wyld: *Nucl. Phys.,* **4**:206 (1957).

Joh 63 Johnson, C. H., F. Pleasanton, and T. A. Carlson: *Phys. Rev.,* **132**:1149 (1963).

Käl 68 Källen, G.: Radiative Corrections in Elementary Particle Physics, in "Springer Tracts in Modern Physics," vol. 46, pp. 67–132, Springer, Berlin, 1968.

Kim 69 Kim, C. W., and H. Primakoff: *Phys. Rev.,* **180**:1502 (1969).

Kir 68 Kirsten, T., O. A. Schaeffer, E. Norton, and R. W. Stoenner: *Phys. Rev. Letters,* **20**:1300 (1968). [See also: T. Kirsten, W. Gentner, and O. A. Schaeffer: *Z. Physik.,* **202**:273 (1967).]

Kir 69 Kirsten, T., and H. W. Müller: *Earth & Planetary Sci. Letters,* **6**:271 (1969).

Kon 66 Konopinski, E. J.: "The Theory of Beta Radioactivity," Oxford, London, 1966.

Laz 70 Lazarus, D. M., and J. S. Greenberg: *Phys. Rev.,* **2D**:45 (1970).

Lee 63 Lee, Y. K., L. Mo, and C. S. Wu: *Phys. Rev. Letters,* **10**:253 (1963).

Mai 68 Maiani, L.: *Phys. Letters,* **26B**:538 (1968).

NBS 52 National Bureau of Standards Applied Mathematics Ser. 13, "Tables for Analysis of Beta Spectra," chap. 2, 1952. (Note that these tables do not contain corrections for finite nuclear size and atomic screening.)

Pal 69 Palathingal, J. C.: *Bull. Am. Phys. Soc.,* **14**:587 (1969).

Ros 65 Rosen, S., and H. Primakoff: Double Beta Decay, in K. Siegbahn (ed.), "Alpha, Beta, and Gamma Ray Spectroscopy," vol. 2, p. 1499, North-Holland, Amsterdam, 1965.

Sch 66 Schopper, H. F.: "Weak Interactions and Nuclear Beta Decay," North-Holland, Amsterdam, 1966.

Wic 69 Wick, G. L., D. C. Robinson, and J. M. Freeman: *Nucl. Phys.,* **A138**:209 (1969).

Wil 70a Wilkinson, D. H.: *Phys. Letters,* **31B**:447 (1970).

Wil 70b Wilkinson, D. H.: *Phys. Rev. Letters,* **24**:1134 (1970).

Wil 70c Wilkinson, D. H., and D. E. Alburger: *Phys. Rev. Letters,* **24**:1134 (1970).

Wu 66 Wu, C. S., and S. A. Moszkowski: "Beta Decay," Interscience, a division of Wiley, New York, 1966.

6

MUON CAPTURE

6.1 INTRODUCTION

Muon capture by a proton

$$\mu^- + p \rightarrow n + \nu_\mu \qquad (6.1)$$

or by a heavier nucleus

$$\mu^- + (Z, N) \rightarrow (Z + 1, N - 1) + \nu_\mu \qquad (6.2)$$

is characterized by two essential features which distinguish it from the somewhat analogous process of orbital electron capture

$$e^- + (Z, N) \rightarrow (Z + 1, N - 1) + \nu_e \qquad (6.3)$$

These are competition with muon decay and the large momentum transfer imparted to the nucleon or nucleus upon capture of the relatively massive muon ($m_\mu \approx 0.1 m_p$). As we shall see, the muon capture rate is proportional to Z^4 (at least for light nuclei). Thus, although capture is less than 10^{-3} times as probable as muon decay in the case of hydrogen, the capture and decay rates are approximately equal for $Z = 6$.

The large momentum transfer implies that forbidden terms in the hadronic matrix element (to borrow a phrase from nuclear beta decay), including terms in the coefficients f_2 and g_3, contribute very substantially to the muon capture rate. Moreover, it is no longer legitimate, as was the case in nuclear beta decay, to replace the arguments of the form factors $f_1(q^2), f_2(q^2), g_1(q^2), g_3(q^2)$ by zero. Instead, one finds that at least $f_1[q_\mu{}^2 = (p_p - p_n)^2]$, $f_2(q_\mu{}^2)$, and $g_3(q_\mu{}^2)$ differ significantly from their values at $q^2 = 0$.

The study of muon capture presents formidable theoretical and experimental difficulties. While in nuclear beta decay the emission of an $e\nu$ pair generally occurs at rather low energy and results in a comparatively gentle disturbance of the nucleus, muon capture is a violent event leading to the generation of a profusion of nuclear excited states and/or breakup of the nucleus. This occurs because the muon rest energy released on capture is much larger than the binding energy of a typical nucleon in a nucleus. Thus, in general, we must employ a closure approximation (first introduced by Primakoff) in order to estimate the total muon capture rate to all accessible nuclear final states.

On the experimental side, we note that the one lepton emerging from muon capture is the virtually undetectable neutrino. Thus we are denied the possibility of observing numerous lepton correlations, polarizations, etc., which were so useful in the elucidation of beta decay. In principle one can glean valuable information from observations of the asymmetry in neutron emission following polarized muon capture and from radiative muon capture:

$$\mu^- + (Z, N) \to (Z - 1, N + 1) + \nu + \gamma \qquad (6.4)$$

However, in practice the experimental and theoretical obstacles are so formidable that no great advances have yet been made from an examination of these last two phenomena.

In the following pages we shall first summarize the theoretical background and then survey the experimental results in order of ascending Z.

6.2 OUTLINE OF THE THEORY OF MUON CAPTURE

The differential transition probability per unit time for muon capture is expressed by the formula

$$dW = \frac{2\pi\delta(\omega + E_i - m_\mu - E_f)\, d^3\mathbf{p}_\nu}{(2\pi)^3} |\mathfrak{A}|^2 \qquad (6.5)$$

where $\mathbf{p}_\nu$ and ω are the neutrino three-momentum and energy, respectively, and E_i and E_f are the initial and final nuclear energies, respectively. In the delta function, which expresses energy conservation, we neglect the atomic binding energy of the muon. Also, in formula (6.5) the neutrino, the muon, and the nuclei are normalized to one per large, arbitrary volume V. The quantity $\mathfrak{A}$ is rather similar in form to the corresponding expression in nuclear beta decay.[1] For example, for muon capture by a proton

$$\mathfrak{A} = \frac{G \cos \Theta}{\sqrt{2}} \int d^3\mathbf{x}\overline{\psi_n(\mathbf{x})}\mathbf{O}\psi_p(\mathbf{x}) \tag{6.6}$$

where

$$\mathbf{O} = f_1(q_\mu{}^2)\gamma^\alpha j_\alpha + if_2(q_\mu{}^2)\sigma^{\alpha\beta}q_{\mu\beta}j_\alpha + g_1(q_\mu{}^2)\gamma^\alpha\gamma^5 j_\alpha + g_3(q_\mu{}^2)\gamma^5 q_\mu{}^\alpha j_\alpha \tag{6.7}$$

and

$$j_\alpha = \overline{\psi_\nu(\mathbf{x})}\gamma_\alpha(1 + \gamma^5)\psi_\mu(\mathbf{x}) \tag{6.8}$$

In (6.7) we have omitted a possible induced tensor contribution in g_2 since, as pointed out in the preceding chapter, there is no evidence for such a term.

Let us evaluate the functions f_1, f_2, g_1, and g_3 at $q^2 = q_\mu{}^2$. We have

$$q_\mu = p_p - p_n \tag{6.9}$$

Now, by conservation of energy,

$$m_\mu + m_p = m_n + \frac{|\mathbf{p}_n|^2}{2m_n} + \omega$$

Thus, ignoring the neutron-proton mass difference,

$$\omega + \frac{|\mathbf{p}_n|^2}{2} = m_\mu \tag{6.10}$$

Also, by conservation of momentum,

$$\mathbf{p}_n + \mathbf{p}_\nu = 0 \tag{6.11}$$

Thus (6.10) becomes $\omega^2/2 + \omega = m_\mu$

or

$$\omega \approx m_\mu\left(1 - \frac{m_\mu}{2}\right) \tag{6.12}$$

Now, from (6.9),

$$q_\mu{}^2 = q_{\mu 0}{}^2 - \mathbf{q}_\mu \cdot \mathbf{q}_\mu$$

$$q_\mu{}^2 = \left(\frac{\mathbf{p}_n{}^2}{2}\right)^2 - \omega^2 = \frac{\omega^4}{4} - \omega^2 \approx -m_\mu{}^2(1 - m_\mu) \cong -0.88m_\mu{}^2 \tag{6.13}$$

According to the conserved vector current hypothesis, $f_1(q^2)$ and $f_2(q^2)$ ought to vary with q^2 in just the same way as the corresponding isovector

[1] See Eq. (5.8).

electromagnetic form factors (cf. Chap. 4). For these we have quite complete information at $q^2 = q_\mu^2$. Thus we can conclude that

$$f_1(q_\mu^2) = 0.972 f_1(0) = 0.972 \tag{6.14}$$

and

$$f_2(q_\mu^2) = 0.972 f_2(0) = 0.972\left(\frac{\mu_p - \mu_n - 1}{2}\right) \tag{6.15}$$

There is no corresponding electromagnetic analogy to g_1 or g_3. However, we expect g_1 to vary very slowly with q^2 (Fuj 59), and it turns out that

$$g_1(q_\mu^2) = 0.999 g_1(0) \approx +1.23 = -C_A \tag{6.16}$$

As for g_3, an estimate of it is available from the Goldberger–Treiman relation. From Eq. (4.81) we have

$$g_3(q_\mu^2) \approx \frac{2C_A}{q_\mu^2 - m_\pi^2}$$

Defining $g_p = m_\mu g_3$ and inserting the value (6.13) for q_μ^2, we find

$$g_p \approx 7|C_A| \tag{6.17}$$

Continuing to discuss the terms and factors in (6.7) and (6.8), we consider the lepton wave functions. The muon is invariably captured from a $1S$ state (hydrogenic orbital). Thus, for muon capture on a proton, we can write

$$\psi_\mu(\mathbf{r}) = \frac{1}{(\pi a_0^{\,3})^{\frac{1}{2}}} e^{-r/a_0} u_\mu = \phi_\mu u_\mu \tag{6.18}$$

where $u_\mu = \begin{pmatrix} \chi_\mu \\ 0 \end{pmatrix}$, with χ_μ a two-component Pauli spinor, and where a_0 is the muon Bohr radius

$$a_0 = \frac{\hbar^2}{m_\mu e^2} \tag{6.19}$$

or, more accurately, $a_0 = \hbar^2/m'e^2$, where m' is the reduced mass of the muon-nucleus atomic system. For muon capture on a nucleus of charge Z, if the charge can be approximated as a point charge,

$$\psi_\mu(\mathbf{r}) = \frac{Z^{3/2}}{(\pi a_0^{\,3})^{1/2}} e^{-Zr/a_0} u_\mu = \phi_{\mu Z} u_\mu \tag{6.20}$$

However, for large Z, the nucleus is so extended and the muon Bohr radius is so small that the muon spends a large fraction of its lifetime within the nuclear radius, so that (6.20) is incorrect. The neutrino wave function may be expressed as a plane wave,

$$\psi_\nu(\mathbf{r}) = \frac{1}{\sqrt{2}} e^{i\mathbf{p}_\nu \cdot \mathbf{r}} \begin{pmatrix} \chi_\nu \\ \boldsymbol{\sigma} \cdot \hat{\mathbf{p}}_\nu \chi_\nu \end{pmatrix} \tag{6.21}$$

where χ_ν is a two-component Pauli spinor normalized to unity.

Expression (6.6) for $\mathfrak{A}$ may be reduced to a convenient form if we take advantage of the fact that the nucleon motions are nonrelativistic even after absorption of a muon. In order to make this reduction, we first express j_α in terms of χ_μ and χ_ν:

$$\begin{aligned} j_0 &= \bar{\psi}_\nu \gamma^0 (1 + \gamma^5) \psi_\mu \\ &= \frac{1}{\sqrt{2}} [\chi_\nu^\dagger (1 - \boldsymbol{\sigma} \cdot \hat{\mathbf{p}}_\nu) \chi_\mu] e^{-i\mathbf{p}_\nu \cdot \mathbf{x}} \phi_\mu \end{aligned} \tag{6.22}$$

$$\begin{aligned} -\mathbf{j} &= -\bar{\psi}_\nu \boldsymbol{\gamma} (1 + \gamma^5) \psi_\mu \\ &= +\frac{1}{\sqrt{2}} [\chi_\nu^\dagger (1 - \boldsymbol{\sigma} \cdot \hat{\mathbf{p}}_\nu) \boldsymbol{\sigma} \chi_\mu] e^{-i\mathbf{p}_\nu \cdot \mathbf{x}} \phi_\mu \end{aligned} \tag{6.23}$$

where we have used (6.18), (6.20), and the usual properties of γ matrices. Now writing

$$\psi_p = \begin{pmatrix} \chi_p \\ \dfrac{\boldsymbol{\sigma}_A \cdot \mathbf{p}_p \chi_p}{2} \end{pmatrix} \qquad \text{and} \qquad \bar{\psi}_n = \left(\chi_n^\dagger, \; -\chi_n^\dagger \frac{\boldsymbol{\sigma}_A \cdot \mathbf{p}_n}{2} \right)$$

where χ_n and χ_p are two-component Pauli wave functions for neutron and proton, respectively, and $\boldsymbol{\sigma}_A$ is the nucleon Pauli spin operator, we find the following for the various terms in O [Eq. (6.7)]:

For the vector term,

$$\begin{aligned} f_1 \bar{\psi}_n \gamma^\alpha j_\alpha \psi_p &= f_1 e^{-i\mathbf{p}_\nu \cdot \mathbf{x}} \phi_\mu \left\{ \frac{1}{\sqrt{2}} [\chi_n^\dagger \chi_\nu^\dagger (1 - \boldsymbol{\sigma} \cdot \hat{\mathbf{p}}) \chi_p \chi_\mu] \right. \\ &\quad \left. + \frac{1}{2\sqrt{2}} [\chi_n^\dagger \chi_\nu^\dagger (1 - \boldsymbol{\sigma} \cdot \hat{\mathbf{p}}_\nu)(\boldsymbol{\sigma}_A \cdot \mathbf{p}_n \boldsymbol{\sigma}_A \cdot \boldsymbol{\sigma} + \boldsymbol{\sigma}_A \cdot \boldsymbol{\sigma} \boldsymbol{\sigma}_A \cdot \mathbf{p}_p) \chi_p \chi_\mu] \right\} \\ &= f_1 e^{-i\mathbf{p}_\nu \cdot \mathbf{x}} \phi_\mu \left\{ \frac{1}{\sqrt{2}} [\chi_n^\dagger \chi_\nu^\dagger (1 - \boldsymbol{\sigma} \cdot \hat{\mathbf{p}}_\nu) \chi_p \chi_\mu] \left(1 + \frac{\omega}{2} \right) \right. \\ &\quad \left. + \frac{i}{2\sqrt{2}} [\chi_n^\dagger \chi_\nu^\dagger (1 - \boldsymbol{\sigma} \cdot \hat{\mathbf{p}}_\nu) \boldsymbol{\sigma}_A \cdot \mathbf{q} \times \boldsymbol{\sigma} \chi_p \chi_\mu] \right\} \end{aligned} \tag{6.24}$$

For the weak magnetism term to lowest order in q,

$$+ if_2 \bar{\psi}_n \sigma^{\alpha\beta}(p_p - p_n)_\beta j_\alpha \psi_p \approx \frac{if_2}{\sqrt{2}} e^{-i\mathbf{p}_\nu \cdot \mathbf{x}} \phi_\mu [\chi_n{}^\dagger \chi_\nu{}^\dagger (1 - \boldsymbol{\sigma} \cdot \hat{\mathbf{p}}_\nu) \boldsymbol{\sigma}_A \cdot \mathbf{q} \times \boldsymbol{\sigma} \chi_p \chi_\mu] \tag{6.25}$$

For the axial vector term,

$$g_1 \bar{\psi}_n \gamma^\alpha \gamma^5 j_\alpha \psi_p = -\frac{g_1}{\sqrt{2}} e^{-i\mathbf{p}_\nu \cdot \mathbf{x}} \phi_\mu [\chi_n{}^\dagger \chi_\nu{}^\dagger (1 - \boldsymbol{\sigma} \cdot \hat{\mathbf{p}}_\nu) \boldsymbol{\sigma} \cdot \boldsymbol{\sigma}_A \chi_p \chi_\mu]$$

$$+ \frac{g_1}{\sqrt{2}} \omega e^{-i\mathbf{p}_\nu \cdot \mathbf{x}} \phi_\mu [\chi_n{}^\dagger \chi_\nu{}^\dagger (1 - \boldsymbol{\sigma} \cdot \hat{\mathbf{p}}_\nu) \boldsymbol{\sigma}_A \cdot \hat{\mathbf{p}}_\nu \chi_p \chi_\mu] \tag{6.26}$$

For the induced pseudoscalar term to lowest order in q,

$$g_3 \bar{\psi}_n \gamma^5 (p_p - p_n)^\alpha j_\alpha \psi_p \approx \frac{1}{2\sqrt{2}} g_3 \omega^2 e^{-i\mathbf{p}_\nu \cdot \mathbf{x}} \phi_\mu [\chi_n{}^\dagger \chi_\nu{}^\dagger (1 - \boldsymbol{\sigma} \cdot \hat{\mathbf{p}}_\nu) \boldsymbol{\sigma}_A \cdot \hat{\mathbf{p}}_\nu \boldsymbol{\sigma} \cdot \hat{\mathbf{p}}_\nu \chi_p \chi_\mu] \tag{6.27}$$

Now making the substitution

$$g_p = -m_\mu g_3$$

and combining terms, we find from (6.24) to (6.27) that (6.5) becomes

$$\mathfrak{A} = \tfrac{1}{2} \int \chi_n{}^\dagger(\mathbf{x}) \chi_\nu{}^\dagger \mathcal{H} \chi_\mu \chi_p(\mathbf{x}) e^{-i\mathbf{p}_\nu \cdot \mathbf{x}} \phi_\mu(\mathbf{x})\, d^3\mathbf{x} \tag{6.28}$$

where the *effective hamiltonian* $\mathcal{H}$ is

$$\mathcal{H} = (1 - \boldsymbol{\sigma} \cdot \hat{\mathbf{p}}_\nu)[G_V + G_A \boldsymbol{\sigma} \cdot \boldsymbol{\sigma}_A + G_P \boldsymbol{\sigma}_A \cdot \hat{\mathbf{p}}_\nu] \tag{6.29}$$

and where

$$G_V = G \cos \Theta f_1(q_\mu{}^2)\left(1 + \frac{\omega}{2}\right) = 1.02 G \cos \Theta \tag{6.30}$$

$$G_A = G \cos \Theta \left[-g_1(q_\mu{}^2) - 0.972 \frac{(\mu_p - \mu_n)}{2} \omega\right] = 1.38\, G \cos \Theta \tag{6.31}$$

$$G_P = G \cos \Theta \left[g_p(q_\mu{}^2) + g_1(q_\mu{}^2) - 0.972(\mu_p - \mu_n) \frac{\omega}{2}\right] = -0.63\, G \cos \Theta \tag{6.32}$$

[The numerical values for G_V, G_A, and G_P are obtained using (6.14) to (6.16).)] Equation (6.28) is valid for muon capture by a proton. In the more general case of muon capture by a nucleus,

$$\mathfrak{A} = \tfrac{1}{2} \sum_A \int \chi'^\dagger \chi_\nu{}^\dagger \mathcal{H}_A \chi_\mu \tau_A \chi e^{-i\mathbf{p}_\nu \cdot \mathbf{x}_A} \phi_\mu(\mathbf{x}_A)\, d^3x \tag{6.33}$$

where $\quad \mathcal{H}_A = (1 - \boldsymbol{\sigma} \cdot \hat{\mathbf{p}}_\nu)[G_V + G_A \boldsymbol{\sigma} \cdot \boldsymbol{\sigma}_A + G_P \boldsymbol{\sigma}_A \cdot \hat{\mathbf{p}}_\nu] \qquad (6.34)$

Now that we have obtained a formal expression for $\mathfrak{A}$, let us return to the expression for dW [Eq. (6.5)]. Writing $E_f = (M_f^2 + \omega^2)^{1/2}$, we have for the total transition probability

$$W = \frac{1}{(2\pi)^2} \int d\Omega_\nu \left[\int \omega^2 \, d\omega \, \delta(\omega + (M_f^2 + \omega^2)^{1/2} - E_i - m_\mu) \right] |\mathfrak{A}|^2$$

$$= \left[\frac{\omega^2}{1 + \dfrac{\omega}{(M_f^2 + \omega^2)^{1/2}}} \right] \int \frac{d\Omega_\nu}{4\pi} \left(\frac{Z^3}{\pi^2 a_0^3} \right)$$

$$\times \, |\chi_\nu^\dagger (1 - \boldsymbol{\sigma} \cdot \hat{\mathbf{p}}_\nu)[G_V \langle 1 \rangle_\mu + (G_A \boldsymbol{\sigma} + G_P \hat{\mathbf{p}}_\nu) \langle \boldsymbol{\sigma}_A \rangle_\mu] \chi_\mu|^2 \qquad (6.35)$$

where

$$\langle 1 \rangle = \frac{\pi^{1/2} a_0^{3/2}}{Z^{3/2}} \langle f | \sum \tau_A^- e^{-i\mathbf{p}_\nu \cdot \mathbf{x}_A} \phi_\mu(\mathbf{x}_A) | i \rangle \qquad (6.36)$$

and

$$\langle \boldsymbol{\sigma}_A \rangle_\mu = \frac{\pi^{1/2} a_0^{3/2}}{Z^{3/2}} \langle f | \sum \tau_A^- e^{-i\mathbf{p}_\nu \cdot \mathbf{x}_A} \phi_\mu(\mathbf{x}_A) \boldsymbol{\sigma}_A | i \rangle \qquad (6.37)$$

These last two quantities would reduce to the allowed beta decay matrix elements $\langle 1 \rangle$ and $\langle \boldsymbol{\sigma} \rangle$ appropriate for the description of positron emission or orbital electron capture if we could set the lepton functions equal to unity.

As it stands, Eq. (6.35) applies only to one particular final nuclear state. However, we may easily generalize (6.35) to account for the total transition probability to all accessible final nuclear states f:

$$W = \sum_f \left[\frac{\omega_2}{1 + \dfrac{\omega}{(M_f^2 + \omega^2)^{1/2}}} \right] \int \frac{d\Omega_\nu}{4\pi} \frac{Z^3}{\pi^2 a_0^3}$$

$$\times \, |\chi_\nu^\dagger (1 - \boldsymbol{\sigma} \cdot \hat{\mathbf{p}}_\nu)[G_V \langle 1 \rangle_\mu + (G_A \boldsymbol{\sigma} + G_P \hat{\mathbf{p}}_\nu) \langle \boldsymbol{\sigma}_A \rangle_\mu] \chi_\mu|^2 \qquad (6.38)$$

Let us assume that the neutrino energies are pretty much the same for all final states, so that the factor

$$\left[\frac{\omega^2}{1 + \omega/(M_f + \omega^2)^{1/2}} \right]$$

can be placed in front of the summation sign. Further, we note that in integration over $d\Omega_\nu$ all terms linear in $\hat{\mathbf{p}}_\nu$ disappear. Thus (6.38) becomes, when averaged over muon polarizations,

$$W = \left[\frac{\omega^2}{1 + \dfrac{\omega}{(M_f^{\,2} + \omega^2)^{1/2}}}\right] \frac{Z^3}{\pi^2 a_0^{\,3}} \int \frac{d\Omega_\nu}{4\pi}$$

$$\times \sum_f [G_V^{\,2}|\langle 1\rangle_\mu|^2 + G_A^{\,2}|\langle \boldsymbol{\sigma}_A\rangle_\mu|^2 + (G_P^{\,2} - 2G_A G_P)|\langle \boldsymbol{\sigma}_A \cdot \hat{\mathbf{p}}_\nu\rangle|^2] \tag{6.39}$$

In (6.39) we are faced with quantities of the form

$$\sum_f |\langle 1\rangle_\mu|^2 \qquad \sum |\langle \sigma_A\rangle_\mu|^2 \qquad \text{and} \qquad \sum |\langle \boldsymbol{\sigma}_A \cdot \hat{\mathbf{p}}_\nu\rangle|^2 \tag{6.40}$$

These may be evaluated if we make the *closure approximation* (Pri 59), in which we replace the sum over *accessible* final states by a sum over *all* final states $|f\rangle$. Such an approximation is a reasonable one (particularly for $A > 12$) because the energy released in muon capture is of order 100 MeV, whereas the matrix elements of (6.40) are largest for low-lying states differing in energy from the ground state by no more than 5 MeV. (It should be noted, of course, that for the proton there are no bound excited states and the closure sum for $\mu^- + p \rightarrow \nu_\mu + n$ reduces to a sum over spins of the final state.)

Since the set $\{|f\rangle\}$ is complete in the closure approximation, we have for any operator A

$$\sum |\langle f|A|i\rangle|^2 = \sum_f \langle i|A^\dagger|f\rangle\langle f|A|i\rangle = \langle i|A^\dagger A|i\rangle$$

Thus the expressions (6.40) become

$$\sum_f |\langle 1\rangle_\mu|^2 = \frac{\pi a_0^{\,3}}{Z} \langle i| \sum_{A,B} \tau_A^{\,+}\tau_B^{\,-} e^{i\mathbf{p}_\nu\cdot(\mathbf{x}_A - \mathbf{x}_B)} \phi_\mu(\mathbf{x}_A)\phi_\mu(\mathbf{x}_B)|i\rangle \tag{6.41}$$

$$\sum_f |\langle \boldsymbol{\sigma}_A\rangle_\mu^{\,2}| = \frac{\pi a_0^{\,3}}{Z} \langle i| \sum_{A,B} \tau_A^{\,+}\tau_B^{\,-} e^{i\mathbf{p}_\nu\cdot(\mathbf{x}_A - \mathbf{x}_B)} \phi_\mu(\mathbf{x}_A)\phi_\mu(\mathbf{x}_B)\boldsymbol{\sigma}_A \cdot \boldsymbol{\sigma}_B|i\rangle \tag{6.42}$$

and

$$\sum_f |\langle \boldsymbol{\sigma}_A \cdot \hat{\mathbf{p}}_\nu\rangle|^2 = \frac{\pi a_0^{\,3}}{Z} \langle i| \sum_{A,B} \tau_A^{\,+}\tau_B^{\,-} e^{i\mathbf{p}_\nu\cdot(\mathbf{x}_A - \mathbf{x}_B)} \phi_\mu(\mathbf{x}_A)\phi_\mu(\mathbf{x}_B)\boldsymbol{\sigma}_A \cdot \hat{\mathbf{p}}_\nu\, \boldsymbol{\sigma}_B \cdot \hat{\mathbf{p}}_\nu|i\rangle \tag{6.43}$$

Now, when we average over all neutrino directions,

$$\int \frac{d\Omega}{4\pi} e^{i\mathbf{p}_\nu\cdot(\mathbf{x}_A - \mathbf{x}_B)} = \frac{\sin \omega(r_A - r_B)}{\omega(r_A - r_B)} \tag{6.44}$$

and $\boldsymbol{\sigma}_A \cdot \hat{\mathbf{p}}_\nu \boldsymbol{\sigma}_B \cdot \hat{\mathbf{p}}_\nu$ may be replaced by $\frac{1}{3}\boldsymbol{\sigma}_A \cdot \boldsymbol{\sigma}_B$. Thus, using (6.41) to (6.44), W becomes, from (6.39),

$$W = \frac{G_V{}^2\omega^2 Z^3}{2\pi^2\left[1 + \dfrac{\omega}{(M_f{}^2 + \omega^2)^{1/2}}\right]a_0{}^3} \langle i | \sum_{AB} \tau_A{}^+ \tau_B{}^- \frac{\sin \omega(r_A - r_B)}{\omega(r_A - r_B)} \phi_\mu(r_A)\phi_\mu(r_B) \times (1 + \eta \boldsymbol{\sigma}_A \cdot \boldsymbol{\sigma}_B) | i \rangle \tag{6.45}$$

where

$$\eta = \frac{G_A{}^2 + \frac{1}{3}(G_P{}^2 - 2G_A G_P)}{G_P{}^2} \approx 1.4$$

At this point it is useful to separate the sum in (6.45) into two parts:

$$\sum_{AB} = \sum_{A=B} + \sum_{A \neq B}$$

For the moment, taking into account only those terms for which $A = B$, we find

$$W_1 = \frac{G_V{}^2\omega^2 Z^3}{2\pi^2\left[1 + \dfrac{\omega}{(M_f{}^2 + \omega^2)^{1/2}}\right]a_0{}^3} \langle i | \sum_A \left(\frac{1 + \tau_{AZ}}{2}\right) \phi_\mu{}^2(r_A)(1 + 3\eta) | i \rangle \tag{6.46}$$

since

$$\boldsymbol{\sigma}_A{}^2 = 3 \qquad \text{and} \qquad \tau_A{}^+ \tau_A{}^- = \frac{1 + \tau_{Az}}{2}$$

If we could ignore those terms in (6.45) for which $A \neq B$, then (6.46) would represent an extremely simple expression for the transition probability. For light nuclei (for example, H^1, H^2, He^3, He^4) the nuclear radius is sufficiently small that we can set $\phi_\mu(r_A) = \phi_\mu(0) = 1$. Then

$$W_1 = \frac{G_V{}^2 Z^4}{2\pi^2}(1 + 3\eta) \frac{(m')^3 \alpha^3 \omega^2}{1 + \dfrac{\omega}{(M_f{}^2 + \omega^2)^{1/2}}} \tag{6.47}$$

since

$$Z = \langle i | \sum_A \left(\frac{1 + \tau_{Az}}{2}\right) | i \rangle$$

For purposes of a rather crude and simple comparison with muon decay, we may replace

$$1 + \frac{\omega}{(M_f{}^2 + \omega^2)^{1/2}}$$

by unity and set $m' = \omega = m_\mu$. Then

$$W_1 \approx \frac{G^2Z^4}{2\pi^2} m_\mu{}^5\alpha^3(1+3\eta)\cos^2\Theta$$

while

$$W(\mu \text{ decay}) = \frac{G^2 m_\mu{}^5}{192\pi^3}$$

Thus, since $\cos\Theta \approx 1$,

$$\frac{W_1}{W(\mu \text{ decay})} \approx \frac{G^2Z^4}{2\pi^2} m_\mu{}^5\alpha^3(1+3\eta)\left(\frac{G^2 m_\mu{}^5}{192\pi^3}\right)^{-1}$$

$$\approx 6 \times 10^{-4} Z^4 \tag{6.48}$$

Since $W(\mu \text{ decay}) = 5 \times 10^5 \text{ s}^{-1}$, we have for this crude estimate

$$W_1(\mu + p \rightarrow n + \nu_\mu) \approx 300 \text{ s}^{-1} \tag{6.49}$$

This is really not so bad since a detailed calculation gives $W = 169 \text{ s}^{-1}$.

A better approximation is obtained for W_1 if we use accurate values for $(1 + \omega/(M_f{}^2 + \omega^2)^{1/2})$, m', and ω. Then we find

$$W_1 = (290 \pm 10) Z_{\text{eff}}{}^4 \left(\frac{\omega}{m_\mu}\right)^2_{\text{eff}} \text{ s}^{-1} \tag{6.50}$$

where Z_{eff}, defined by

$$\left(\frac{Z_{\text{eff}}}{Z}\right)^4 = \frac{\langle i | \sum_A \frac{1}{2}(1+\tau_z{}^A)\phi_\mu{}^2(r_A) | i \rangle}{Z}$$

is written to emphasize that for large Z we cannot set $\phi_\mu{}^2(r_A) = 1$ in (6.46), and also

$$\left(\frac{\omega}{m_\mu}\right)^2_{\text{eff}} = \left(\frac{\omega}{m_\mu}\right)^2 \left[1 + \frac{\omega}{(M_f{}^2 + \omega^2)^{1/2}}\right] \left(1 + \frac{m_\mu}{M_i}\right)^{-3} \tag{6.51}$$

For the light nuclei H^1, H^2, He^3, and He^4, $Z_{\text{eff}} = Z$, of course, and we have the values of Table 6.1.

Table 6.1

Nucleus	$(\omega/m_\mu)_{\text{eff}}{}^2$
H^1	0.58
H^2	0.65
He^3	0.78
He^4	0.50

To complete the calculation of the muon capture rate, we still need to evaluate the *interference* terms in the sum of (6.45), i.e., those for which $A \neq B$. This is a difficult and lengthy task involving the evaluation of nucleon *pair-correlation* integrals (Pri 59). We shall not repeat any of the details here; we shall merely point out that the contributions of these interference terms can be understood crudely and qualitatively to *diminish* the total rate on account of the Pauli principle as follows: A proton which captures a muon must turn into a neutron, and this it cannot do if the neutron state it would fill is already occupied by an existing neutron. This diminution must be proportional to the neutron number and the nucleon spatial overlap volume $\approx d^3$, where

$$d = \frac{\hbar}{m_\pi c} \approx \frac{0.7}{m_\mu}$$

Primakoff has, in fact, shown that

$$\boxed{W_{\text{total}} = W_1 + W_{\text{interference}} = (290\ \text{s}^{-1}) Z_{\text{eff}}^4 \left(\frac{\omega}{m_\mu}\right)_{\text{eff}}^2 \left(1 - \frac{\delta N}{2A}\right)} \tag{6.52}$$

where $$\delta \approx \left(\frac{d}{r_0}\right)^3 \qquad \text{with } r_0 = \frac{R}{A^{1/3}} \approx 1.2 \times 10^{-13}\ \text{cm}$$

Formula (6.52) represents the final theoretical result which must be compared with experiment.

6.3 COMPARISON OF THE THEORY WITH EXPERIMENTAL RESULTS

6.3.1 Hydrogen (H^1)

Only for muon capture on protons is it possible to investigate the basic questions relevant to weak interactions that are completely free of the complications of nuclear structure. Unfortunately, the capture rate for hydrogen is low, and so it is necessary to work at high hydrogen concentrations. The early experiments were done with liquid H_2 (Hil 62, Rot 64), but more recently muon capture has been investigated in H_2 gas at high pressures (Qua 67). It may be shown[1] that in the gas the atom $p\mu^-$ is formed, whereas in the liquid $p\mu^-$ mesoatoms exchange muons with H_2 molecules so that the muons are mainly in mesomolecules $pp\mu^-$. The situation in the latter case is therefore complicated by the necessity of estimating the muon density at either proton in the molecule.

[1] See the useful review, Mesonic Atoms, by E. H. S. Burhop, in E. H. S. Burhop (ed.), "High Energy Physics," vol. 3, Academic, New York, 1969.

Considering the atom, we note that the hfs splitting $\Delta\nu$ of the ground state, only 1,420 Mc/s for ordinary atomic hydrogen, is a factor $(m_\mu/m_e)^2$ larger for $p\mu^-$. This splitting is comparable to thermal energies, and so $p\mu^-$ atoms formed in the upper ($F = 1$) hfs level quickly go to the lower ($F = 0$) level by collision-induced transitions. At the high gas densities necessary for practical experiments, the lifetime of the triplet state is only some tens of nanoseconds, and so practically all $p\mu^-$ captures occur from the singlet state.

Now, the muon capture rate depends on $\boldsymbol{\sigma} \cdot \boldsymbol{\sigma}_A$, which has different expectation values in the $F = 1$ and $F = 0$ states:

$$\langle \boldsymbol{\sigma} \cdot \boldsymbol{\sigma}_A \rangle_{F=1} = +1 \qquad (6.53a)$$

$$\langle \boldsymbol{\sigma} \cdot \boldsymbol{\sigma}_A \rangle_{F=0} = -3 \qquad (6.53b)$$

Thus we expect a difference in the capture rates for these two hfs levels. This can easily be calculated starting from (6.28) and (6.29) rather than (6.52). We have for a sum over final spins

$$W = \frac{1}{\pi^2 a_0^{\,3}} \left[\frac{\omega^2}{1 + \omega/(1 + \omega^2)^{1/2}} \right] \int \frac{d\Omega_\nu}{4\pi} \times \sum_{\substack{\text{final}\\ \text{spins}}} \chi_\mu^{\dagger} \chi_p^{\dagger} [G_V + (G_A \boldsymbol{\sigma} + G_P \hat{\mathbf{p}}_\nu) \cdot \boldsymbol{\sigma}_A](1 - \boldsymbol{\sigma} \cdot \hat{\mathbf{p}}_\nu)^2 \times [G_V + (G_A \boldsymbol{\sigma} + G_P \hat{\mathbf{p}}_\nu) \cdot \boldsymbol{\sigma}_A] \chi_\mu \chi_p \qquad (6.54)$$

obtained by invoking the closure principle. The sum in (6.54) reduces to

$$\sum = 2[G_V^{\,2} + G_A^{\,2}(\boldsymbol{\sigma} \cdot \boldsymbol{\sigma}_A)^2 + 2G_A G_V \boldsymbol{\sigma} \cdot \boldsymbol{\sigma}_A - 2G_P G_V \boldsymbol{\sigma}_A \cdot \hat{\mathbf{p}}_\nu \boldsymbol{\sigma} \cdot \hat{\mathbf{p}}_\nu - G_A G_P(\boldsymbol{\sigma}_A \cdot \hat{\mathbf{p}}_\nu \boldsymbol{\sigma} \cdot \hat{\mathbf{p}}_\nu \boldsymbol{\sigma} \cdot \boldsymbol{\sigma}_A + \boldsymbol{\sigma} \cdot \boldsymbol{\sigma}_A \boldsymbol{\sigma} \cdot \hat{\mathbf{p}}_\nu \boldsymbol{\sigma}_A \cdot \hat{\mathbf{p}}_\nu) + G_P^{\,2}(\boldsymbol{\sigma}_A \cdot \hat{\mathbf{p}}_\nu)^2] + \text{terms linear in } \hat{\mathbf{p}}_\nu$$

Now, when we average over neutrino directions, the terms linear in $\hat{\mathbf{p}}_\nu$ vanish and $\boldsymbol{\sigma}_A \cdot \hat{\mathbf{p}}_\nu \boldsymbol{\sigma} \cdot \hat{\mathbf{p}}_\nu = \frac{1}{3}\boldsymbol{\sigma}_A \cdot \boldsymbol{\sigma}$. Also, $(\boldsymbol{\sigma} \cdot \boldsymbol{\sigma}_A)^2 = 3 - 2\boldsymbol{\sigma} \cdot \boldsymbol{\sigma}_A$. Thus

$$\begin{aligned} \int \frac{d\Omega_\nu}{4\pi} \sum_{\text{spins}} &= 2(G_V^{\,2} + 3G_A^{\,2} + G_P^{\,2} - 2G_A G_P) \\ &\quad - 2(2G_A^{\,2} - 2G_A G_V + \tfrac{2}{3}G_P G_V - \tfrac{4}{3}G_A G_P)\boldsymbol{\sigma} \cdot \boldsymbol{\sigma}_A \\ &= 2G_V^{\,2}(1 + 3\eta)\left(1 - \frac{\langle \boldsymbol{\sigma} \cdot \boldsymbol{\sigma}_A \rangle \xi}{1 + 3\eta}\right) \end{aligned} \qquad (6.55)$$

where

$$\xi = 2\,\frac{G_A(G_A - \frac{2}{3}G_P) - G_V(G_A - \frac{1}{3}G_P)}{G_V^{\,2}} \qquad (6.56)$$

Using (6.53a) and (6.53b) and inserting numerical values for the various coefficients, we obtain

$$W(F=0) = 636\ \mathrm{s}^{-1} \qquad (6.57)$$

$$W(F=1) = 13\ \mathrm{s}^{-1} \qquad (6.58)$$

These results are consistent with the total rate for muon capture in hydrogen averaged over muon spins: i.e., from Eq. (6.52) and Table 6.1 we find

$$W_{\text{total}} = 169\ \mathrm{s}^{-1} = \tfrac{3}{4}W(F=1) + \tfrac{1}{4}W(F=0) \qquad (6.59)$$

Equation (6.57) may also be compared with the experimental result obtained in gaseous hydrogen at high pressure:

$$W_{\text{exp}}(F=0) = (640 \pm 70)\ \mathrm{s}^{-1} \qquad (6.60)$$

The agreement is satisfactory and the result (6.60) may be interpreted in terms of limits on g_p:

$$g_p = (8.7 \pm 3.7)|C_A| \qquad (6.61)$$

compared with the theoretical value $g_p = 7|C_A|$ [from Eq. (6.17)].

6.3.2 Muon Capture in He^3

The process

$$\mu^- + \mathrm{He}^3 \to \mathrm{H}^3 + \nu \qquad (6.62)$$

is a very favorable one for study for the following reasons:

1 The final nucleus H^3 has no observed excited states.
2 It is easy to detect H^3 recoils (which have 1.9 MeV energy) especially since He^3 gas has very good scintillation properties.
3 The vector form factors can be determined from electron scattering experiments on He^3.
4 The effective axial coupling for $He^3 \to H^3$ beta decay is known from the *ft* value, and this can be extrapolated to the muon capture momentum transfer q_μ.

The experimental results are

$$W_{\text{exp}}(\mathrm{He}^3 + \mu^- \to \mathrm{H}^3 + \nu) = (1410 \pm 140)\ \mathrm{s}^{-1}\ddagger \qquad (6.63a)$$

$$= (1505 \pm 50)\ \mathrm{s}^{-1}\S \qquad (6.63b)$$

$$= (1450 \pm 75)\ \mathrm{s}^{-1}\P \qquad (6.63c)$$

‡ Fal 62, 63.
§ Ede 64.
¶ Aue 63.

The theoretical values (based on the general formula 6.52) are

$$W_{\text{theo}}(\text{He}^3 + \mu^- \rightarrow \text{H}^3 + \nu) = (1{,}530 \pm 150)\ \text{s}^{-1}\ddagger \qquad (6.64\text{a})$$

$$= (1{,}450 \pm 150)\ \text{s}^{-1}\S \qquad (6.64\text{b})$$

Thus (6.63) and (6.64) are in excellent agreement.

6.3.3 Muon Capture in He^4

In a recent experiment by M. Block et al. (Blo 68), the total muon capture rate for

$$\mu^- + \text{He}^4 \rightarrow 3n + p + \nu$$

$$\rightarrow 2n + d + \nu$$

$$\rightarrow n + t + \nu$$

was compared with the muon decay rate, with the result

$$W_{\text{total}}(\mu^-, \text{He}^4 \text{ capture}) = (364 \pm 46)\ \text{s}^{-1} \qquad (6.65)$$

An earlier experiment (Biz 64) gave the result $(336 \pm 46)\ \text{s}^{-1}$; thus, a weighted average gives

$$W_{\text{total, exp}}(\text{He}^4) = (359 \pm 39)\ \text{s}^{-1} \qquad (6.66)$$

The theory for He^4, developed along the lines sketched in the previous section, and assuming $g_p = 8|C_A|$, gives $(324 \pm 65)\ \text{s}^{-1}$ (Gou 64) or $(345 \pm 110)\ \text{s}^{-1}$ (Cai 63). The experimental and theoretical results are therefore in agreement, but this is not a particularly sensitive test of the value of g_p, for if $g_p = 16|C_A|$ were assumed, we would obtain the theoretical result $(290 \pm 60)\ \text{s}^{-1}$ (Gou 64), which is still in agreement with experiment.

6.3.4 Total Muon Capture Rates in Other Nuclei

In C^{12} the situation is somewhat similar to He^3 in that capture goes mostly to the ground state of B^{12}, which then beta-decays back to C^{12}. However, the experimental results (Rey 63, Mai 64) have so large a statistical spread that one can only conclude that the vector and axial couplings f_1 and g_1 are about the same in beta decay and muon capture. The accuracy is not good enough to determine anything about f_2 and g_3.

‡ Fuj 59.
§ Wol 62.

The capture rates for other complex nuclei (ranging from lithium to uranium) are in reasonable agreement with the Primakoff formula (6.51), but, again, it is impossible to conclude anything definite from these results in general, except that the VA theory seems correct for muon capture.[1] It should be noted that for certain complex nuclei (such as the doubly magic O^{16} and Ca^{40}) the nuclear structure calculations are somewhat more straightforward than in other cases. In the case of Ca^{40}, in particular, one may assume that the conserved vector current hypothesis and the Goldberger–Treiman relation are valid and compute the capture rate for definite nuclear models. The results do not quite agree with experiment, and there has been some speculation (Bou 67) that the discrepancy arises from the existence of a second-class term in g_2.

6.4 RADIATIVE MUON CAPTURE

The process

$$\mu^- + (Z, N) \to (Z - 1, N + 1) + \nu + \gamma \tag{6.67}$$

is somewhat analogous to inner *bremsstrahlung* from electron capture in nuclear beta decay. The rate is roughly 10^{-4} times the rate for ordinary muon capture. Therefore, on account of the low rate, experiments to detect radiative muon capture are extremely difficult. It can be shown that the photon energy spectrum depends rather sensitively on g_p (Roo 65). However, preliminary experimental results for radiative muon capture in O^{16} and Ca^{40} are simply too crude to give any useful definite values for g_p.

We also expect an asymmetry α_γ in the intensity distribution of γ rays emitted in radiative capture of polarized muons because of parity violation:

$$I(\theta) = 1 + \alpha_\gamma \mathcal{T} \cos\theta \tag{6.68}$$

Here $\mathcal{T}$ is the muon polarization and θ is the angle between muon spin and photon momentum. It is possible to show that if the electromagnetic radiation is coupled only to the muon current, then α_γ is equal to the photon circular polarization, and for the V − A theory the photons are all right-circularly polarized with $\alpha_\gamma = +1$ (Cut 57). Recently Di Lella et al. (DiL 71) measured α_γ for photon energies between 57 and 75 MeV for radiative muon capture by Ca^{40}; their result is $\alpha_\gamma \le -0.32 \pm 0.48$, which is inconsistent with the VA prediction. However, this result must be regarded as rather preliminary and imprecise, and it is clear that further work must be done.

[1] For a summary of these results, see C. Rubbia, Weak Interaction Physics, in E. H. S. Burhop (ed.), "High Energy Physics," vol. 3, p. 308, Academic, New York, 1969.

6.5 ASYMMETRY OF NEUTRONS FROM CAPTURE OF POLARIZED MUONS

It may be shown directly from (6.29) that when polarized muons are captured by unpolarized protons, the neutrons are emitted in an angular distribution of the form

$$1 + \mathcal{P}\alpha \cos \theta \qquad (6.69)$$

where $\mathcal{P}$ is the muon polarization, θ is the angle between neutron emission and muon spin, and α is the asymmetry parameter

$$\alpha = \frac{G_V{}^2 - 2G_A{}^2 + (G_A - G_P)^2}{G_V{}^2 + 2G_A{}^2 + (G_A - G_P)^2} \qquad (6.70)$$

Unfortunately, such a calculation ignores the hyperfine effect, which couples the muon and proton spins. In fact, we know that for μ capture in hydrogen at high densities, practically all the capture events proceed from the $F = 0$ state, where there is no net muon polarization. Thus, in order to observe an asymmetry it is necessary to study polarized muon capture in the even-even nuclei such as Si^{28}, S^{32}, and Ca^{40} where no hyperfine effect exists. However, in these cases the asymmetry is complicated by nuclear structure effects and by interaction of the outgoing neutron with the residual nucleus. The theoretical analysis is difficult (Pri 59) and contains some uncertain approximations. Measurements of the asymmetry on Si^{28}, S^{32}, and Ca^{40} (Sun 68) yield values which do not agree with the theory and are difficult to understand. Thus, at present, it seems impossible to conclude anything very useful about g_p from neutron asymmetry measurements.

In conclusion, we may say that the results for muon capture are in reasonable agreement with the V-A law, the CVC hypothesis, and the Goldberger–Treiman relation, but the experimental and theoretical difficulties have so far prevented really definitive tests of these theories.

REFERENCES

Aue 63 Auerbach, L. B., R. J. Esterling, R. E. Hill, D. A. Jenkins, J. T. Lach, and N. H. Lipiman: *Phys. Rev. Letters,* **11**:23 (1963).

Biz 64 Bizzarri, R., et al.: *Nuovo Cimento,* **33**:1497 (1964).

Blo 68 Block, M. M., T. Kikuchi, S. Koetke, C. R. Sun, R. Walker, G. Culligan, V. L. Telegdi, and R. Winston: *Nuovo Cimento,* **55A**:501 (1968).

Bou 67 Bouyssy, A., M. Castaignet, and N. V. Mau: *Phys. Letters,* **25B**:533 (1967).

Cai 63 Caine, C. A., and P. S. H. Jones: *Nucl. Phys.*, **44**:177 (1963).

Cut 57 Cutkosky, R. E.: *Phys. Rev.*, **107**:330 (1957).

DiL 71 Di Lella, L., T. Hammerman, and L. Rosenstein: *Phys. Rev. Letters*, **27**:830 (1971).

Ede 64 Edelstein, R. M., D. Clay, J. W. Keuffel, and R. L. Wagner, Jr.: *Internat. Conf. Fundamental Aspects Weak Interactions*, Brookhaven National Laboratory, Upton, N.Y., 303 (1964).

Fal 62, 63 Falomkin, A. I., A. Filippov, M. M. Kutynkin, B. Pontecorvo, Y. A. Scherbokov, R. M. Sulyer, V. M. Tsupko-Sitnikov, and D. A. Zaimidoroga: *Phys. Letters*, **1**:318 (1962); *ibid.*, **3**:229 (1963); *ibid.*, **6**:100 (1963).

Fuj 59 Fujii, A., and H. Primakoff: *Nuovo Cimento*, **12**:327 (1959).

Gou 64 Goulard, B., G. Goulard, and H. Primakoff: *Phys. Rev.*, **133B**:186 (1964).

Hil 62 Hildebrand, R. H., and J. H. Doede: Invited paper A3, *Proc. Internatl. Conf. High-Energy Physics, CERN, Geneva* (1962).

Mai 64 Maier, E. J., R. M. Edelstein, and R. T. Siegel: *Phys. Rev.*, **133B**:663 (1964).

Pri 59 Primakoff, H.: *Rev. Mod. Phys.*, **31**:802 (1959).

Qua 67 Quaranta, A. A., et al.: *Phys. Letters*, **25B**:429 (1967).

Rey 63 Reynolds, G. T., D. B. Scarl, R. A. Swasson, J. R. Waters, and R. A. Zdanis: *Phys. Rev.*, **129**:1790 (1963).

Roo 65 Rood, H. R., and H. A. Tolhoek: *Nucl. Phys.*, **70**:658 (1965).

Rot 64 Rothberg, J. E., E. W. Anderson, E. J. Bleser, L. M. Lederman, P. L. Meyer, J. L. Rosen, and I. T. Wang: *Phys. Rev.*, **132**:2664 (1964).

Sun 68 Sundelin, R. M., R. M. Edelstein, A. Suzuki, and K. Takahashi: *Phys. Rev. Letters*, **20**:1198 and 1201 (1968).

Wol 62 Wolfenstein, L.: *Proc. Internatl. Conf. High-Energy Physics, CERN, Geneva*, 821 (1962).

7

WEAK INTERACTIONS AND UNITARY SYMMETRY; CABIBBO'S HYPOTHESIS

7.1 INTRODUCTION

In Chap. 4 we began to discuss the general properties of the hadronic weak currents $\mathfrak{J}_0$ and $\mathfrak{J}_1$. We saw how Lorentz invariance, time-reversal and G-parity invariance, and the CVC and PCAC hypotheses provide useful restrictions on the possible forms of matrix elements of $\mathfrak{J}_0$ and $\mathfrak{J}_1$ between various hadron states. However, many fundamental questions were ignored in Chap. 4. We now take up the thread of that chapter once again, with the goal of seeing how $\mathfrak{J}_0$ and $\mathfrak{J}_1$ are related to one another and how they fit into the basic scheme of hadron classification: SU_3. In the present chapter we limit ourselves to a discussion of general principles, including the fundamental hypothesis of Cabibbo and the current commutation relations.[1] Specific applications to various weak processes are given in succeeding chapters. Before we plunge into a formal treatment, however, it is appropriate to recollect the relevant empirical facts about hadrons and their weak decays which we seek to explain.

[1] Comprehensive discussions of SU_3 symmetry and hadron interactions are given in many texts, monographs, and reviews. See, for example, Gel 64a, Lip 65, and Car 66.

7.2 SURVEY OF EMPIRICAL FACTS

7.2.1 Hadron Multiplet Structure

An examination of the masses and quantum numbers of the hadrons reveals a number of striking regularities. Let us consider first the the mesons with $J^P = 0^-$, namely, the particles K^+, K^0, $\pi^\pm$, π^0, η^0, $\bar{K}^0$, and K^-. These mesons can be depicted as points in a two-dimensional diagram (Fig. 7.1), where the abscissa is the z component of isotopic spin I_3 and the ordinate is hypercharge Y (hypercharge equals strangeness S plus baryon number B). Figure 7.1 displays four isomultiplets in this *octet* of mesons: the isosinglet η^0, the two isodoublets (K^+, K^0) and $(\bar{K}^0, K^-)$, and the isotriplet (π^+, π^0, π^-). We note that the mass differences between various members of a given isodoublet or isotriplet are always very small. As we know, this is because the strong interaction is isospin-invariant. Presumably, the small mass differences within an isomultiplet are caused by the electromagnetic interaction, which does not conserve isospin.

As a second example we consider the 1^- mesons (which are unstable against decay by strong interaction). These particles also form an octet (Fig. 7.2) with the same isomultiplet structure as the 0^- mesons. Higher meson resonances are also grouped in similar octets (and singlets), which are distinguished from one another by spin, parity, and G parity.

Turning next to the baryons, we consider first the nucleons and metastable hyperons p, n, $\Sigma^\pm$, Σ^0, Λ^0, Ξ^0, Ξ^-, which have $J^P = \frac{1}{2}^+$ and also form an octet. Again, the isomultiplet structure forms a now familiar pattern (Fig. 7.3), and

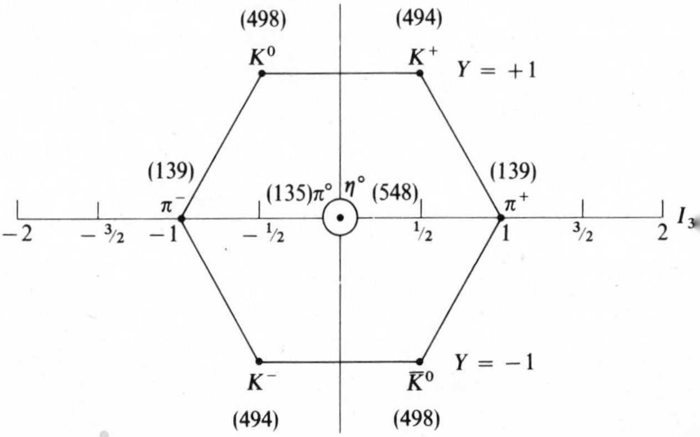

FIGURE 7.1
The 0^- meson octet. Masses in MeV/c^2 are given in parentheses.

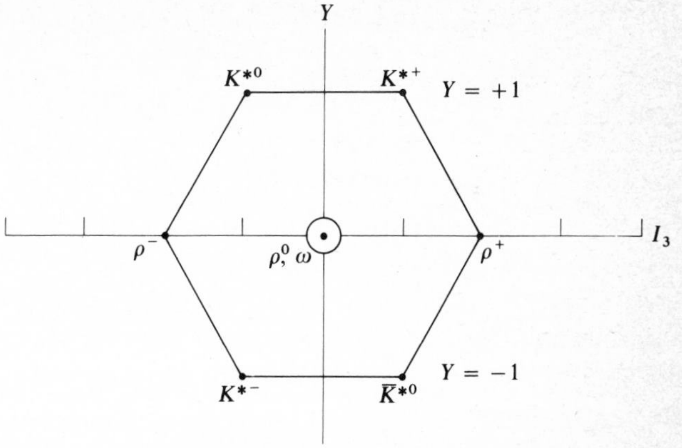

FIGURE 7.2
The 1^- meson octet. Mass splittings between various members of a given isomultiplet are obscured by large resonance decay widths.

the mass differences between members of a given isomultiplet are small. The $J^P = \frac{3}{2}^+$ baryon resonances also form a well-defined multiplet (Fig. 7.4), but this time it is a decuplet rather than an octet.

Quite obviously the multiplet patterns shown in Figs. 7.1 to 7.4 result from the (approximate) invariance of the strong interaction under some symmetry group larger than the SU_2 isospin group. As the reader is undoubtedly well aware, and as we shall discuss in more detail below, it is the group SU_3 which

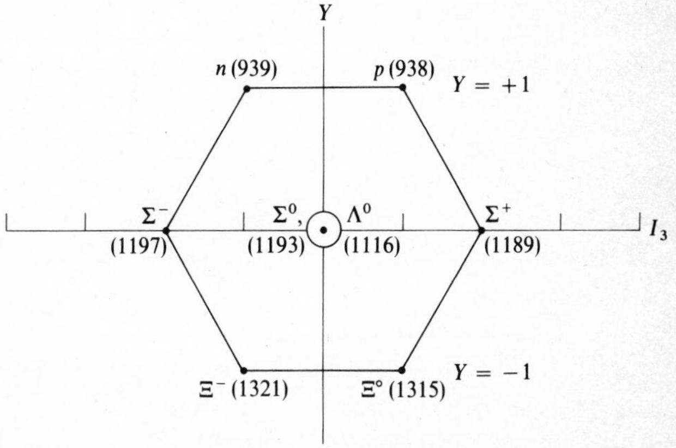

FIGURE 7.3
The $\frac{1}{2}^+$ "stable" baryon octet.

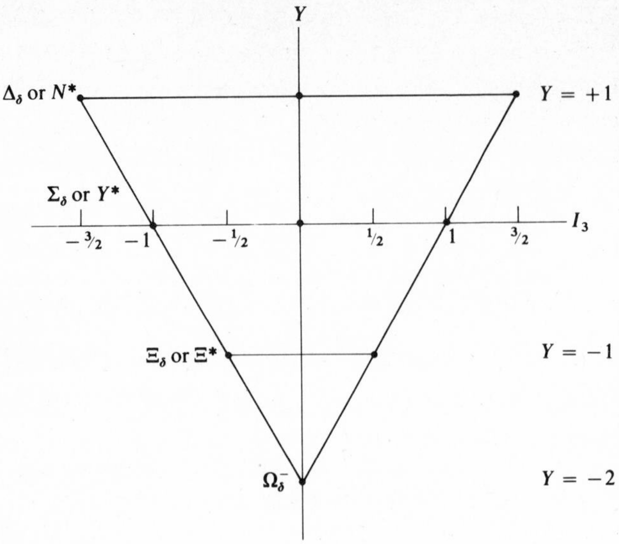

FIGURE 7.4
The $\frac{3}{2}^+$ baryon resonance decuplet.

provides a very natural explanation for these aforementioned regularities. In addition, there are a number of striking empirical selection rules governing the weak decays of hadrons, which must be explained by any theory of weak interactions incorporating SU_3 symmetry. These are now presented.

7.2.2 Selection Rules Governing Weak Decays of Hadrons

7.2.2.1 $|\Delta S| \leq 1$ There are no observed weak decays for which $|\Delta S| > 1$. For example, the transitions $\Xi^0 \to p\pi^-$ and $\Xi^- \to n\pi^-$, for which $|\Delta S| = 2$, have never been observed. The limits on $|\Delta S| = 2$ transitions for all the leptonic and nonleptonic decays are $\Xi(\Delta S = 2)/\Xi(\text{all}) \leq 10^{-3}$.

7.2.2.2 For $\Delta S = 0$ decays, $|\Delta I_3| = 1$ Actually, we have already discussed this selection rule in some detail in Chap. 4. It will be recalled that the $\Delta S = 0$ weak hadronic vector currents $\mathcal{V}_0$ and $\mathcal{V}_0^\dagger$ together with the isovector portion of the hadronic electromagnetic current, form an isotriplet vector (CVC hypothesis). The $\Delta S = 0$ weak hadronic axial vector currents $\mathcal{A}_0$ and $\mathcal{A}_0^\dagger$ also satisfy the rule $|\Delta I_3| = 1$.

7.2.2.3 For $|\Delta S| = 1$ decays, $\Delta S = \Delta Q$ This rule applies to leptonic and semileptonic $|\Delta S| = 1$ transitions, for example, the $|\Delta S| = 1$ K-meson and hyperon decays, and refers to the change in S and Q (strangeness and electric charge) of the transforming hadron. It predicts, for example, that decays such as

$$\Sigma^+ \to n l^+ \nu_l$$

and

$$K^+ \to \pi^+ \pi^+ e^- \bar{\nu}_e$$

are forbidden. Experimentally, one finds (as of 1970)

$K^+ \to \pi^+ \pi^- e^+ \nu_e$	$\Delta S = \Delta Q$	264 world events
$K^+ \to \pi^+ \pi^+ e^- \bar{\nu}_e$	$\Delta S = -\Delta Q$	0 world events

$$\frac{\Gamma(\Sigma^+ \to n e^+ \nu_e)}{\Gamma(\Sigma^- \to n e^- \nu_e)} \approx 0.4 \times 10^{-2} \leq 1.6\% \text{ with } 90\% \text{ confidence}$$

$$\frac{\Gamma(\Sigma^+ \to n \mu^+ \nu_\mu)}{\Gamma(\Sigma^- \to n \mu^- \bar{\nu}_\mu)} \approx 5 \times 10^{-2} \leq 9\% \text{ with } 90\% \text{ confidence}$$

Another test of the $\Delta S = \Delta Q$ rule, in connection with K_{l3} decays, is discussed in detail in Chap. 10.

7.2.2.4 For $|\Delta S| = 1$ transitions, $|\Delta I| = \frac{1}{2}$ Whereas the strangeness non-changing weak hadronic current transforms like an isovector, the $|\Delta S| = 1$ weak hadronic current transforms like an isospinor. It will be noted that in a $|\Delta S| = 1$ weak transition the total isospin cannot be conserved, a conclusion which follows even without knowledge of the $|\Delta I| = \frac{1}{2}$ rule. For, we have by definition

$$I_3 = Q - \frac{S + B}{2} \tag{7.1}$$

where B is the baryon number, which is conserved. Thus,

$$\Delta I_3 = \Delta Q - \frac{\Delta S}{2} \tag{7.2}$$

Since Q and S change by ± 1, we must have

$$|\Delta I_3| = \tfrac{1}{2} \text{ or } \tfrac{3}{2}$$

Therefore $|\Delta I| = \frac{1}{2}, \frac{3}{2}, \frac{5}{2}, \ldots$ are possible a priori. However, if $|\Delta I| = \frac{1}{2}$ only, then $\Delta I_3 = \pm \frac{1}{2}$ and Eq. (7.2) implies $\Delta S = \Delta Q$. Thus the $|\Delta I| = \frac{1}{2}$ rule is more general than the $\Delta S = \Delta Q$ rule, and the former implies the latter.

Examples of the $|\Delta I| = \frac{1}{2}$ rule are found in the decays $K \to 2\pi$, $K \to 3\pi$, $K \to \pi l \nu$ and in hyperon nonleptonic decays. Here we shall consider merely $K \to 2\pi$, deferring detailed discussions of the remaining cases to Chaps. 9 to 11. In the decays $K_S{}^0 \to \pi^+\pi^-$, $K_S{}^0 \to \pi^0\pi^0$, and $K^+ \to \pi^+\pi^0$, the two-pion final state always has zero angular momentum and is thus spatially symmetric. Since pions are bosons, this implies that the isospin part of the two-pion wave function must also be symmetric. Now $I(\pi) = 1$, and so a priori the possible isospin of two pions is $I = 2$, 1, or 0. Of these, the $I = 2$ and 0 states are symmetric with respect to pion exchange, while the $I = 1$ state is antisymmetric and is therefore excluded. Hence, only an $I = 0$ final state is possible if $|\Delta I| = \frac{1}{2}$. Now, noting that

$$\left| I = 0, I_3 = 0 \right\rangle = \frac{1}{\sqrt{3}} \left| \pi_1{}^+\pi_2{}^- + \pi_1{}^-\pi_2{}^+ - \pi_1{}^0\pi_2{}^0 \right\rangle \tag{7.3}$$

and taking into account that the probability for $K_s{}^0 \to \pi^+\pi^-$ is proportional to

$$\tfrac{1}{3}|\pi_1{}^+\pi_2{}^-|^2 + \tfrac{1}{3}|\pi_1{}^-\pi_2{}^+|^2 = \tfrac{2}{3}$$

while the probability for $K_s{}^0 \to \pi^0\pi^0$ is proportional to

$$\tfrac{1}{3}|\pi_1{}^0\pi_2{}^0|^2 = \tfrac{1}{3}$$

we find that the $|\Delta I| = \frac{1}{2}$ rule implies

$$\frac{\Gamma(K_s{}^0 \to \pi^+\pi^-)}{\Gamma(K_s{}^0 \to \pi^0\pi^0)} = 2 \tag{7.4}$$

This result is in doubtful agreement with experiment (see Chap. 11).

However, in the case $K^+ \to \pi^+\pi^0$, the final two-pion state has $I_z = +1$ and therefore $I = 0$ is excluded. Thus I_{final} must be 2 in definite violation of the $|\Delta I| = \frac{1}{2}$ rule. Of course, the $K^+ \to \pi^+\pi^0$ decay rate is very slow, compared, for example, with the rates for $K_s{}^0 \to \pi^+\pi^-$, $\pi^0\pi^0$. Let us denote the $|\Delta I| = \frac{1}{2}$, $\frac{3}{2}$, and $\frac{5}{2}$ transition amplitudes by a_1, a_3, and a_5, respectively. Also, we define the amplitudes

$$A_0 = \langle (2\pi)_I{}^S{}_{=0} | H_{wk} | K^0 \rangle \tag{7.5}$$

and

$$A_2 = \langle (2\pi)_I{}^S{}_{=2} | H_{wk} | K^0 \rangle \tag{7.6}$$

where the superscript S means *stationary*, i.e., no final-state interaction between the pions. Then we have

$$a_1 = A_0 e^{i\delta_0}$$

and

$$a_3 + a_5 = A_2 e^{i\delta_2}$$

where δ_0 and δ_2 are the 2π final-state phase shifts for $I = 0$ and $I = 2$, respectively. One can then show that

$$\frac{\Gamma(K^+ \to \pi^+\pi^0)}{\Gamma(K_s{}^0 \to \pi^+\pi^-) + \Gamma(K_s{}^0 \to 2\pi^0)} = \frac{3}{4} \frac{|a_3 - 2a_5/3|^2}{A_0{}^2 + (\text{Re } A_2)^2} \tag{7.7}$$

If we assume $|a_3| \gg |a_5|$, the experimental value of this ratio allows us to conclude that

$$\left|\frac{A_2}{A_0}\right| \approx 4.5\% \tag{7.8}$$

It is interesting to speculate on the cause of this rather small violation of the $|\Delta I| = \frac{1}{2}$ rule. Possibly it is due to electromagnetic effects. However, in this case one might expect a violation of the order of $\alpha/2\pi \approx 0.2$ percent. On the other hand, since no one has shown how to calculate such electromagnetic effects precisely, the question remains open.[1] In spite of the disturbing, if small, violation of the $|\Delta I| = \frac{1}{2}$ rule in $K^+ \to \pi^+\pi^0$ decay and other small violations in $K_{\pi 3}$ decay, in discussions which follow we shall assume that the rule is valid.

7.3 SU_2

We turn now to a discussion of the symmetry groups of hadron states. We shall be concerned with the groups SU_2 (e.g., the group of isospin transformations under which the strong interaction is invariant) and SU_3 (the group under which the strong interaction is approximately invariant and which explains among other things the multiplet structure of Figs. 7.1 to 7.4).

In general, SU_n is the name given to the group of unitary transformations U on an n-dimensional complex vector space for which

$$\det U = +1 \tag{7.9}$$

We can always express a unitary operator U in terms of a hermitian operator H as follows:

$$U = e^{iH} \tag{7.10}$$

Moreover, we can express H in terms of certain standard hermitian operators F_i, called *generators* of the group SU_n, as follows:[2]

$$H = \alpha^i F_i \tag{7.11}$$

[1] See also Chap. 9.
[2] The repeated-index-summation convention is assumed here.

The numerical parameters α^i characterize H and thus U. We have

$$U = e^{i\alpha^k F_k} \tag{7.12}$$

and (7.9) and (7.12) imply

$$\text{Tr}(F_k) = 0 \tag{7.13}$$

In general, there are n^2 independent $n \times n$ matrices. With the imposition of condition (7.13), however, the number of linearly independent, traceless, hermitian generators F_k is $n^2 - 1$. Is it possible to give a simple prescription for finding such matrices F_k? One method (not unique) is to start by forming the $n - 1$ diagonal matrices

$$\begin{pmatrix} 1 & & & & \\ & -1 & & & \\ & & 0 & & \\ & & & \ddots & \\ & & & & 0 \end{pmatrix}, \quad \frac{1}{\sqrt{3}}\begin{pmatrix} 1 & & & & & \\ & 1 & & & & \\ & & -2 & & & \\ & & & 0 & & \\ & & & & \ddots & \\ & & & & & 0 \end{pmatrix}, \ldots,$$

$$\sqrt{\frac{2}{n(n-1)}}\begin{pmatrix} 1 & & & & & \\ & 1 & & & & \\ & & 1 & & & \\ & & & \ddots & & \\ & & & & 1 & \\ & & & & & -(n-1) \end{pmatrix}$$

Then we form the $(n^2 - n)/2$ off-diagonal matrices with 1 in a given off-diagonal position, 1 in the transposed position, and zeros elsewhere. Also, we form the $(n^2 - n)/2$ off-diagonal matrices with $-i$ in a given off-diagonal position, $+i$ in the transposed position, and zeros elsewhere.

This whole set of $n^2 - 1$ independent $n \times n$ hermitian matrices will be called $\lambda_k (k = 1, \ldots, n^2 - 1)$ with the conditions, obviously satisfied,

$$\text{Tr}\,\lambda_k = 0 \tag{7.14}$$

$$\text{Tr}\,\lambda_i \lambda_k = 2\delta_{ik} \tag{7.15}$$

One representation of the F_k is obtained by putting

$$F_k = \tfrac{1}{2}\lambda_k \tag{7.16}$$

Setting $n = 2$, we have three λ matrices:

$$\lambda_1 = \begin{pmatrix} 0 & 1 \\ 1 & 0 \end{pmatrix} \qquad \lambda_2 = \begin{pmatrix} 0 & -i \\ i & 0 \end{pmatrix} \qquad \lambda_3 = \begin{pmatrix} 1 & 0 \\ 0 & -1 \end{pmatrix}$$

These are just the familiar Pauli matrices, and the $F_k = \frac{1}{2}\lambda_k$ are the three-spin-$\frac{1}{2}$ operators which satisfy the commutation relations

$$[F_i, F_j] = i\varepsilon_{ijk} F_k \qquad (7.17)$$

Relations (7.17) can be satisfied not only by the 2×2 matrices F (the "fundamental" representation, spin $\frac{1}{2}$) but also by matrices of dimension

$$(2I + 1) \times (2I + 1)$$

(corresponding to spin $I = 1, \frac{3}{2}, 2, \frac{5}{2}, \ldots$).

In the fundamental representation we can distinguish two kinds of vectors:

1 Covariant, or column, spinors $\psi = \begin{pmatrix} \psi_1 \\ \psi_2 \end{pmatrix}$, which transform under a unitary transformation as follows:

$$\psi_i' = U_i{}^j \psi_j$$

2 Contravariant, or row, spinors $\bar{\phi} = (\bar{\phi}^1, \bar{\phi}^2)$, which transform as

$$\bar{\phi}^{j\prime} = \bar{\phi}^i U_i{}^{j\dagger}$$

A covariant two-spinor is transformed to a contravariant two-spinor and vice versa with the aid of the completely antisymmetric unit tensors ε_{ij} and ε^{ij}. Thus, for example,

$$\phi_i = \varepsilon_{ij} \bar{\phi}^j$$

transforms as a covariant spinor and has the form

$$\phi_i = \begin{pmatrix} \bar{\phi}^2 \\ -\bar{\phi}^1 \end{pmatrix} \qquad (7.18)$$

Higher representations of SU_2 are systematically built up from the fundamental 2×2 representation by forming tensor products of the basic spinors ψ_i, $\bar{\phi}^j$, or ϕ_i and then by symmetrizing and antisymmetrizing. Here we utilize a basic theorem which states that once a multispinor has been broken up into its different symmetry parts it has been decomposed uniquely into irreducible representations of the special unitary group (SU_n).

For example, consider the tensor product of two covariant spinors ψ_i, η_j:

$$\psi_i \eta_j = \tfrac{1}{2}(\psi_i \eta_j + \psi_j \eta_i) + \tfrac{1}{2}(\psi_i \eta_j - \psi_j \eta_i) \qquad (7.19)$$

On the left-hand side of (7.19) we have four distinct elements since i, j independently can take on the values one or two. On the right-hand side the first (symmetric) term in parentheses has three independent components and transforms as a spin-one object. The second (antisymmetric) term in parentheses is nonzero only if $i \neq j$. It thus consists of only one component and transforms as a spin-zero object. We obtain the familiar result

$$
\begin{array}{lll}
J = 1 \begin{cases} J_3 = +1 \\ J_3 = 0 \\ J_3 = -1 \end{cases} & \begin{array}{l} \psi_1 \eta_1 \\ \dfrac{1}{\sqrt{2}}(\psi_1 \eta_2 + \psi_2 \eta_1) \\ \psi_2 \eta_2 \end{array} \\
J = 0 \;\; J_3 = 0 & \dfrac{1}{\sqrt{2}}(\psi_1 \eta_2 - \psi_2 \eta_1)
\end{array}
\tag{7.20}
$$

as in the theory of angular momentum. Alternatively, in the theory of isotopic spin we may combine

$$
\begin{aligned}
\psi_i \phi_j &= \tfrac{1}{2}(\psi_i \phi_j + \psi_j \phi_i) + \tfrac{1}{2}(\psi_i \phi_j - \psi_j \phi_i) \\
&= \tfrac{1}{2}(\psi_i \varepsilon_{jk} \bar{\phi}^k + \psi_j \varepsilon_{ik} \bar{\phi}^k) + \tfrac{1}{2}(\psi_i \varepsilon_{jk} \bar{\phi}^k - \psi_j \varepsilon_{ik} \bar{\phi}^k)
\end{aligned}
\tag{7.21}
$$

Here ψ represents, e.g., the nucleon isodoublet, and ϕ the antinucleon isodoublet; that is, $\bar{\phi} = (\bar{p}\bar{n})$ $\phi = \begin{pmatrix} \bar{n} \\ -\bar{p} \end{pmatrix}$, and $\psi = \begin{pmatrix} p \\ n \end{pmatrix}$. We thus obtain an isospin-one object and an isospin-zero object:

$$
\begin{array}{lll}
I = 1 \begin{cases} I_3 = 1 \\ I_3 = 0 \\ I_3 = -1 \end{cases} & \begin{array}{l} p\bar{n} \\ \dfrac{1}{\sqrt{2}}(p\bar{p} - n\bar{n}) \\ n\bar{p} \end{array} \\
I = 0 \;\; I_3 = 0 & \dfrac{1}{\sqrt{2}}(p\bar{p} + n\bar{n})
\end{array}
\tag{7.22}
$$

Here note carefully the signs in the $I_3 = 0$ components. The *pi* mesons (π^+, π^0, π^-) possess the same isospin transformation properties as the three $I = 1$ states given in (7.22).

A convenient way to visualize the irreducible representations of SU_2 is to recall that any given representation consists of a set of eigenstates ($2I + 1$ of them for the I multiplet) and that each of these eigenstates can be depicted as a point on the so-called *root diagram*. In the case of SU_2 this is simply a one-dimensional array of points along I_3. In general, we can imagine root diagrams for each group SU_n where the number of dimensions in the diagram corresponds to the number of operators F_k which can be simultaneously diagonalized. (In the case SU_2 the number of dimensions is, of course, only one.)

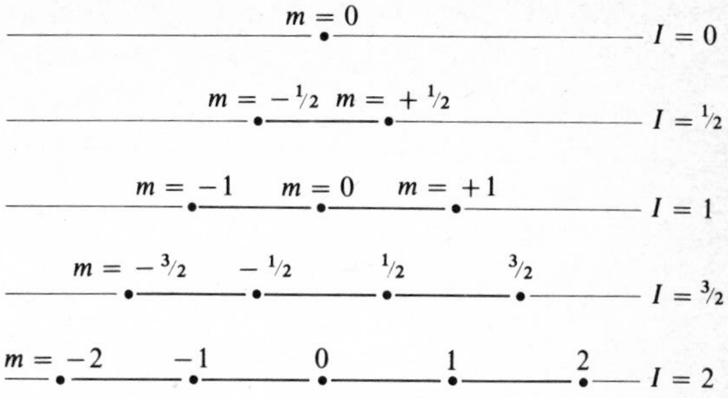

FIGURE 7.5
Root diagrams for SU_2 multiplets $I = 0, \frac{1}{2}, 1, \frac{3}{2}, 2$.

When we take the tensor product of two irreducible representations, we must somehow combine their root diagrams. Of course, this is an extremely simple procedure for SU_2. For example, consider the tensor product of two spin-one representations *a* and *b*. The original root diagrams are as shown in Fig. 7.6. To find the new diagrams, we just superpose *b* on each state of *a*. The result is Fig. 7.7 (which is not a two-dimensional root diagram but merely an illustration). The new irreducible representations consist of a spin-zero object (one state), a spin-one object (three states) and a spin-two object (five states). Algebraically we obtain these by antisymmetrizing and removing lower-rank tensors such as the trace, as follows:

1 Start with $T_{ij} = \psi_i \eta_j$; $i, j = 1, 2, 3$.

2 Form $A_k = \frac{1}{2}\psi_i \eta_j \varepsilon_{ijk} = \frac{1}{2}(\psi_i \eta_j - \psi_j \eta_i)$. A_k is an antisymmetric tensor of rank unity with three components.

3 Remove A_k from T_{ij}:

$$
\begin{aligned}
T_{ij} &= T_{ij} - \tfrac{1}{2}\psi_i \eta_j \varepsilon_{ijk} + \tfrac{1}{2}\psi_i \eta_j \varepsilon_{ijk} \\
&= \tfrac{1}{2}(\psi_i \eta_j + \psi_j \eta_i) + \tfrac{1}{2}(\psi_i \eta_j - \psi_j \eta_i)
\end{aligned}
$$

4 Form $S = \operatorname{tr} T_{ij} = T_{kk}\delta_{ij}$. S is a symmetric tensor of zero rank with one component.

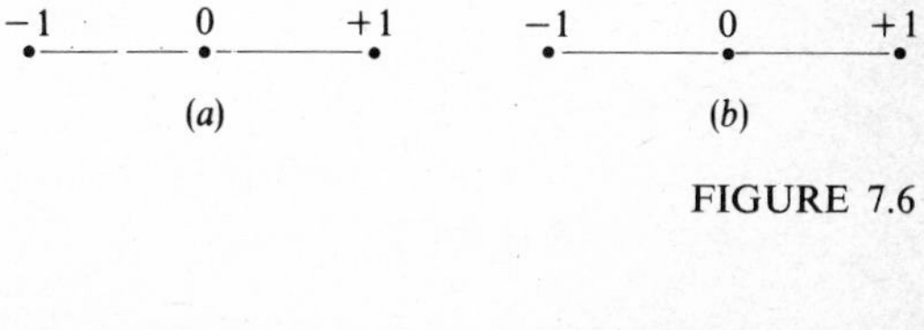

FIGURE 7.6

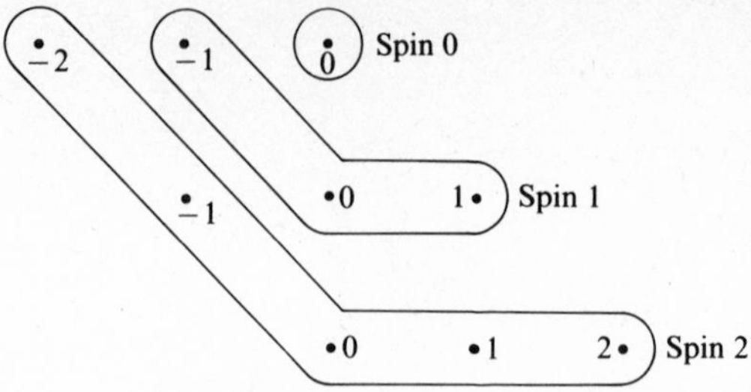

FIGURE 7.7

5 Remove S from T_{ij} as follows:

$$T_{ij} = [\tfrac{1}{2}(\psi_i \eta_j + \psi_i \eta_j) - \tfrac{1}{3} T_{kk} \delta_{ij}] + A_k + S$$
$$= M_{ij} + A_k + S$$

Thus, finally M_{ij} is an irreducible tensor of rank two, and has five independent components.

7.4 SU_3

For $n = 3$, we have $3^2 - 1 = 8$ λ matrices:

$$\lambda_1 = \begin{pmatrix} 0 & 1 & 0 \\ 1 & 0 & 0 \\ 0 & 0 & 0 \end{pmatrix} \qquad \lambda_2 = \begin{pmatrix} 0 & -i & 0 \\ i & 0 & 0 \\ 0 & 0 & 0 \end{pmatrix} \qquad \lambda_3 = \begin{pmatrix} 1 & 0 & 0 \\ 0 & -1 & 0 \\ 0 & 0 & 0 \end{pmatrix}$$

$$\lambda_4 = \begin{pmatrix} 0 & 0 & 1 \\ 0 & 0 & 0 \\ 1 & 0 & 0 \end{pmatrix} \qquad \lambda_5 = \begin{pmatrix} 0 & 0 & -i \\ 0 & 0 & 0 \\ i & 0 & 0 \end{pmatrix} \qquad \lambda_6 = \begin{pmatrix} 0 & 0 & 0 \\ 0 & 0 & 1 \\ 0 & 1 & 0 \end{pmatrix}$$

$$\lambda_7 = \begin{pmatrix} 0 & 0 & 0 \\ 0 & 0 & -i \\ 0 & i & 0 \end{pmatrix} \qquad \lambda_8 = \sqrt{\tfrac{1}{3}} \begin{pmatrix} 1 & 0 & 0 \\ 0 & 1 & 0 \\ 0 & 0 & -2 \end{pmatrix}$$

We sometimes write also

$$\lambda_0 = \sqrt{\tfrac{2}{3}} \begin{pmatrix} 1 & 0 & 0 \\ 0 & 1 & 0 \\ 0 & 0 & 1 \end{pmatrix}$$

Taking $F_k = \frac{1}{2}\lambda_k$ as before, simple manipulations lead us to the commutation relations

$$[F_i, F_j] = i f_{ij}{}^k F_k \qquad (7.23)$$

where the nonzero SU_3 *structure constants* $f_{ij}{}^k$ are given in Table 7.1. It is easy to show that these structure constants satisfy the following identities:

$$f_{ij}{}^k = -f_{ji}{}^k \qquad (7.24)$$

and

$$f_{ij}{}^{l}f_{lk}{}^{m} + f_{ki}{}^{l}f_{lj}{}^{m} + f_{jk}{}^{l}f_{li}{}^{m} = 0 \tag{7.25}$$

We may also verify that the following equation holds:[1]

$$\{F_i, F_j\} = \tfrac{1}{3}\delta_{ij} + d_{ijk}F_k \tag{7.26}$$

where the coefficients d_{ijk} are totally symmetric and are given in Table 7.1.

[1] $\{F_i, F_j\} \equiv F_iF_j + F_jF_i$.

Table 7.1 SU_3: ANTISYMMETRIC STRUCTURE CONSTANTS $f_{ij}{}^{k}$ AND SYMMETRIC CONSTANTS d_{ijk}

i	j	k	$f_{ij}{}^{k}$	i	j	k	d_{ijk}
1	2	3	1	1	1	8	$\frac{1}{\sqrt{3}}$
1	4	7	$\frac{1}{2}$	1	4	6	$\frac{1}{2}$
1	5	6	$-\frac{1}{2}$	1	5	7	$\frac{1}{2}$
2	4	6	$\frac{1}{2}$	2	2	8	$\frac{1}{\sqrt{3}}$
2	5	7	$\frac{1}{2}$	2	4	7	$-\frac{1}{2}$
3	4	5	$\frac{1}{2}$	2	5	6	$\frac{1}{2}$
3	6	7	$-\frac{1}{2}$	3	3	8	$\frac{1}{\sqrt{3}}$
4	5	8	$\frac{\sqrt{3}}{2}$	3	4	4	$\frac{1}{2}$
6	7	8	$\frac{\sqrt{3}}{2}$	3	5	5	$\frac{1}{2}$
				3	6	6	$-\frac{1}{2}$
				3	7	7	$-\frac{1}{2}$
				4	4	8	$-\frac{1}{2\sqrt{3}}$
				5	5	8	$-\frac{1}{2\sqrt{3}}$
				6	6	8	$-\frac{1}{2\sqrt{3}}$
				7	7	8	$-\frac{1}{2\sqrt{3}}$
				8	8	8	$-\frac{1}{\sqrt{3}}$

We shall also find it very useful to define the following operators:

$$I_\pm = F_1 \pm iF_2 \tag{7.27}$$

$$I_3 = F_3 \tag{7.28}$$

$$U_\pm = F_6 \pm iF_7 \tag{7.29}$$

$$U_3 = -\tfrac{1}{2}F_3 + \sqrt{\tfrac{3}{4}}\,F_8 \tag{7.30}$$

$$V_\pm = F_4 \mp iF_5 \tag{7.31}$$

and

$$V_3 = -\tfrac{1}{2}F_3 - \sqrt{\tfrac{3}{4}}\,F_8 \tag{7.32}$$

Only eight of these nine operators are independent since $I_3 + U_3 + V_3 = 0$. Obviously $I_\pm$ and I_3 refer to isospin (I spin). Similarly, Eqs. (7.29) and (7.30) define the U-spin operators and (7.31) and (7.32) define the V-spin operators. The commutation relations (7.23) may be written in terms of I-, U-, and V-spin operators (Table 7.2). We see from Table 7.2 that the I-, U-, V-spin operators generate three noncommuting SU_2 subgroups in SU_3.

The key assumption which links the theory of SU_3 to the physics of hadrons, is that F_8 is proportional to hypercharge:

$$F_8 = \frac{\sqrt{3}}{2}\,Y \tag{7.33}$$

Since the two generators F_3 and $F_8 = \sqrt{3}\,Y/2$ are simultaneously diagonalized in SU_3, the root diagrams for SU_3 are two-dimensional. Choosing $F_3 = I_3$ and

Table 7.2 SU_3 COMMUTATION RELATIONS FOR I, U, V SPIN OPERATORS

$[I_+, I_-] = 2I_3$	$[U_+, U_-] = 2U_3$	$[V_+, V_-] = 2V_3$
$[I_+, I_3] = -I_+$	$[U_+, U_3] = -U_+$	$[V_+, V_3] = -V_+$
$[I_-, I_3] = +I_-$	$[U_-, U_3] = +U_-$	$[V_-, V_3] = +V_-$
$[I_3, U_3] = 0$	$[I_+, U_3] = \tfrac{1}{2}I_+$	$[I_-, U_3] = -\tfrac{1}{2}I_-$
$[I_3, U_+] = -\tfrac{1}{2}U_+$	$[I_+, U_+] = V_-$	$[I_-, U_+] = 0$
$[I_3, U_-] = \tfrac{1}{2}U_-$	$[I_+, U_-] = 0$	$[I_-, U_-] = -V_+$
$[I_3, V_3] = 0$	$[I_+, V_3] = \tfrac{1}{2}I_+$	$[I_-, V_3] = -\tfrac{1}{2}I_-$
$[I_3, V_+] = -\tfrac{1}{2}U_+$	$[I_+, V_+] = -U_-$	$[I_-, V_+] = 0$
$[I_3, V_-] = +\tfrac{1}{2}V_-$	$[I_+, V_-] = 0$	$[I_-, V_-] = U_+$
$[U_3, V_3] = 0$		
$[U_3, V_+] = -\tfrac{1}{2}V_+$		
$[U_3, V_-] = \tfrac{1}{2}V_-$		
$[U_+, V_3] = \tfrac{1}{2}U_+$		
$[U_+, V_+] = I_-$		
$[U_+, V_-] = 0$		
$[U_-, V_3] = -\tfrac{1}{2}U_-$		
$[U_-, V_+] = 0$		
$[U_-, V_-] = -I_+$		

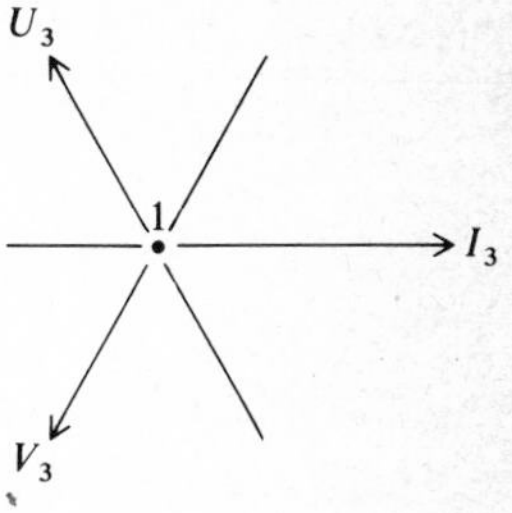

FIGURE 7.8
Root diagram for SU_3, showing the simplest representation, the singlet, or 1.

F_8 as abscissa and ordinate, respectively, we see from Eqs. (7.30) and (7.32) that the U_3 and V_3 axes make angles of 120 and 240°, respectively, with respect to the I_3 axis. Each eigenstate of a given representation is depicted by a point which must fall on the vertices of a lattice which satisfies the condition I_3, U_3, V_3 integral or half-integral. Also each point must belong simultaneously to SU_2 multiplets in I, U, and V. The simplest representation is just a singlet called the 1 and associated with a single point at the origin of the root diagram (Fig. 7.8).

The next higher representations, the 3 and $\bar{3}$, each consist of three independent states, as shown in Fig. 7.9.[1] The 3 is described by a covariant column spinor ψ_i:

$$\psi_i = \begin{pmatrix} p \\ n \\ \lambda \end{pmatrix} \qquad \psi_i' = U_i^{\ j}\psi_j \tag{7.34}$$

while the $\bar{3}$ is described by a contravariant spinor ϕ^j:[2]

$$\phi^j = (\bar{p}, \bar{n}, \bar{\lambda}) \qquad \phi^{j\prime} = \phi^i\, U_i^{\ j+} \tag{7.35}$$

[1] The phases of the states p, n, λ; $\bar{p}$, $\bar{n}$, $\bar{\lambda}$ can be confusing. Referring to the root diagram of Fig. 7.9*b* for the $\bar{3}$ and the commutation relations of Table 7.2, we start with state $\bar{p}$ and go once around the triangle in the counterclockwise direction by applying the operators I_+, U_+, V_+ in succession:

$$(V_+ U_+ I_+)\bar{p} = (U_+ V_+ I_+ - I_- I_+)\bar{p} = -I_- I_+ \bar{p}$$

since $U_+ V_+ I_+ \bar{p} = 0$. However $-I_- I_+ \bar{p} = -I_+ I_- \bar{p} + 2I_3 \bar{p}$. Thus, because $I_- \bar{p} = 0$,

$$(V_+ U_+ I_+)\bar{p} = 2I_3 \bar{p} = -\bar{p}$$

Therefore in going once around, we do not recover $\bar{p}$ but rather its negative. In the case of the 3 representation, a clockwise circuit of the triangle involves applying $I_+ U_+ V_+$ to p and results in no phase change:

$$(I_+ U_+ V_+)p = +p$$

[2] Here and in what follows we omit the bar over the contravariant spinor ϕ.

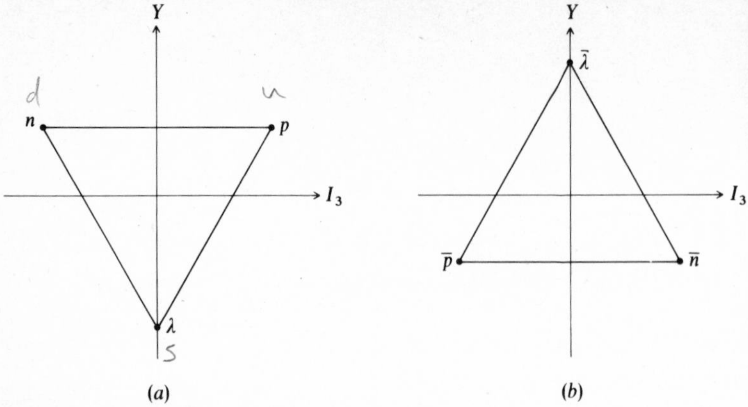

FIGURE 7.9
Root diagrams for 3 and $\bar{3}$ representations. (*a*) The 3; (*b*) the $\bar{3}$.

It is important to note that the symbols $p, n, \lambda, \bar{p}, \bar{n}, \bar{\lambda}$ do *not* refer to physical proton, neutron, ... states. In fact, by inspection we can write down the quantum numbers I_3, U_3, V_3 and Y for the 3 and $\bar{3}$ states (see Table 7.3), and we find that the hypercharge of each state is $\frac{1}{3}$ integral. Thus these states do not correspond to any observed particles since these always have integral hypercharge. The 3, ($\bar{3}$) states can be associated with quark (antiquark) states, hypothetical particles with spin $\frac{1}{2}$ and with the assumed quantum numbers for strangeness, baryon number B, and electric charge Q given in Table 7.3.

Higher representations are constructed from the 3 and $\bar{3}$ by utilization of the invariant tensors $\delta_i{}^j$, ε^{ijk}, and ε_{ijk} to break down an arbitrary tensor into its

Table 7.3 QUANTUM NUMBERS OF QUARK AND ANTIQUARK STATES

		I_3	U_3	V_3	Y	S	B	Q
	p	$\frac{1}{2}$	0	$-\frac{1}{2}$	$\frac{1}{3}$	0	$\frac{1}{3}$	$\frac{2}{3}$
Quarks	n	$-\frac{1}{2}$	$\frac{1}{2}$	0	$\frac{1}{3}$	0	$\frac{1}{3}$	$-\frac{1}{3}$
	λ	0	$-\frac{1}{2}$	$\frac{1}{2}$	$-\frac{2}{3}$	-1	$\frac{1}{3}$	$-\frac{1}{3}$
	$\bar{p}$	$-\frac{1}{2}$	0	$\frac{1}{2}$	$-\frac{1}{3}$	0	$-\frac{1}{3}$	$-\frac{2}{3}$
Antiquarks	$\bar{n}$	$\frac{1}{2}$	$-\frac{1}{2}$	0	$-\frac{1}{3}$	0	$-\frac{1}{3}$	$\frac{1}{3}$
	$\bar{\lambda}$	0	$\frac{1}{2}$	$-\frac{1}{2}$	$\frac{2}{3}$	1	$-\frac{1}{3}$	$\frac{1}{3}$

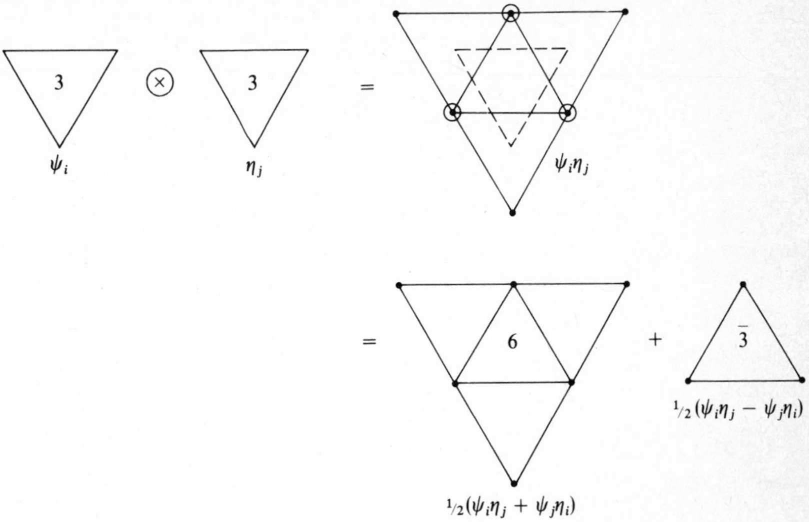

FIGURE 7.10
The tensor product $3 \otimes 3$.

irreducible parts. For example, consider the tensor $\psi_i \eta_j$ composed of two covariant three-spinors. This tensor has nine components and it is reducible:

$$\psi_i \eta_j = (\psi_i \eta_j - \tfrac{1}{2}\varepsilon^{ijk}\psi_i \eta_j) + \tfrac{1}{2}\varepsilon^{ijk}\psi_i \eta_j$$
$$= \tfrac{1}{2}(\psi_i \eta_j + \psi_j \eta_i) + \tfrac{1}{2}(\psi_i \eta_j - \psi_j \eta_i) \qquad (7.36)$$

or symbolically

$$3 \otimes 3 = 6 + \bar{3} \qquad (7.37)$$

The symmetric tensor 6 on the right-hand side of (7.36) has six independent elements, while the antisymmetric tensor $\bar{3}$ has three independent elements. This is illustrated in Fig. 7.10 where we obtain the root diagram for $3 \otimes 3$ by placing the center of the triangle for ψ_i at each corner of the η_j triangle. Wherever the corners of two ψ_i triangles touch, there is degeneracy, labeled by the symbol $\odot$.

As another example, consider the mixed reducible tensor $\psi_i \phi^j$ with nine components:

$$\psi_i \phi^j = (\psi_i \phi^j - \tfrac{1}{3}\,\delta_i^{\,j}\psi_k \phi^k) + \tfrac{1}{3}\,\delta_i^{\,j}\psi_k \phi^k \qquad (7.38)$$

In Eq. (7.38) we have separated out the trace of $\psi_i \phi^j$; the last quantity is a tensor of zero rank with one component (singlet). The remaining irreducible traceless second-rank tensor has eight independent components (octet). Symbolically we have

$$3 \otimes \bar{3} = 8 + 1 \tag{7.39}$$

This is represented geometrically in Fig. 7.11.

According to the quark model of hadron structure (which has great intuitive appeal but may not correspond to reality), a meson is composed of a quark–antiquark pair. (See, e.g., Gel 64b). Thus, the octet of Eq. (7.38) or (7.39) should correspond to a meson octet (Fig. 7.1 or 7.2) when the 3 and $\bar{3}$ correspond to triplets of quarks and antiquarks, respectively. Indeed, the isospin and hypercharge quantum numbers of the octet diagrams of Figs. 7.1, 7.2, and 7.11 are identical, which is strongly suggestive. We write the 0^- meson octet tensor in matrix form and make the $Q\bar{Q}$ composition of the various mesons explicit:

$$M = \psi_i \phi^j - \tfrac{1}{3}\delta_i^{\,j}\psi_k \phi^k$$

$$= \begin{pmatrix} \frac{2}{3}p\bar{p} - \frac{1}{3}n\bar{n} - \frac{1}{3}\lambda\bar{\lambda} & p\bar{n} & p\bar{\lambda} \\ n\bar{p} & -\frac{1}{3}p\bar{p} + \frac{2}{3}n\bar{n} - \frac{1}{3}\lambda\bar{\lambda} & n\bar{\lambda} \\ \lambda\bar{p} & \lambda\bar{n} & -\frac{1}{3}p\bar{p} - \frac{1}{3}n\bar{n} + \frac{2}{3}\lambda\bar{\lambda} \end{pmatrix}$$

$$= \begin{pmatrix} \frac{\eta^0}{\sqrt{6}} + \frac{\pi^0}{\sqrt{2}} & \pi^+ & K^+ \\ \pi^- & +\frac{\eta^0}{\sqrt{6}} - \frac{\pi^0}{\sqrt{2}} & K^0 \\ K^- & \bar{K}^0 & -\frac{2\eta^0}{\sqrt{6}} \end{pmatrix} \tag{7.40}$$

Of course, the quark model may not be correct; even so, the last matrix on the right-hand side of (7.40) must correspond to the 0^- meson octet.

It will be noted that under an arbitrary infinitesimal SU_3 transformation U the eight generators F_i transform analogously to the eight independent states of the meson octet. Thus we obtain the so-called *adjoint*, or *generator octet*, representation, which is conveniently expressed by the matrix

$$F = \begin{pmatrix} \frac{1}{\sqrt{2}}\left(F_3 + \frac{1}{\sqrt{3}}F_8\right) & I_+ = F_1 + iF_2 & V_- = F_4 + iF_5 \\ I_- = F_1 - iF_2 & \frac{1}{\sqrt{2}}\left(-F_3 + \frac{1}{\sqrt{3}}F_8\right) & U_+ = F_6 + iF_7 \\ V_+ = F_4 - iF_5 & U_- = F_6 - iF_7 & -\frac{2}{\sqrt{6}}F_8 \end{pmatrix} \tag{7.41}$$

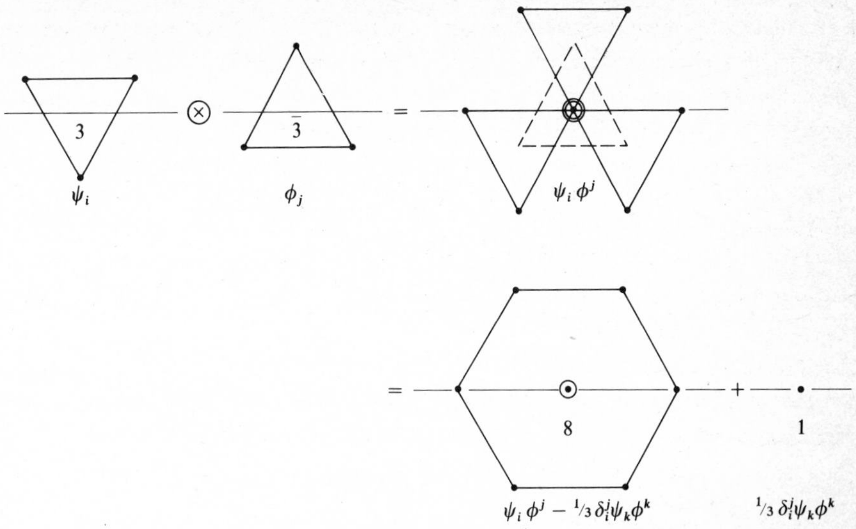

FIGURE 7.11
The tensor product $3 \otimes \bar{3}$.

According to the quark model, baryons are assumed to be composed of three quarks QQQ, and antibaryons of three antiquarks $\bar{Q}\bar{Q}\bar{Q}$. Here the situation is much more complex and obscure than in the quark model of mesons. It may seem reasonable enough to have a strong attractive force between Q and $\bar{Q}$ leading to meson states, but it is difficult to understand why the configuration QQQ should be more stable than that of four or five quarks, etc. If we assume the QQQ configuration, however, we can make considerable progress in understanding the baryons. Neglecting spin for the moment, the algebraic problem is clearly to find the irreducible representations of the covariant tensor $T_{ijk} = \psi_i \phi_j \eta_k$:

$$\begin{aligned} 3 \otimes 3 \otimes 3 &= 3 \otimes (3 \otimes 3) \\ &= 3 \otimes (6 + \bar{3}) \\ &= (3 \otimes 6) + (3 \otimes \bar{3}) \\ &= (10 + 8) + (8 + 1) \qquad (7.42) \end{aligned}$$

The singlet is completely antisymmetric in ijk, the decuplet is completely symmetric, and the octets both have mixed symmetry.

If we now allow for the assumption that quarks have spin $\frac{1}{2}$, the three representation must be enlarged to a six, which represents the three possible quark states, each one of which has two possible spin orientations. We are now dealing with the group SU_6 instead of SU_3, and for three quarks we need to find the irreducible representations of $6 \otimes 6 \otimes 6$. It can be shown that

$$\begin{aligned} 6 \otimes 6 \otimes 6 &= 6 \otimes (6 \otimes 6) \\ &= 6 \otimes (21 + 15) \\ &= (6 \otimes 21) + (6 \otimes 15) \\ 6 \otimes 6 \otimes 6 &= (56 + 70) + (70 + 20) \end{aligned} \tag{7.43}$$

The 56 can be broken down to an SU_3 10 with spin $\frac{3}{2}$ ($2J + 1 = 4$, thus 40 states in all) and an SU_3 8 with spin $\frac{1}{2}$ ($2J + 1 = 2$, thus 16 states in all). Of course, the 10 is identified with the $J^p = \frac{3}{2}^+$ decuplet of Fig. 7.4, while the 8 corresponds to the "stable" baryon $\frac{1}{2}^+$ octet of Fig. 7.3. The baryon octet is conveniently written as a matrix:

$$B = \begin{pmatrix} \dfrac{\Sigma^0}{\sqrt{2}} + \dfrac{\Lambda^0}{\sqrt{6}} & \Sigma^+ & p \\ \Sigma^- & -\dfrac{\Sigma^0}{\sqrt{2}} + \dfrac{\Lambda^0}{\sqrt{6}} & n \\ \Xi^- & \Xi^0 & -\dfrac{2\Lambda^0}{\sqrt{6}} \end{pmatrix} \tag{7.44}$$

which is a valid expression independent of the quark model.

One can easily show that under charge conjugation the meson or baryon octet matrix transforms to a new matrix in which all the rows and columns are transposed and each particle state is replaced by its charge conjugate. Thus from Eq. (7.40),

$$M_c = \begin{pmatrix} \dfrac{\overline{\eta^0}}{\sqrt{6}} + \dfrac{\overline{\pi^0}}{\sqrt{2}} & \overline{\pi^-} & \overline{K^-} \\ \overline{\pi^+} & \dfrac{\overline{\eta^0}}{\sqrt{6}} - \dfrac{\overline{\pi^0}}{\sqrt{2}} & \overline{\overline{K^0}} = K^0 \\ \overline{K^+} & \overline{K^0} & -\dfrac{\overline{2\eta^0}}{\sqrt{6}} \end{pmatrix} \tag{7.45}$$

However, since $\overline{\pi^+} = \pi^-$, $\overline{K^-} = K^+$, and so on, we find that $M_c = M$; in other words, the meson octet is *self-conjugate*. This is not true of the baryon octet, however:

$$B_c = \begin{pmatrix} \dfrac{\overline{\Lambda^0}}{\sqrt{6}} + \dfrac{\overline{\Sigma^0}}{\sqrt{2}} & \overline{\Sigma^-} & \overline{\Xi^-} \\ \overline{\Sigma^+} & \dfrac{\overline{\Lambda^0}}{\sqrt{6}} - \dfrac{\overline{\Sigma^0}}{\sqrt{2}} & \overline{\Xi^0} \\ \bar{p} & \bar{n} & -\dfrac{\overline{2\Lambda^0}}{\sqrt{6}} \end{pmatrix} \tag{7.46}$$

and $B_c \neq B$.

7.5 THE HIERARCHY OF INTERACTIONS

At this point, it is very useful to think in terms of a hierarchy of elementary particle interactions:

1 Superstrong
2 Medium strong
3 Electromagnetic
4 Weak

We shall find it instructive to consider the various hadron states as one after another of these interactions is *turned on*. At first we imagine that all interactions save for the superstrong are *switched off*. The superstrong interaction is assumed to be SU_3-invariant, so that in this circumstance all the particles in a given SU_3 multiplet are mass-degenerate. Thus, for example, all the 0^- mesons would have the same mass, all the $\frac{1}{2}^+$ baryons would have another mass, etc.

The medium-strong interaction breaks SU_3, conserves isospin, and has a coupling strength about 10 percent that of the superstrong interaction. When it is turned on, mass splittings are introduced between one isomultiplet and another, within a given SU_3 multiplet. Thus, for example, within the $\frac{1}{2}^+$ baryon octet, the p and n now have the same mass, while $\Sigma^\pm$, Σ^0 have another common mass, etc. So far the mass splitting within an SU_3 multiplet depends on hypercharge only. A reasonable account of the mass splittings, at least for the $\frac{1}{2}^+$ baryon octet and $\frac{3}{2}^+$ baryon decuplet, is given by assuming that the medium-strong interaction transforms as F_8 under SU_3 transformations [Gell-Mann–Okubo mass formula (Gel 61, Oku 62)].

Next we turn on the electromagnetic interaction. It violates SU_3 and breaks isospin invariance; thus it introduces mass splittings between members of a given isomultiplet within a given SU_3 multiplet. However, it conserves hypercharge, and so in electromagnetic transitions there is no strangeness change.

Let us try to find how the hadronic electromagnetic current transforms under SU_3. This current is a four-vector:

$$j_{\mathrm{EM}}{}^{\lambda}(x) \qquad \lambda = 0, 1, 2, 3$$

and its zeroth component is the charge density. Thus the charge operator $\hat{Q}$ is given by the equation

$$\hat{Q} = \int j_{\mathrm{EM}}{}^{0}(x)\, d^3x$$

We also know that the expectation value of $\hat{Q}$ (namely, Q) for a hadron is related to isospin and hypercharge by the formula

$$Q = \langle I_3 + \tfrac{1}{2}Y \rangle$$

Thus we put

$$\hat{Q} = F_3 + \frac{1}{\sqrt{3}} F_8$$

and the hadronic EM current $j_{\mathrm{EM}}{}^{\lambda}(x)$ is written as

$$j_{\mathrm{EM}}{}^{\lambda}(x) = j_3{}^{\lambda}(x) + \frac{1}{\sqrt{3}} j_8{}^{\lambda}(x) \qquad (7.47)$$

where the $j_i{}^{\lambda}(x)$ are members of an octet of vector currents, with

$$F_i = \int j_i{}^{0}(x)\, d^3x \qquad (7.48)$$

The form (7.47) is the sum of an I-spin vector and an I-spin scalar, and, in fact, it transforms as a U-spin scalar. A number of useful predictions about baryon magnetic moments, electromagnetic decays of vector (1^-) mesons, etc., follow immediately from the assumption of form (7.47) for the hadronic electromagnetic current (see, for example, Oak 63, Cab 61, Col 60, and Ros 64).

7.6 CABIBBO'S HYPOTHESIS

Finally we consider the weak interaction of hadrons. As we know, this interaction satisfies the selection rules

$$|\Delta S| \le 1 \qquad (7.49a)$$

$$|\Delta I_3| = 1 \qquad \text{for } \Delta S = 0 \qquad (7.49b)$$

$$\Delta S = \Delta Q \qquad \text{for } |\Delta S| = 1 \qquad (7.49c)$$

$$|\Delta I| = \tfrac{1}{2} \qquad \text{for } |\Delta S| = 1 \text{ (approximately true)} \qquad (7.49d)$$

Furthermore, we know that the $\Delta S = 0$ weak hadronic vector currents $\mathcal{V}_0$, $\mathcal{V}_0^\dagger$ and the isovector portion of the hadronic electromagnetic current form a single isotriplet (CVC hypothesis). It is therefore natural to assume that $\mathcal{V}_0^\lambda$, $\mathcal{V}_0^{\lambda\dagger}$, and j_{EM}^λ belong to the same octet of vector current operators $j_i(x)$. In fact, we *tentatively* assume

$$\mathcal{V}_0^\lambda(x) = j_1^\lambda(x) - ij_2^\lambda(x) \tag{7.50}$$

$$\mathcal{V}_0^{\lambda\dagger}(x) = j_1^\lambda(x) + ij_2^\lambda(x) \tag{7.51}$$

as well as

$$j_{\text{EM}}^\lambda(x) = j_3^\lambda(x) + \frac{1}{\sqrt{3}} j_8^\lambda(x)$$

Let us consider next the $|\Delta S| = 1$ weak hadronic vector currents $\mathcal{V}_1^\lambda(x)$ and $\mathcal{V}_1^{\lambda\dagger}(x)$ in the light of the selection rules (7.49a), (7.49c), and (7.49d). It is helpful to focus attention on one particular decay as an instructive example:

$$K^+ \to \pi^0 l^+ \nu_l \tag{7.52}$$

for which the operative hadronic weak current is $\mathcal{V}_1^\lambda(x)$. Here $\Delta S = \Delta Q = -1$ and $I_i = \frac{1}{2}$, $I_f = 1$, $\Delta I = +\frac{1}{2}$. Let us consider the V-spin triplet of which K^+ is the member with $V_3 = -1$; we attempt to find the other members of this triplet. Obviously the $V_3 = +1$ member is K^-. We shall call the $V_3 = 0$ member π_V; it is obtained by writing

$$\pi_V = \frac{1}{\sqrt{2}} V_+ K^+ \tag{7.53}$$

where V_+ is the V-spin raising operator. Here, it is helpful to pretend that the quark model is valid. Then we have

$$K^+ = p\bar{\lambda}$$

$$V_+ K^+ = V_+ p\bar{\lambda} = (V_+ p)\bar{\lambda} + p(V_+ \bar{\lambda}) = \lambda\bar{\lambda} - p\bar{p} \tag{7.54}$$

Thus

$$\pi_V = \frac{1}{\sqrt{2}}(\lambda\bar{\lambda} - p\bar{p}) = -\tfrac{1}{2}\pi^0 - \frac{\sqrt{3}}{2}\eta^0 \tag{7.55}$$

Similarly, we can obtain an expression for the V-spin singlet η_V which lies at the same position in the root diagram as π_V (at the origin):

$$\eta_V = -\frac{\sqrt{3}}{2}\pi^0 + \tfrac{1}{2}\eta^0 \tag{7.56}$$

Furthermore, we can express π^0 and η^0 in terms of π_V and η_V. In particular, for π^0

$$\pi^0 = -\frac{\pi_V}{2} - \frac{\sqrt{3}}{2}\eta_V \tag{7.57}$$

Now, returning to the transition $K^+ \to \pi^0 l^+ \nu_l$, it seems very reasonable to regard the $|\Delta S| = 1$ weak hadronic vector current $\mathcal{V}_1(x)$ as an operator which transforms like a V-spin raising operator and converts K^+ to the π_V component of π_0.

From Eq. (7.31), we see that with this assumption, $\mathcal{V}_1$ transforms like a vector current of the form $j_4(x) - ij_5(x)$. We therefore tentatively assume that $\mathcal{V}_1(x)$ and $\mathcal{V}_1{}^\dagger(x)$ are associated with the same octet of vector current operators as the $\Delta S = 0$ weak hadronic vector currents and the hadronic electromagnetic current:

$$\mathcal{V}_1(x) = j_4(x) - ij_5(x) \tag{7.58}$$

$$\mathcal{V}_1{}^\dagger(x) = j_4(x) + ij_5(x) \tag{7.59}$$

In order to account for the axial vector hadronic weak currents, we assume tentatively that these are associated with a related octet of axial current operators g_i:

$$\mathcal{A}_0{}^\lambda(x) = g_1{}^\lambda(x) - ig_2{}^\lambda(x) \tag{7.60}$$

$$\mathcal{A}_1{}^\lambda(x) = g_4{}^\lambda(x) - ig_5{}^\lambda(x) \tag{7.61}$$

It still remains to account for the strengths of these various currents relative to the leptonic weak currents and, in particular, for the fact (noted in earlier discussions) that the $|\Delta S| = 1$ weak decays generally have smaller amplitudes than the $|\Delta S| = 0$ decays. In order to do this, let us consider the leptonic weak currents

$$j_e{}^\lambda(x) = v_e{}^\lambda(x) + a_e{}^\lambda(x)$$

$$j_\mu{}^\lambda(x) = v_\mu{}^\lambda(x) + a_\mu{}^\lambda(x)$$

Using an obvious notation for electron, muon, and neutrino field operators, we write

$$v^\lambda(x) = v_e{}^\lambda(x) + v_\mu{}^\lambda(x) = \bar{e}(x)\gamma^\lambda\nu_e(x) + \bar{\mu}(x)\gamma^\lambda\nu_\mu(x)$$

and
$$a^\lambda(x) = a_e{}^\lambda(x) + a_\mu{}^\lambda(x) = \bar{e}(x)\gamma^\lambda\gamma^5\nu_e(x) + \bar{\mu}(x)\gamma^\lambda\gamma^5\nu_\mu(x) \tag{7.62}$$

We also define a new set of operators L

$$L_+(t) = \int v^0(x)\, d^3x = \int v^0(\mathbf{x}, t)\, d^3x \tag{7.63}$$

and
$$L_+{}^5(t) = \int a^0(\mathbf{x}, t)\, d^3x \tag{7.64}$$

and their hermitian conjugates

$$L_- = (L_+)^\dagger \qquad L_-^5 = (L_+^5)^\dagger \tag{7.65}$$

Furthermore, we define

$$L_3 = \tfrac{1}{2}\int d^3x\{e^\dagger e + \mu^\dagger\mu - \nu_e^\dagger\nu_e - \nu_\mu^\dagger\nu_\mu\} \tag{7.66}$$

and

$$L_3^5 = \tfrac{1}{2}\int d^3x\{e^\dagger\gamma^5 e + \mu^\dagger\gamma^5\mu - \nu_e^\dagger\gamma^5\nu_e - \nu_\mu^\dagger\gamma^5\nu_\mu\} \tag{7.67}$$

Then, by using the lepton field anticommutation relations, we can prove that

$$[L_+, L_-] = 2L_3 \tag{7.68}$$

$$[L_+^5, L_-^5] = 2L_3 \tag{7.69}$$

$$[L_+, L_-^5] = 2L_3 \tag{7.70}$$

$$[L_+^5, L_-] = 2L_3^5 \tag{7.71}$$

$$[L_+, L_+^5] = [L_-, L_-^5] = 0 \tag{7.72}$$

For example, consider (7.69). From (7.64) we have

$$[L_+^5, L_-^5] = \iint d^3x\, d^3x'[a^0(\mathbf{x}, t), a^{0\dagger}(\mathbf{x}', t)][a^0(\mathbf{x}, t), a^{0\dagger}(\mathbf{x}', t)]$$

$$= [e_\alpha^\dagger(\mathbf{x}, t)\gamma_{\alpha\beta}{}^5\nu_{e\beta}(\mathbf{x}, t)\nu_{e\phi}^\dagger(\mathbf{x}', t)\gamma_{\phi\sigma}{}^5 e_\sigma(\mathbf{x}', t)$$

$$- \nu_{e\phi}^\dagger(\mathbf{x}', t)\gamma_{\phi\sigma}{}^5 e_\sigma(\mathbf{x}', t)e_\alpha^\dagger(\mathbf{x}, t)\gamma_{\alpha\beta}{}^5\nu_{e\beta}(\mathbf{x}, t)] + \text{muon term}$$

Now,

$$\{\nu_{e\beta}(\mathbf{x}, t), \nu_{e\phi}^\dagger(\mathbf{x}', t)\} = \delta_{\beta\phi}\,\delta^3(\mathbf{x} - \mathbf{x}')$$

and

$$\{e_\sigma(\mathbf{x}', t), e_\alpha^\dagger(\mathbf{x}, t)\} = \delta_{\sigma\alpha}\,\delta^3(\mathbf{x} - \mathbf{x}')$$

are the anticommutation relations.[1] Therefore,

$$[a^0(\mathbf{x}, t), a^{0\dagger}(\mathbf{x}', t)] = [e_\alpha^\dagger(\mathbf{x}, t)\gamma_{\alpha\beta}{}^5\gamma_{\beta\sigma}{}^5 e_\sigma(\mathbf{x}', t)\,\delta^3(\mathbf{x} - \mathbf{x}')$$

$$- \nu_{e\phi}^\dagger(\mathbf{x}', t)\gamma_{\phi\sigma}{}^5\gamma_{\sigma\beta}{}^5\nu_{e\beta}(\mathbf{x}, t)\,\delta^3(\mathbf{x} - \mathbf{x}')] + \text{muon term}$$

Since $(\gamma^5)^2 = 1$, we have

$$[L_+^5, L_-^5] = 2L_3$$

Similar arguments are used to derive the other commutation relations. It is convenient to define the cartesian components

$$L_\pm = L_1 \pm iL_2 \tag{7.73}$$

and

$$L_\pm^5 = L_1^5 \pm iL_2^5 \tag{7.74}$$

[1] See, e.g., J. Bjorken and S. Drell, "Relativistic Quantum Fields," McGraw-Hill, New York, 1965.

Then, using the commutation relations (7.68) to (7.72), we obtain

$$[L_i, L_j] = i\varepsilon_{ijk} L_k \tag{7.75a}$$

$$[L_i^5, L_j^5] = i\varepsilon_{ijk} L_k \tag{7.75b}$$

$$[L_i, L_j^5] = [L_i^5, L_j] = i\varepsilon_{ijk} L_k^5 \tag{7.75c}$$

Equations (7.75) tell us that the six operators L_i and L_i^5 form an $SU_2 \otimes SU_2$ algebra.

Next we turn to the hadronic weak current to see what restrictions can be imposed on it by the requirement of *universality* of the weak interaction. In the context of the current-current lagrangian, *universality* must mean, in some sense, equal strengths for the various pieces of the weak current. We know that the $\Delta S = 0$ and $|\Delta S| = 1$ parts of the hadron current do *not* have the same strength, but perhaps we can gain some insight by requiring the lepton current and the *total* hadronic current to have the same strength.

What does the requirement of equal strength mean for two different operators operating on different spaces? Presumably, a reasonable definition is that they have the same spectrum of eigenvalues. Actually, the currents themselves are highly singular operators, and so, instead of comparing their eigenvalues directly, we prefer to compare the spectra of eigenvalues of the corresponding *charges* associated with these currents. Note from Eqs. (7.75) that according to this definition of universality, the vector and axial lepton currents are equal in strength because $2L_3$ and $2L_3^5$ are both associated with the same spectrum of eigenvalues: all positive and negative integers and zero.

As we have seen, the various parts of the hadronic weak current are identified with various components of an octet of vector current operators j_i and an octet of axial vector current operators g_i:

$$\mathcal{V}_0^\lambda = j_1^\lambda - ij_2^\lambda$$

$$\mathcal{V}_0^{\lambda\dagger} = j_1^\lambda + ij_2^\lambda$$

$$\mathcal{V}_1^\lambda = j_4^\lambda - ij_5^\lambda$$

$$\mathcal{V}_1^{\lambda\dagger} = j_4^\lambda + ij_5^\lambda$$

$$\mathcal{A}_0^\lambda = g_1^\lambda - ig_2^\lambda$$

$$\mathcal{A}_1^\lambda = g_4^\lambda - ig_5^\lambda$$

and corresponding to each of the currents j_i and g_i we have the charges F_i, F_i^5:

$$\begin{aligned} F_i &= \int j_i^0(\mathbf{x}, t)\, d^3x \qquad i = 1, \ldots, 8 \\ F_i^5 &= \int g_i^0(\mathbf{x}, t)\, d^3x \end{aligned} \tag{7.76}$$

The following commutation relations hold because the F_1, F_2, F_3 are the generators of the I-spin SU_2 subgroup:

$$[F_1 + F_2, F_1 - iF_2] = 2F_3$$
$$[F_1 + iF_2, F_3] = -(F_1 + iF_2)$$
$$[F_1 - iF_2, F_3] = F_1 - iF_2$$

which are the same relations we obtained for the L_i [Eq. (7.75a)].

Thus it would seem that $\mathcal{V}_0$ alone has the same strength as the leptonic vector current. However, we want the total hadronic weak vector current, *including* a contribution from $\mathcal{V}_1$, to have the same strength as the leptonic vector current. To achieve this we make an SU_3 transformation R (a *rotation*) on the charge associated with $\mathcal{V}_0$:

$$W = R(F_1 - iF_2)R^{-1} \tag{7.77}$$

This transformation does not affect the eigenvalues of the charge operator. We choose

$$R = R(\Theta) = e^{-2i\Theta F_7}$$

Then, since

$$[F_7, F_1] = \frac{i}{2} F_4$$

$$[F_7, F_2] = \frac{i}{2} F_5$$

$$[F_7, F_4] = -\frac{i}{2} F_1$$

and

$$[F_7, F_5] = -\frac{i}{2} F_2$$

we easily find that

$$d^2W(\Theta)/d\Theta = -W(\Theta)$$
$$W'(\Theta)|_{\Theta=0} = +(F_4 - iF_5)$$

and

$$W(0) = F_1 - iF_2$$

Thus

$$W(\Theta) = \cos\Theta(F_1 - iF_2) + \sin\Theta(F_4 - iF_5) \tag{7.78}$$

which has the same eigenvalues as $F_1 - iF_2$. We assume that exactly the same rotation about the F_7 axis applies for the axial current, and so we finally obtain for the hadronic weak current

$$\begin{aligned} \mathfrak{J}^\lambda &= \mathfrak{J}_0{}^\lambda + \mathfrak{J}_1{}^\lambda \\ &= \cos\Theta[(j_1{}^\lambda - ij_2{}^\lambda) + (g_1{}^\lambda - ig_2{}^\lambda)] + \sin\Theta[(j_4{}^\lambda - ij_5{}^\lambda) + (g_4{}^\lambda - ig_5{}^\lambda)] \end{aligned} \tag{7.79}$$

The specific value of Θ, Cabibbo's angle, can be found only by experiment. As we have noted, it is

$$\Theta \approx 15^0 \tag{7.80}$$

Equation (7.79) summarizes the content of Cabibbo's important and fundamental hypothesis concerning the SU_3 transformation properties of the hadronic weak current (see Cab 63).

7.7 $SU_3 \otimes SU_3$ AND CURRENT ALGEBRA

To complete this chapter we consider a further, more speculative analogy between the hadronic and leptonic currents and their associated charges. First we recall the $SU_2 \times SU_2$ commutation relations for the lepton charges:

$$\begin{aligned} [L_i, L_j] &= [L_i^5, L_j^5] = i\varepsilon_{ijk} L_k \\ [L_i, L_j^5] &= [L_i^5, L_j] = i\varepsilon_{ijk} L_k^5 \end{aligned} \tag{7.81}$$

Recall also that for the charges associated with the vector octet of hadronic currents,

$$[F_i, F_j] = if_{ij}{}^k F_k \tag{7.82}$$

Moreover, the charges F_i^5 associated with the axial octet also transform as an octet:

$$\begin{aligned} [F_i, F_j^5] &= if_{ij}{}^k F_k^5 \\ [F_i^5, F_j] &= if_{ij}{}^k F_k^5 \end{aligned} \tag{7.83}$$

In order to complete the analogy we postulate a *new* commutation relation

$$[F_i^5, F_j^5] = if_{ij}{}^k F_k \tag{7.84}$$

The relations

$$\begin{aligned} [F_i, F_j] &= if_{ij}{}^k F_k \\ [F_i, F_j^5] = [F_i^5, F_j] &= if_{ij}{}^k F_k^5 \\ [F_i^5, F_j^5] &= if_{ij}{}^k F_k \end{aligned} \tag{7.85}$$

give an $SU_3 \times SU_3$ algebra as compared with the $SU_2 \times SU_2$ algebra of leptonic charges. These are *equal-time* commutation relations in all cases. Finally we make the important assumption that each of the hadronic currents transforms under this algebra in the same way as the corresponding charge:

$$[F_i, j_j^\lambda] = if_{ij}{}^k j_k^\lambda \tag{7.86}$$

$$[F_i, g_j^\lambda] = if_{ij}{}^k g_k^\lambda \tag{7.87}$$

$$[F_i^5, j_j^\lambda] = if_{ij}{}^k g_k^\lambda \tag{7.88}$$

$$[F_i^5, g_j^\lambda] = if_{ij}{}^k j_k^\lambda \tag{7.89}$$

The commutation relations (7.85) to (7.89) lead to a number of important theoretical results, which will be discussed in subsequent chapters.[1] Among these are the relation of Adler and Weisberger (Adl 65, Wei 65), who independently derived an expression for the axial vector coupling constant of neutron beta decay in terms of pion-nucleon scattering cross sections with the aid of PCAC and Eq. (7.84). Also, the current commutation relations yield useful predictions concerning high-energy neutrino interactions and semileptonic and nonleptonic strange particle decays.

REFERENCES

Adl 65 Adler, S. L.: *Phys. Rev. Letters,* **14**:1051 (1965); *Phys. Rev.,* **140**:B736 (1965).

Adl 68 Adler, S. L., and R. F. Dashen: "Current Algebras and Applications to Particle Physics," Benjamin, New York, 1968.

Cab 61 Cabibbo, N., and R. Gatto: *Nuovo Cimento,* **21**:872 (1961).

Cab 63 Cabibbo, N.: *Phys. Rev. Letters,* **10**:531 (1963).

Car 66 Carruthers, P.: "Introduction to Unitary Symmetry," Wiley, New York, 1966.

Col 60 Coleman, S., and S. Glashow: *Phys. Rev. Letters,* **6**:1 (1960).

Gel 61 Gell–Mann, M.: *California Institute of Technology Rept.* CTSL-20, 1961.

Gel 64a Gell–Mann, M., and Y. Ne'eman: "The Eightfold Way," Benjamin, New York, 1964.

Gel 64b Gell–Mann, M.: *Phys. Letters,* **8**:214 (1964).

Lip 65 Lipkin, H. J.: "Lie Groups for Pedestrians," North-Holland, Amsterdam, 1965.

Oak 63 Oakes, R. J.: *Phys. Rev.,* **132**:2349 (1963).

Oku 62 Okubo, S.: *Progr. Theoret. Phys.,* **27**:949 (1962).

Oku 67 Okubo, S., R. E. Marshak, and V. S. Mathur: *Phys. Rev. Letters,* **19**:407 (1967).

Ros 64 Rosen, S.: *Phys. Rev. Letters,* **11**:100 (1964).

Wei 65 Weisberger, W. I.: *Phys. Rev. Letters,* **14**:1047 (1965); *Phys. Rev.,* **143**:1302 (1966).

[1] See, for example, Adl 68 for a general survey of the field of current algebra.

8

BARYON SEMILEPTONIC DECAYS

8.1 THE AXIAL VECTOR COUPLING CONSTANT OF NEUTRON BETA DECAY; THE ADLER–WEISBERGER RELATION

The PCAC hypothesis and the assumed commutation relation for the *chiral* charges

$$[I_+{}^5, I_-{}^5] = 2I_3 \tag{8.1}$$

which was obtained in Chap. 7, have been employed independently by Adler and Weisberger to derive the numerical value of the axial vector coupling constant in neutron beta decay, and their results are in good agreement with experiment. We outline now the method of Adler (Adl 65) for solution of this problem, without presenting all the details.

To begin, we consider the matrix element of both sides of Eq. (8.1), taken between proton plane-wave states of momentum q_1 and q_2:

$$\begin{aligned}\langle p(q_2)|\,[I_+{}^5, I_-{}^5]\,|p(q_1)\rangle &= \langle p(q_2)|\,2I_3\,|p(q_1)\rangle \\ &= \langle p(q_2)|p(q_1)\rangle \\ &= (2\pi)^3\delta^3(\mathbf{q}_2 - \mathbf{q}_1)2q_0\end{aligned} \tag{8.2}$$

The final expression on the right-hand side of (8.2) is obtained using the proton spinor normalization

$$\bar{u}u = 2m_p$$

which is equivalent to

$$u^\dagger u = 2q_0$$

where q_0 is the energy of the proton.

Now let us consider the left-hand side of (8.2). Inserting a complete set of states $|\alpha\rangle$ and summing, with $\Sigma\ |\alpha\rangle\langle\alpha|/2q_0 = 1$, we obtain

$$\langle p(q_2)|\,[I_+{}^5, I_-{}^5]\,|p(q_1)\rangle$$

$$= \sum_\alpha \frac{\langle p(q_2)|I_+{}^5|\alpha\rangle\langle\alpha|I_-{}^5|p(q_1)\rangle}{2q_0} - \sum_\alpha \frac{\langle p(q_2)|I_-{}^5|\alpha\rangle\langle\alpha|I_+{}^5|p(q_1)\rangle}{2q_0} \tag{8.3}$$

It is convenient to consider separately those states α which correspond, on the one hand, to a neutron, and, on the other hand, to all other possible states coupled to the proton by $I_\pm{}^5$. Thus

$$\sum_\alpha = \sum_n + \sum_{\alpha \neq n}$$

Confining our attention to the neutron states for the moment, we have

$$\sum_n = \sum_{\text{spin}} \int \frac{d^3\mathbf{k}}{(2\pi)^3}$$

where $\mathbf{k}$ is the momentum of the neutron. Thus

$$\sum_n = \sum_{\text{spin}} \int \frac{d^3\mathbf{k}}{(2\pi)^3} \frac{1}{2q_0} \int d^3\mathbf{x} \int d^3\mathbf{x}' \langle p(q_2)|\mathcal{A}_0{}^{0\dagger}(\mathbf{x}, t)|n(\mathbf{k})\rangle\langle n(\mathbf{k})|\mathcal{A}_0{}^0(\mathbf{x}', t)|\,p(q_1)\rangle \tag{8.4}$$

which is obtained from the definition of $I_\pm{}^5$, and where we note that the second sum on the right-hand side of (8.3) is zero. Now,

$$\int d^3\mathbf{x}\langle p(q_2)|\mathcal{A}_0{}^{0\dagger}(\mathbf{x}, t)|n(\mathbf{k})\rangle = (2\pi)^3\delta^3(\mathbf{q}_2 - \mathbf{k})|\,C_A|\bar{u}(q_2)\gamma^0\gamma^5 u(\mathbf{k})$$

if we neglect momentum transfer-dependent terms in f_2, g_3. Thus

$$\sum_n = \frac{1}{2q_0} \int d^3\mathbf{k}(2\pi)^3\delta^3(\mathbf{q}_2 - \mathbf{k})\delta^3(\mathbf{k} - \mathbf{q}_1)C_A{}^2\bar{u}(q_2)\gamma^0\gamma^5(\not{k} + m_n)\gamma^0\gamma^5 u(q_1) \tag{8.5}$$

Now making use of the δ functions in (8.5) which yield $q_1 = q_2 = k = q$, and writing

$$\begin{aligned}
\gamma^0\gamma^5(\not{k} + m_n)\gamma^0\gamma^5 u(q_1) &= \gamma^0(\not{k} - m_n)\gamma^0 u(q_1)\\
&= (\gamma^0 q_0 + \boldsymbol{\gamma}\cdot\mathbf{q} - m_p)u(q_1)\\
&= (2\gamma^0 q_0 - m_p)(uq_1)
\end{aligned}$$

we find

$$\sum_n = 2C_A{}^2 q_0\left(1 - \frac{m_n{}^2}{q_0{}^2}\right)(2\pi)^3\delta^3(\mathbf{q}_2 - \mathbf{q}_1) \tag{8.6}$$

Next we consider the sum $\sum_{\alpha \neq n}$, and here the PCAC hypothesis enters. Let us consider the matrix element of $\partial_\mu A^{\mu\dagger}(x)$ taken between a physical π^- state and vacuum, and recall from Eq. (4.77) that this is

$$\langle 0 | \partial_\mu \mathcal{A}_0{}^{\mu\dagger}(x) | \pi^- \rangle = f_\pi m_\pi{}^2 e^{-iq_1 \cdot x} \tag{8.7}$$

where f_π is the constant of pion decay. Now we also consider the renormalized π^- field operator, which satisfies[1]

$$\begin{aligned}\langle 0 | \phi_\pi{}^-(x) | \pi^- \rangle &= \frac{1}{\sqrt{2}} \langle 0 | \phi_1(x) + i\phi_2(x) | \pi^- \rangle \\ &= e^{-iq \cdot x}\end{aligned} \tag{8.8}$$

Comparing Eqs. (8.7) and (8.8) we have

$$\partial_\mu \mathcal{A}_0{}^{\mu\dagger}(x) = a(\phi_1 + i\phi_2) \tag{8.9}$$

where

$$a = \frac{f_\pi m_\pi{}^2}{\sqrt{2}}$$

Moreover, according to the Goldberger-Treiman relation [Eqs. (4.79 and 4.80)], we have

$$f_\pi \approx \frac{\sqrt{2} m_p |C_A|}{g_{NN\pi}}$$

where $g_{NN\pi}$ is the pion-nucleon coupling constant. Thus

$$a = \frac{m_p m_\pi{}^2 |C_A|}{g_{NN\pi}}$$

and

$$\partial_\mu \mathcal{A}_0{}^{\mu\dagger}(x) = \frac{m_p m_\pi{}^2 |C_A|}{g_{NN\pi}} (\phi_1 + i\phi_2) \tag{8.10}$$

which is a very useful way of writing the PCAC hypothesis.

In order to apply Eq. (8.10) to evaluation of the sum $\sum_{\alpha \neq n}$, let us consider the equation

$$[I_+{}^5, H] = i\frac{dI_+{}^5}{dt} \tag{8.11}$$

[1] $\phi_\pi{}^-$ is expressed, as is usual, in terms of hermitian scalar fields ϕ_1 and ϕ_2. See, for example, Bjo 65.

(with H the total hamiltonian), which is valid in the Heisenberg picture. We take the matrix element of both sides of (8.11) between $|p(q_2)\rangle$ and $|\alpha\rangle$ states:

$$\langle p(q_2)|[I_+{}^5, H]|\alpha\rangle = i\int \langle p(q_2)\left|\frac{d\mathcal{A}_0{}^{0\dagger}(x)}{dt}\right|\alpha\rangle\, d^3\mathbf{x} \tag{8.12}$$

However,

$$\int \frac{d\mathcal{A}_0{}^{0\dagger}(x)}{dt}\, d^3\mathbf{x} = \int \partial_\mu \mathcal{A}_0{}^{\mu\dagger}(x)\, d^3\mathbf{x}$$

since

$$\int \nabla \cdot \mathcal{A}_0{}^\dagger\, d^3\mathbf{x} = 0$$

Therefore we obtain from (8.12)

$$\langle p(q_2)|I_+{}^5|\alpha\rangle = \frac{i}{q_{0\alpha} - q_0}\int \langle p(q_2)|\partial_\mu \mathcal{A}_0{}^{\mu\dagger}(x)|\alpha\rangle\, d^3\mathbf{x} \tag{8.13}$$

and, similarly,

$$\langle \alpha|I_-{}^5|p(q_1)\rangle = \frac{-i}{q_{0\alpha} - q_0}\int \langle \alpha|\partial_\mu \mathcal{A}_0{}^{\mu}(x)|p(q_1)\rangle\, d^3\mathbf{x} \tag{8.14}$$

Now, utilizing (8.10) and an analogous expression for $\partial_\mu \mathcal{A}_0{}^\mu(x)$ and taking into account (8.13) and (8.14), we obtain

$$\sum_{\alpha\neq n} = \sum_{\alpha\neq n} \frac{1}{2q_0}\frac{1}{(q_0 - q_{0\alpha})^2}\left[\frac{m_\pi{}^2 m_p |C_A|^2}{g_{NN\pi}}\right]^2$$
$$\times \int d^3\mathbf{x}\int d^3\mathbf{x}'[\langle p(q_2)|\phi_1(\mathbf{x}) + i\phi_2(\mathbf{x})|\alpha\rangle\langle\alpha|\phi_1(\mathbf{x}') - i\phi_2(\mathbf{x}')|p(q_1)\rangle$$
$$- \langle p(q_2)|\phi_1(\mathbf{x}) - i\phi_2(\mathbf{x})|\alpha\rangle\langle\alpha|\phi_1(\mathbf{x}') + i\phi_2(\mathbf{x}')|p(q_1)\rangle] \tag{8.15}$$

It can be seen that the integrand of (8.15) involves matrix elements between proton states and intermediate states $|\alpha\rangle$ consisting of proton plus pion. It thus turns out that in the limit as $q_0 \to \infty$, (8.15) can be expressed conveniently in terms of the cross sections for $\pi^\pm$-proton scattering.

The details of this analysis by Adler and Weisberger are omitted, and at this point we merely present Adler's result. Combining (8.15) with (8.6) and (8.2), we are led to the conclusion that

$$1 - \frac{1}{C_A{}^2} = \frac{4m_n{}^2}{\pi g_{NN\pi}{}^2}\int_{m_n+m_\pi}^{\infty} \frac{W\, dW}{W^2 - m_n{}^2}[\sigma_0{}^+(w) - \sigma_0{}^-(w)] \tag{8.16}$$

where $\sigma_0{}^\pm(W)$ are the total cross sections for $\pi^\pm$ on protons at CM energy W and in these cross sections the pions are assumed to have zero mass. Since experimental values of σ^+ and σ^- refer, of course, to finite-mass pions, corrections are necessary to obtain $\sigma_0{}^{\pm}(W)$. These can be estimated in terms of known pion-nucleon scattering phase shifts. The result of Adler's calculation, involving numerical integration of the pion-nucleon cross sections, is

$$|C_A|_{\text{theo}} = 1.24 \tag{8.17}$$

in excellent agreement with the experimental value

$$|C_A|_{\text{exp}} = 1.23 \pm 0.01 \tag{8.18}$$

8.2 BARYON SEMILEPTONIC DECAYS AND THE CABIBBO HYPOTHESIS

Let us now consider the totality of baryon semileptonic decays:

$$\left.\begin{aligned} n &\to p l^- \bar{\nu} \\ \Sigma^- &\to \Sigma^0 l^- \bar{\nu} \\ \Sigma^- &\to \Lambda^0 l^- \bar{\nu} \\ \Sigma^+ &\to \Lambda^0 l^+ \nu \\ \Xi^- &\to \Xi^0 l^- \bar{\nu} \end{aligned}\right\} \quad \Delta S = 0$$

and

$$\left.\begin{aligned} \Sigma^- &\to n l^- \bar{\nu} \\ \Lambda^0 &\to p l^- \bar{\nu} \\ \Xi^- &\to \Lambda^0 l^- \bar{\nu} \\ \Xi^0 &\to \Sigma^+ l^- \bar{\nu} \\ \Xi^- &\to \Sigma^0 l^- \bar{\nu} \end{aligned}\right\} \quad |\Delta S| = 1$$

Our goal is to calculate the vector and axial vector coupling constants for all these decays, with the aid of Cabibbo's hypothesis (Cab 63). We shall assume that the decays take place in the zero momentum transfer limit where weak magnetism, induced pseudoscalar, etc., corrections can be neglected. In addition, we assume perfect SU_3 symmetry and the Adler–Weisberger result (or, alternatively, the experimental value of $|C_A|$ for neutron decay) as a starting point.

We know that the hadronic portion of the matrix element for the $\Delta S = 0$ decays takes the form

$$M_\lambda{}^0 = \cos\Theta\langle b'|(j_1 \pm i j_2)_\lambda + (g_1 \pm i g_2)_\lambda|b\rangle \tag{8.19}$$

while the $|\Delta S| = 1$ decays are associated with the form

$$M_\lambda{}^1 = \sin\Theta\langle b'|(j_4 \pm i j_5)_\lambda + (g_4 \pm i g_5)_\lambda|b\rangle \tag{8.20}$$

Here the symbols b, b' refer to the initial and final baryon, respectively, while the j's and g's refer to the vector and axial vector octets of current operators, respectively. Each operator j, g contains sums of products of annihilation operators B for the initial baryon and creation operators $\bar{B}$ for the final baryon. The annihilation operators B transform under SU_3 as members of the baryon octet

$$B_i^{\,j} = \begin{pmatrix} \frac{1}{\sqrt{2}}\left(\Sigma^0 + \frac{1}{\sqrt{3}}\Lambda^0\right) & \Sigma^+ & p \\ \Sigma^- & \frac{1}{\sqrt{2}}\left(-\Sigma^0 + \frac{1}{\sqrt{3}}\Lambda^0\right) & n \\ \Xi^- & \Xi^0 & -\frac{2}{\sqrt{6}}\Lambda^0 \end{pmatrix} \tag{8.21}$$

while the creation operators $\bar{B}$ transform as members of the antibaryon octet

$$\bar{B}_i^{\,j} = \begin{pmatrix} \frac{1}{\sqrt{2}}\left(\bar{\Sigma}^0 + \frac{1}{\sqrt{3}}\bar{\Lambda}^0\right) & \bar{\Sigma}^- & \bar{\Xi}^- \\ \bar{\Sigma}^+ & \frac{1}{\sqrt{2}}\left(-\bar{\Sigma}^0 + \frac{1}{\sqrt{3}}\bar{\Lambda}^0\right) & \bar{\Xi}^0 \\ \bar{p} & \bar{n} & -\frac{2}{\sqrt{6}}\bar{\Lambda}^0 \end{pmatrix} \tag{8.22}$$

In Eqs. (8.21) and (8.22) the Latin indices on the left-hand side refer to the row and column of the matrix; thus $B_1^{\,2} = \Sigma^+$, $\bar{B}_1^{\,3} = \bar{\Xi}^-$, and so on.

Now when one takes the tensor product of two octets, it can be shown that the resulting 64 terms fall into irreducible representations, as given by the right-hand side of the following equation:

$$8 \otimes 8 = 27 + 10 + \overline{10} + 8_S + 8_A + 1 \tag{8.23}$$

The symbols S and A on the right-hand side of (8.23) refer to the fact that one of the resulting octets is symmetric while the other is antisymmetric. In particular, the symmetric and antisymmetric combinations of B, $\bar{B}$ octets are

$$S_i^{\,n} = \bar{B}_i^{\,k} B_k^{\,n} + \bar{B}_k^{\,n} B_i^{\,k} - \tfrac{2}{3}\delta_i^{\,n} \bar{B}_m^{\,k} B_k^{\,m} \tag{8.24}$$

and

$$A_i^{\,n} = \bar{B}_k^{\,n} B_i^{\,k} - \bar{B}_i^{\,k} B_k^{\,n} \tag{8.25}$$

respectively. Since, according to Cabibbo's hypothesis, the j and g operators transform as members of V and A octets, and since they are made up of products of $\bar{B}$ and B octets, it follows that operators $j_i{}^n$ and $g_i{}^n$ are independently expressible as linear combinations of the $S_i{}^n$ and $A_i{}^n$. In fact, we can easily show that the vector operators j are made up exclusively of the antisymmetric $A_i{}^n$. This follows, for example, because the $\Delta S = 0$ weak hadronic vector current transforms like the isovector lowering operator I_-. Such an operator does not connect an $I = 1$ state with an $I = 0$ state. Therefore, for example, the vector contribution to the $\Delta S = 0$ transition $\Sigma^+ \to \Lambda e^+ \nu$ must be zero. (This is similar to the $\Delta T = 0$ selection rule already obtained in Chap. 5 for pure Fermi allowed beta decays.) However, one may easily verify from (8.21), (8.22), (8.24), and (8.25) that

$$S_1{}^2 = \left\{\frac{2\bar{\Lambda}\Sigma^+}{\sqrt{6}} + 2\frac{\bar{\Sigma}^-\Lambda}{\sqrt{6}} + \bar{n}p + \bar{\Xi}^-\Xi^0\right\} \tag{8.26}$$

$$A_1{}^2 = \{\sqrt{2}\bar{\Sigma}^-\Sigma^0 + \sqrt{2}\bar{\Sigma}^0\Sigma^+ - \bar{\Xi}^-\Xi^0 + \bar{n}p\} \tag{8.27}$$

$$S_1{}^3 = \left\{\frac{\bar{\Xi}^-\Sigma^0}{\sqrt{2}} + \frac{\bar{\Sigma}^0 p}{\sqrt{2}} - \frac{\bar{\Lambda}p}{\sqrt{6}} - \frac{\bar{\Xi}^-\Lambda^0}{\sqrt{6}} + \bar{\Sigma}^- n - \bar{\Xi}^0\Sigma^+\right\} \tag{8.28}$$

and

$$A_1{}^3 = \left\{\bar{\Xi}^0\Sigma^+ - \bar{\Sigma}^- n - \frac{\bar{\Xi}^-\Sigma^0}{\sqrt{2}} - \frac{\bar{\Sigma}^0 p}{\sqrt{2}} + 3\frac{\bar{\Xi}^-\Lambda}{\sqrt{6}} - \frac{3\bar{\Lambda}^0 p}{\sqrt{6}}\right\} \tag{8.29}$$

As Eq. (8.26) shows, $S_1{}^2$ contains a contribution from the combination $\bar{\Lambda}\Sigma^+$. Therefore, the $\Delta S = 0$ vector operators $j_i{}^j$ must be proportional to $A_i{}^j$ alone and not contain a term in $S_i{}^j$. On the other hand, the axial octet operators $g_i{}^j$ can contain contributions from $S_i{}^j$ as well as $A_i{}^j$. Now, employing Eqs. (8.26) to (8.29), we find that the $\Delta S = 0$ matrix element of (8.19) becomes

$$\begin{aligned} M_\lambda{}^0(b', b) = \cos\Theta\{&Ds_1{}^2(b', b)\bar{u}(b')\gamma_\lambda\gamma^5 u(b) \\ &+ Fa_1{}^2(b', b)\bar{u}(b')\gamma_\lambda\gamma^5 u(b) + a_1{}^2(b', b)\bar{u}(b')\gamma_\lambda u(b)\} \end{aligned} \tag{8.30}$$

while for $|\Delta S| = 1$ we have

$$\begin{aligned} M_\lambda{}^1(b', b) = \sin\Theta\{&Ds_1{}^3(b', b)\bar{u}(b')\gamma_\lambda\gamma^5 u(b) \\ &+ Fa_1{}^3(b', b)\bar{u}(b')\gamma_\lambda\gamma^5 u(b) + a_1{}^3(b', b)\bar{u}(b')\gamma_\lambda u(b)\} \end{aligned} \tag{8.31}$$

where the $a_i{}^j(b', b)$ and $s_i{}^j(b', b)$ are the numerical coefficients multiplying the corresponding combinations $\bar{B}$, B in the $A_i{}^j$ and $S_i{}^j$, respectively, of (8.26) to (8.29), and D and F are constants to be determined. Now, inserting the values for $a_i{}^j(b', b)$ and $b_i{}^j(b', b)$ from (8.26) and (8.29), we arrive at the predictions given in Table 8.1 for the coupling constants of baryon semileptonic decays at zero momentum transfer, all assuming perfect SU_3 symmetry.

In actual fact, of course, perfect SU_3 symmetry does not hold in the real world. For example, the masses of the various baryons in the $\frac{1}{2}^+$ octet are not all the same. What effect does the breaking of SU_3 symmetry have on the preceding argument? The strong interaction may be described by the hamiltonian

$$H_{\text{strong}} = H_0 + \lambda H_M$$

where H_0 represents the superstrong interaction (SU_3 invariant) and H_M represents the medium-strong interaction (which transforms like F_8, thus breaking SU_3 invariance). Parameter λ characterizes the coupling strength of the medium-strong interaction relative to the superstrong interaction: $\lambda \approx 0.1$.

An important theorem due to Ademollo and Gatto (Ade 64) shows that, to first order in λ, the $|\Delta S| = 1$ weak hadronic vector current remains unrenormalized. (So does the $|\Delta S| = 0$ vector current, of course. However, no such result can be derived for the axial current.) Therefore Eqs. (8.30) and (8.31) remain valid even when SU_3 symmetry is not perfect, at least to first order in λ. A convenient proof of the Ademollo–Gatto theorem relies on charge commutation relations and concerns the decays $K \to \pi l \nu$, although it can be generalized to the baryon octet also. This proof is discussed in Chap. 10. For the moment we merely accept its result, which tells us that Table 8.1 should give a reasonably good account of semileptonic decays of baryons even though SU_3 symmetry is not perfect.

Table 8.1 COUPLING CONSTANTS FOR BARYON SEMILEPTONIC DECAYS‡

		Axial vector coupling $g_1(0)$	Vector coupling $f_1(0)$
$\lvert\Delta S\rvert = 0$			
(1) $n \to pe^-\bar{\nu}$	$\cos\Theta$	$D+F$	1
(2) $\Sigma^- \to \Sigma^0 l^-\bar{\nu}$	$\cos\Theta$	$2^{1/2}F$	$2^{1/2}$
(3) $\Sigma^- \to \Lambda^0 l^-\bar{\nu}$	$\cos\Theta$	$(\frac{2}{3})^{1/2}D$	0
(4) $\Sigma^+ \to \Lambda^0 l^+\nu$	$\cos\Theta$	$(\frac{2}{3})^{1/2}D$	0
(5) $\Xi^- \to \Sigma^0 l^-\bar{\nu}$	$\cos\Theta$	$D-F$	-1
$\lvert\Delta S\rvert = 1$			
(6) $\Sigma^- \to nl^-\bar{\nu}$	$\sin\Theta$	$D-F$	-1
(7) $\Lambda^0 \to pl^-\bar{\nu}$	$\sin\Theta$	$-\frac{1}{6^{1/2}}(D+3F)$	$-\frac{3}{6^{1/2}}$
(8) $\Xi^- \to \Lambda^0 l^-\bar{\nu}$	$\sin\Theta$	$-\frac{1}{6^{1/2}}(D-3F)$	$\frac{3}{6^{1/2}}$
(9) $\Xi^0 \to \Sigma^+ l^-\bar{\nu}$	$\sin\Theta$	$D+F$	1
(10) $\Xi^- \to \Sigma^0 l^-\bar{\nu}$	$\sin\Theta$	$\frac{1}{2^{1/2}}(D+F)$	$\frac{1}{2^{1/2}}$

‡ As predicted from Cabibbo's hypothesis, assuming perfect SU_3 symmetry.

Next we consider the coefficients D and F which characterize the axial coupling constants in Table 8.1. We know that $D + F = 1.23$ is required either from the Adler–Weisberger relation or from the experimental value of the axial coupling constant in neutron decay. It is of interest to see how the separate coefficients D and F may be related theoretically to a generalized Goldberger–Treiman relation and the SU_3 theory of trilinear baryon-baryon-meson couplings.

Let us recall our "elementary" derivation of the Goldberger–Treiman relation of Chap. 4, which made use of the single-pion exchange diagram in Fig. 8.1. Nothing prevents us from drawing analogous *one-kaon* exchange diagrams for the strangeness-violating hyperon semileptonic decays, for example,

$$\Lambda^0 \to pe^-\bar{\nu} \qquad \text{(Fig. 8.2.)}$$

and

$$\Sigma^- \to ne^-\bar{\nu} \qquad \text{(Fig. 8.3)}$$

These diagrams all have certain common features. In each case the amplitude may be written

Amplitude $\approx$ meson-baryon-baryon vertex

$\times$ meson propagator $\times$ meson-lepton vertex

Thus, for Fig. 8.1

$$\text{Amplitude} \propto g_{\pi np} \times \text{propagator} \times \frac{G}{\sqrt{2}} f_\pi q_\nu L^\nu$$

while for Fig. 8.2

$$\text{Amplitude} \propto g_{K\Lambda p} \times \text{propagator} \times \frac{G}{\sqrt{2}} f_K q_\nu L^\nu$$

and for Fig. 8.3

$$\text{Amplitude} \propto g_{K\Sigma n} \times \text{propagator} \times \frac{G}{\sqrt{2}} f_K q_\nu L^\nu$$

where L^ν is the lepton current and the $g_{\pi np}$, $g_{K\Lambda p}$, and $g_{K\Sigma n}$ are meson-baryon-baryon coupling constants. The latter are related by SU_3 symmetry, as we shall see below.

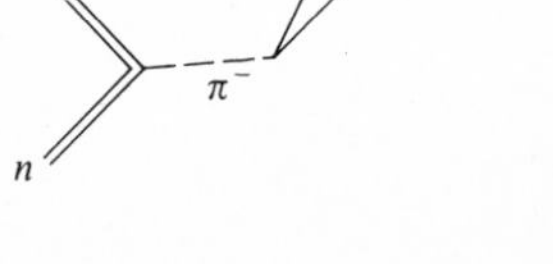

FIGURE 8.1

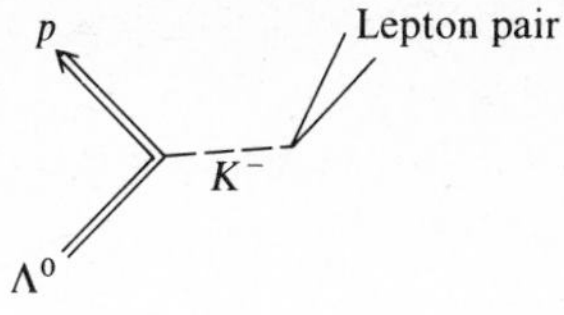

FIGURE 8.2

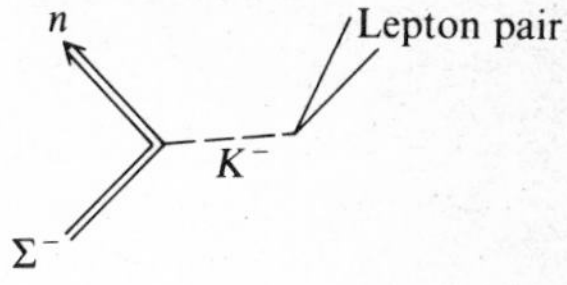

FIGURE 8.3

Just as we derived a Goldberger–Treiman relation

$$-f_\pi = -f\cos\Theta = \frac{m_n + m_p}{\sqrt{2}g_{\pi np}} C_A(n \to p) \tag{8.32}$$

for the decay $n \to pe^-\bar{\nu}$, we can derive similar relations for the decays $\Lambda \to pe^-\bar{\nu}$, $\Sigma^- \to ne^-\bar{\nu}$:

$$-f_K = -f\sin\Theta = \frac{m_n + m_p}{\sqrt{2}g_{K\Lambda P}} C_A(\Lambda \to p) \tag{8.33}$$

$$-f_K = -f\sin\Theta = \frac{m_\Sigma + m_n}{\sqrt{2}g_{\Sigma Kn}} C_A(\Sigma \to n) \tag{8.34}$$

Equation (8.32) is valid in the limit of zero pion mass, and similarly (8.33) and (8.34) are derived with the (rather unjustified) assumption that the K mass is negligible. In the limit of perfect SU_3 symmetry, where hyperon-nucleon mass differences can be neglected, we find from (8.32) to (8.34)

$$C_A(\Lambda \to p) = \frac{g_{K\Lambda p}}{g_{\pi pn}} C_A(n \to p)\tan\Theta \tag{8.35}$$

and

$$C_A(\Sigma \to n) = \frac{g_{\Sigma Kn}}{g_{\pi pn}} C_A(n \to p)\tan\Theta \tag{8.36}$$

All that remains is to find the explicit relationship between $g_{K\Lambda p}$, $g_{\Sigma Kn}$, and $g_{\pi np}$, which can be done by writing down the most general SU_3-invariant baryon-baryon-0^- meson trilinear coupling of gradient form. This lagrangian is

$$\mathfrak{L} = \mathfrak{L}_f + \mathfrak{L}_d \tag{8.37}$$

where

$$\mathfrak{L}_f = gf\,\mathrm{tr}\,[\bar{B}BM - B\bar{B}M] \tag{8.38}$$

and

$$\mathfrak{L}_d = gd\,\mathrm{tr}\,[\bar{B}BM + B\bar{B}M] \tag{8.39}$$

Here g, f, and d are constants and B, $\bar{B}$, and M are the octet matrices for baryons, antibaryons, and the 0^- mesons. (We suppress spatial indices, spatial derivative operators, and γ matrices in this formulation.) Writing out the matrices explicitly and calculating the traces, we find

$$\begin{aligned}\mathcal{L}_f = gf[&\pi^0(\bar{p}p - \bar{n}n + 2\bar{\Sigma}^+\Sigma^+ - 2\bar{\Sigma}^-\Sigma^- + \bar{\Xi}^0\Xi^0 - \bar{\Xi}^-\Xi^-)\\ &+ \pi^+(\sqrt{2}\bar{p}n - \sqrt{2}\bar{\Xi}^0\Xi^- - 2\bar{\Sigma}^+\Sigma^0 + 2\bar{\Sigma}^0\Sigma^-) + \text{h.c.}\\ &+ K^+(-\sqrt{3}\bar{p}\Lambda^0 + \sqrt{3}\bar{\Lambda}^0\Xi^- - \bar{p}\Sigma^0 - \sqrt{2}\bar{n}\Sigma^- + \bar{\Sigma}^0\Xi^- + \sqrt{2}\bar{\Sigma}^+\Xi^0) + \text{h.c.}\\ &+ K^0(-\sqrt{3}\bar{n}\Lambda^0 + \sqrt{3}\bar{\Lambda}^0\Sigma^0 + \bar{n}\Sigma^0 - \sqrt{2}\bar{p}\Sigma^+ - \bar{\Sigma}^0\Xi^0 + 2\sqrt{2}\bar{\Sigma}^-\Xi^-) + \text{h.c.}\\ &+ \sqrt{3}\eta^0(\bar{p}p + \bar{n}n - \bar{\Xi}^0\Xi^0 - \bar{\Xi}^-\Xi^-)]\end{aligned} \tag{8.40}$$

$$\begin{aligned}\mathcal{L}_d = gd\Bigg[&\pi^0\left(\bar{p}p - \bar{n}n + \frac{2}{\sqrt{3}}\bar{\Sigma}^0\Lambda^0 + \frac{2}{\sqrt{3}}\bar{\Lambda}^0\Sigma^0 - \bar{\Xi}^0\Xi^0 + \bar{\Xi}^-\Xi^-\right)\\ &+ \pi^+\left(\sqrt{2}\bar{p}n + \frac{2}{\sqrt{3}}\bar{\Sigma}^+\Lambda^0 + \frac{2}{\sqrt{3}}\bar{\Lambda}^0\Sigma^- - \sqrt{2}\bar{\Xi}^0\Xi^-\right) + \text{h.c.}\\ &+ K^+\left(-\frac{1}{\sqrt{3}}\bar{p}\Lambda^0 + \bar{p}\Sigma^0 + \sqrt{2}\bar{n}\Sigma^- - \frac{1}{\sqrt{3}}\bar{\Lambda}^0\Sigma^- + \bar{\Sigma}^0\Xi^- + \sqrt{2}\bar{\Sigma}^+\Xi^0\right) + \text{h.c.}\\ &+ K^0\left(-\frac{1}{\sqrt{3}}\bar{n}\Lambda^0 - \bar{n}\Sigma^0 + \sqrt{2}\bar{p}\Sigma^+ - \frac{1}{\sqrt{3}}\bar{\Lambda}\Xi^0 - \bar{\Sigma}^0\Xi^0 + \sqrt{2}\bar{\Sigma}^-\Xi^-\right) + \text{h.c.}\\ &+ \frac{\eta^0}{\sqrt{3}}(-\bar{p}p - \bar{n}n - 2\bar{\Lambda}^0\Lambda^0 + 2\bar{\Sigma}^+\Sigma^+ + 2\bar{\Sigma}^0\Sigma^0 + 2\bar{\Sigma}^-\Sigma^- - \bar{\Xi}^0\Xi^0 - \bar{\Xi}^-\Xi^-)\Bigg]\end{aligned} \tag{8.41}$$

From (8.40) and (8.41) it is evident that

$$g_{\pi np} = g\sqrt{2}(f + d) \tag{8.42}$$

$$g_{K\Sigma n} = g\sqrt{2}(d - f) \tag{8.43}$$

and

$$g_{K\Lambda p} = -\frac{g}{\sqrt{3}}(d + 3f) \tag{8.44}$$

Now, inserting (8.42) to (8.44) in (8.35) and (8.36), we obtain

$$\frac{C_A(\Lambda \to p)}{C_A(n \to p)} = -\frac{(d + 3f)}{\sqrt{6}(d + f)}\tan\Theta \tag{8.45}$$

and

$$\frac{C_A(\Sigma \to n)}{C_A(u \to p)} = \frac{(d - f)}{(d + f)}\tan\Theta \tag{8.46}$$

From Table 8.2, however, we obtain

$$\frac{C_A(\Lambda \to p)}{C_A(n \to p)} = -\frac{1}{\sqrt{6}} \frac{(D + 3F)}{(D + F)} \tan \Theta \tag{8.47}$$

and

$$\frac{C_A(\Sigma \to n)}{C_A(u \to p)} = \frac{D - F}{D + F} \tan \Theta \tag{8.48}$$

Further relations between the other axial coupling constants of Table 8.2 and $C_A(n \to p)$ are also readily obtained. Quite obviously, comparison of (8.47) and (8.48) with (8.45) and (8.46) leads us to the conclusion that $d = D$, $f = F$.

8.3 EXPERIMENTAL RESULTS FOR BARYON SEMILEPTONIC DECAYS

The prediction of Cabibbo's theory for the VA coupling constants in baryon semileptonic decays, as given in Table 8.1, may be tested by observations of the following: total decay rates; electron, muon, and final baryon momentum spectra; electron-neutrino angular correlations; asymmetry in the emission of

Table 8.2 BARYON SEMILEPTONIC DECAYS: COMPARISON OF EXPERIMENT WITH CABIBBO'S THEORY‡

	Experimental data		Theoretical predictions	
Decay	Branching ratio	g_1/f_1§	Branching ratio	g_1/f_1§
$n \to pe^-\bar{\nu}_e$		1.23		1.23
$\Sigma^- \to \Lambda e^-\bar{\nu}_e$	$(0.59 \pm 0.06) \times 10^{-4}$	$(f_1 = 0)$	0.62×10^{-4}	$(f_1 = 0)$
$\Sigma^+ \to \Lambda e^+\nu_e$	$(0.19 \pm 0.04) \times 10^{-4}$	$(f_1 = 0)$	0.19×10^{-4}	$f_1 = 0$
$\Lambda \to pe^-\bar{\nu}_e$	$(0.78 \pm 0.09) \times 10^{-3}$	$(0.75^{+0.18}_{-0.15})$	0.86×10^{-3}	0.74
$\Lambda \to p\mu^-\bar{\nu}_\mu$	$(1.35 \pm 0.60) \times 10^{-4}$		1.41×10^{-4}	0.74
$\Sigma^- \to ne^-\bar{\nu}_e$	$(1.10 \pm 0.05) \times 10^{-3}$	$(0.29^{+0.16}_{-0.14})$¶	1.01×10^{-3}	−0.24
$\Sigma^- \to n\mu^-\bar{\nu}_\mu$	$(0.45 \pm 0.05) \times 10^{-3}$		0.48×10^{-3}	−0.24
$\Xi^- \to \Lambda e^-\nu_e$	$(0.90^{+0.71}_{-0.43}) \times 10^{-3}$		0.54×10^{-3}	0.25
$\Xi^- \to \Sigma^0 e^-\bar{\nu}_e$	$(0.62^{+0.20}_{-0.30}) \times 10^{-3}$		0.08×10^{-3}	1.23

‡ Assuming $\Theta = 0.235 \pm 0.006$, $F = 0.49 \pm 0.02$, and $D = 0.74 \pm 0.02$.
§ From Table 8.1 with the above values of F and D.
¶ Absolute value only.

electrons by polarized baryons; and polarization of final baryons. The formulas for these quantities have been worked out in detail (Wil 68). In this section we shall not reproduce all these formulas or attempt to give a complete survey of the experimental results. Instead we shall content ourselves with a presentation of some of the data by way of example and conclude with a summary of the current situation for baryon semileptonic decays.

8.3.1 $\Sigma^\pm \to \Lambda e^\pm \nu_e$

We begin with a consideration of the decays $\Sigma^\pm \to \Lambda e^\pm \nu_e$. The Cabibbo hypothesis predicts that the vector coupling constant $f_1(0)$ should be zero, as indicated in lines 3 and 4 of Table 8.1. This can be tested in several ways:

1 $e\nu$ correlation in the decay of unpolarized Σ. Neglecting the electron mass and $(m_\Sigma - m_\Lambda)/m_\Sigma$ (and disregarding an extremely small weak magnetism contribution and a possible second-class current contribution), the $e\nu$ correlation is given by the formula

$$W(\cos\theta_{e\nu}) = |f_1|^2(1-x) + |g_1|^2(1+x) \tag{8.49}$$

where $$x = \tfrac{1}{2}(1 - \cos\theta_{e\nu})$$

Experimental results (Bar 67, Eis 68) are consistent with $f_1 = 0$ (see Fig. 8.4).

2 Λ energy spectrum. Here again the results confirm $f_1 = 0$ (Bar 67, Eis 68).

3 Polarization of the Λ particle. As we shall discuss in detail in Chap. 10, the Λ polarization can be detected by observation of the angular anisotropy in emission of protons in the decay $\Lambda \to p\pi^-$. Once more, the observations confirm that $f_1 = 0$. According to Table 8.1, the matrix elements for $\Sigma^\pm \to \Lambda e^\pm \nu$ ought to be the same (neglecting second-class currents), and so the branching ratio $\Gamma(\Sigma^+ \to \Lambda e^+ \nu)/\Gamma(\Sigma^- \to \Lambda e^- \bar{\nu})$ ought to be equal to the ratio of the phase space factors, which is 0.61. Experimentally,

$$\frac{\Gamma(\Sigma^+ \to \Lambda e^+ \nu)}{\Gamma(\Sigma^- \to \Lambda e^- \bar{\nu})} = 0.71 \pm 0.15 \tag{8.50}$$

in good agreement. However, even if a second-class term did exist, it could not be detected at present. Finally the world averages for the following branching ratios are given as data which test the Cabibbo hypothesis:

$$\frac{\Gamma(\Sigma^+ \to \Lambda e^+ \nu)}{\Gamma(\Sigma^+ \to \text{all})} = (0.19 \pm 0.04) \times 10^{-4} \tag{8.51}$$

$$\frac{\Gamma(\Sigma^- \to \Lambda e^- \bar{\nu})}{\Gamma(\Sigma^- \to \text{all})} = (0.59 \pm 0.06) \times 10^{-4} \tag{8.52}$$

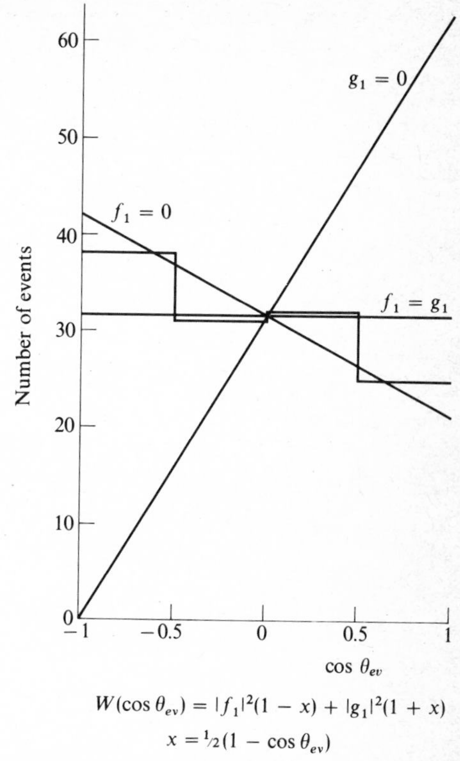

FIGURE 8.4
ev angular correlation in $\Sigma^{\pm} \rightarrow \Lambda e^{\pm} \nu$ decays. (*From H. Filthuth, Hyperon Decays, Proc. Topical Conf. Weak Interactions, CERN, Geneva, 1969, p. 131.*)

$$W(\cos\theta_{ev}) = |f_1|^2(1-x) + |g_1|^2(1+x)$$
$$x = {}^1\!/_2(1 - \cos\theta_{ev})$$

8.3.2 $\Sigma^- \rightarrow ne^-\bar{\nu}_e$, $\Sigma^- \rightarrow n\mu^-\bar{\nu}_\mu$

We have already noted that the $\Delta S = \Delta Q$ rule prohibits decays of the form $\Sigma^+ \rightarrow n\mu^+\nu_\mu$, $\Sigma^+ \rightarrow ne^+\nu_e$. Actually a total of less than five $\Sigma^+ \rightarrow n\mu^+\nu_\mu$ decays have been observed throughout the world at the time of writing, while one $\Sigma^+ \rightarrow ne^+\nu_e$ event seems definitely established. Exactly what the significance of these events is in terms of the $\Delta S = \Delta Q$ rule is not known. As for the decays $\Sigma^- \rightarrow nl^-\bar{\nu}$, the ratio $|g_1/f_1|$ may be established from observations of the charged lepton momentum spectrum (the low momentum portion of which is sensitive to $|g_1/f_1|$) and also from observations of the *ev* correlation. These observations lead to the conclusion

$$\left|\frac{g_1}{f_1}\right| = 0.36^{+0.18}_{-0.15}$$

(see Eis 67, Fra 68, Day 68).

In principle one can even determine the sign of g_1/f_1 from the lepton spectrum shape in the decay of unpolarized Σ's if this is measured with sufficient accuracy. Another method is to measure the electron decay asymmetry from polarized Σ's (Ger 68). Unfortunately the results of these experiments are not sufficiently precise to yield the sign of g_1/f_1. Electron-muon universality requires that the branching ratio of $\Sigma^- \to n\mu^-\bar{\nu}_\mu/\Sigma^- \to ne^-\bar{\nu}_e$ be 0.45 (from the phase-space ratios).

Experimentally, one finds

$$\frac{\Gamma(\Sigma^- \to n\mu^-\bar{\nu}_\mu)}{\Gamma(\Sigma^- \to ue^-\bar{\nu}_e)} = 0.41 \pm 0.06 \qquad (8.53)$$

in good agreement. Finally one finds for a world average

$$\frac{\Gamma(\Sigma^- \to ue^-\bar{\nu}_e)}{\Gamma(\Sigma^- \to \text{all})} = (1.10 \pm 0.05) \times 10^{-3} \qquad (8.54)$$

and

$$\frac{\Gamma(\Sigma^- \to u\mu^-\bar{\nu}_\mu)}{\Gamma(\Sigma^- \to \text{all})} = (0.45 \pm 0.05) \times 10^{-3} \qquad (8.55)$$

8.3.3 $\Lambda \to p\mu^-\bar{\nu}_\mu$, $\Lambda \to pe^-\bar{\nu}_e$

The very small number of $\Lambda \to p\mu^-\bar{\nu}_\mu$ events (five at time of writing) establish that

$$\frac{\Gamma(\Lambda \to p\mu^-\bar{\nu}_\mu)}{\Gamma(\Lambda \to \text{all})} = (0.14 \pm 0.06) \times 10^{-3} \qquad (8.56)$$

As for the more easily observed decay $\Lambda \to pe^-\bar{\nu}_e$, an examination of the e^- momentum spectrum (Can 71) yields the ratio

$$\left|\frac{g_1}{f_1}\right| = 0.75\,^{+0.18}_{-0.15}$$

From the electron asymmetry in the decay of polarized Λ (Bar 65), we determine that $g_1/f_1 < 0$. Finally the branching ratio

$$\frac{\Gamma(\Lambda \to pe^-\bar{\nu}_e)}{\Gamma(\Lambda \to \text{all})} = (0.78 \pm 0.09) \times 10^{-3} \qquad (8.57)$$

is measured (Can 71).

In summary, the experimental results for baryon semileptonic decays, though still quite fragmentary, are in very good agreement with the Cabibbo hypothesis, the predictions of which are set forth in Table 8.1. A fit to all the existing data yields the parameters Θ, F, and D (see Table 8.2). One finds (Fil 69) that

$$\Theta = 0.235 \pm 0.006 \qquad (8.58)$$

$$F = 0.49 \pm 0.02 \qquad (8.59)$$

$$D = 0.74 \pm 0.02 \qquad (8.60)$$

The value for Θ is in reasonable agreement with that obtained from a comparison of $K^{\pm} \to \mu^{\pm} \nu$ versus $\pi^{\pm} \to \mu^{\pm} \nu$ decays and $K^{+} \to \pi^{0} e \nu$ versus $\pi^{+} \to \pi^{0} e \nu$ decays (see Chap. 10).[1] The values of F and D satisfy

$$F + D = 1.23 \pm 0.04 \qquad (8.61)$$

as is obviously required to satisfy the known results of neutron decay, and

$$\frac{F}{F + D} = 0.40 \pm 0.05 \qquad (8.62)$$

The last result is in good agreement with the ratio $f/(f + d)$ obtained from determination of the meson-baryon-baryon coupling constants through measurement of meson-nucleon forward-scattering cross sections (Kim 67, Cha 68, Nie 68):

$$0.37 \leq \frac{f}{f + d} \leq 0.41 \qquad (8.63)$$

REFERENCES

Ade 64 Ademollo, M., and R. Gatto: *Phys. Letters,* **13**:264 (1964).

Adl 65 Adler, S. L.: *Phys. Rev.,* **140B**:736 (1965).

Bar 65 Barlow, J., I. M. Blair, G. Conforto, M. I. Ferrero, C. Rubbia, J. C. Sens, P. J. Duke, and A. K. Mann: *Phys. Letters,* **18**:64 (1965).

Bar 67 Barash, N., T. B. Day, R. G. Glasser, B. Kehoe, R. Knop, B. Sechi-Zorn, and G. A. Snow: *Phys. Rev. Letters,* **19**:181 (1967).

Bjo 65 Bjorken, J. D., and S. D. Drell: "Relativistic Quantum Fields," McGraw-Hill, New York, 1965.

Cab 63 Cabibbo, N.: *Phys. Rev. Letters,* **10**:531 (1963).

Can 71 Canter, J., J. Cole, J. Lee-Franzini, R. J. Loveless, and P. Franzini: *Phys. Rev. Letters,* **26**:868 (1971).

Cha 68 Chan, C. H., and F. T. Meiere: *Phys. Rev. Letters,* **20**:568 (1968).

Day 68 Day, T. B., R. G. Glasser, R. E. Knop, B. Sechi-Zorn, and G. A. Snow: Contribution 375, *Internatl. Conf. High-Energy Physics, Vienna* (1968).

Eis 67 Eisele, F., R. Engelmann, H. Filthuth, W. Föhlisch, V. Hepp, E. Kluge, E. Leitner, P. Lexa, P. Mokry, W. Presser, H. Schneider, L. M. Stevenson, and G. Zech: *Z. Physik.,* **205**:409 (1967).

Eis 68 Eisele, F., R. Engelmann, H. Filthuth, W. Föhlisch, V. Hepp, E. Leitner, V. Linke, W. Presser, H. Schneider, M. L. Stevenson, and G. Zech: Contribution 569, *Internatl. Conf. High-Energy Physics, Vienna* (1968).

Fil 69 Filthuth, H.: Hyperon Decays, *Proc. Topical Conf. Weak Interactions, CERN, Geneva,* 131 (1969).

Fra 68 Franzini, P., C. Baltay, R. Newman, H. Narton, N. Yeh, J. A. Cole, J. Lee-Franzini, R. Loveless, and J. McFadyen: Contribution 262, *Internatl. Conf. High-Energy Physics, Vienna* (1968).

[1] The small differences between the values of Θ derived by these various means are attributed to departures from perfect SU_3 symmetry.

Ger 68 Gershwin, L. K., M. Alston-Garnjost, R. O. Bangester, A. Barbaro-Galtieri, F. T. Solmitz, and R. D. Tripp: *Phys. Rev. Letters,* **20**:1270 (1968).

Kim 67 Kim, J. K.: *Phys. Rev. Letters,* **19**:1079 (1967).

Nie 68 Nieh, H. T.: *Phys. Rev. Letters,* **20**:1254 (1968).

Wil 68 Willis, W. J., and J. Thompson: in R. Cool and R. E. Marshak (eds.), "Advances in Particle Physics," Wiley, New York, 1968.

9
HYPERON NONLEPTONIC DECAYS

9.1 GENERAL REMARKS

The nonleptonic hyperon decays (see Table 9.1)

$$\begin{array}{lll} \Lambda^0 \to p\pi^- & \Sigma^+ \to p\pi^0 & \Xi^0 \to \Lambda^0\pi^0 \\ \Lambda^0 \to n\pi^0 & \Sigma^+ \to n\pi^+ & \Xi^- \to \Lambda^0\pi^- \\ & \Sigma^- \to n\pi^- & \end{array} \tag{9.1}$$

may be described by the $\mathfrak{J}_0\mathfrak{J}_1^\dagger$ or $\mathfrak{J}_1\mathfrak{J}_0^\dagger$ portion of the weak lagrangian. However, we gain little by writing down a formal expression for the amplitude $\mathcal{M}$ for $B \to B'\pi$ in terms of a product of the currents,

$$\mathcal{M} = \frac{G}{\sqrt{2}} \langle B'\pi | \mathfrak{J}_0\mathfrak{J}_1^\dagger \quad \text{or} \quad \mathfrak{J}_1\mathfrak{J}_0^\dagger | B \rangle \tag{9.2}$$

since our ignorance of the strong interactions prevents us from translating (9.2) into a concrete formula which can be used to predict transition rates, angular correlations, etc.

Nevertheless, even though we are unable to utilize (9.2) for practical calculations of hyperon nonleptonic decays, we can impose useful restrictions

Table 9.1 HYPERON NONLEPTONIC DECAYS

Initial hyperon	Λ^0		Σ^+		Σ^-	Ξ^0	Ξ^-	Ω^- ‡
Mass, MeV/c^2	$1{,}115.59 \pm 0.06$		$1{,}189.47 \pm 0.08$		$1{,}197.37 \pm 0.07$	$1{,}314.7 \pm 0.7$	$1{,}321.3 \pm 0.17$	$1{,}672.5 \pm 0.5$
Mean life, 10^{-10} s	2.517 ± 0.024		0.800 ± 0.006		1.489 ± 0.022	3.03 ± 0.18	1.660 ± 0.037	$1.3 {+0.4 \atop -0.3}$
I, I_3	0, 0		1, 1		1, −1	$\frac{1}{2}, \frac{1}{2}$	$\frac{1}{2}, -\frac{1}{2}$	0, 0
J^P	$\frac{1}{2}^+$		$\frac{1}{2}^+$		$\frac{1}{2}^+$	$\frac{1}{2}^+$ §	$\frac{1}{2}^+$ §	$\frac{3}{2}^+$
Decay mode	$p\pi^-$	$n\pi^0$	$p\pi^0$	$n\pi^+$	$n\pi^-$	$\Lambda^0\pi^0$	$\Lambda^0\pi^-$	Λ^0K^-, $\Xi^0\pi^-$, $\Xi^-\pi^0$
Branching ratio, %	64	36	51.7	48.3	100	100	100	
α	0.645 ± 0.016	0.649 ± 0.046	-0.991	0.066 ± 0.016	-0.069 ± 0.008	-0.35 ± 0.08	-0.40 ± 0.03	
ϕ	$(-6.3 \pm 3.5)^\circ$		$(22 \pm 90)^\circ$	$(167 \pm 20)^\circ$	$(10 \pm 15)^\circ$	$(25 \pm 21)^\circ$	$(-4 \pm 8)^\circ$	
γ	0.76		0.12	-0.97	0.98	0.85	0.91	

$$\alpha = \mathrm{Re}\,\frac{s^*p}{|s|^2 + |p|^2} \qquad \gamma = \frac{|s|^2 - |p|^2}{|s|^2 + |p|^2} = (1 - \alpha^2)^{1/2} \cos \phi$$

‡ Ω^- belongs to the $\frac{3}{2}^+$ baryon resonance decuplet but decays by weak interaction.
§ The intrinsic parities of Ξ^0, Ξ^- have not been determined experimentally. They are assumed to be positive.

on the amplitude from the requirements of symmetry and invariance alone. Thus, even if the weak lagrangian is *not* of the current-current form, the mere assumptions of proper Lorentz invariance and validity of the $|\Delta I| = \frac{1}{2}$ rule and that the amplitude can be expressed in terms of an effective hamiltonian H_W, where H_W has definite and simple SU_3 transformation properties, allow us to reduce the problem of hyperon nonleptonic decays to manageable proportions.

9.2 FORM OF THE DECAY AMPLITUDE; RESTRICTIONS DUE TO PROPER LORENTZ INVARIANCE

In each of the decays (9.1) we deal with an initial and final baryon with spin $\frac{1}{2}$ and also a final spin-zero pion. The most general proper-Lorentz-invariant amplitude is then

$$\mathcal{M} = \bar{u}_f(A + B\gamma^5)u_i \tag{9.3}$$

where u_i and u_f are Dirac four-spinors describing the initial and final baryons, respectively, and A and B are constants whose values cannot be calculated until we have a more complete theory of hadron interactions than exists presently.

One might think that (9.3) is not the most general form of the amplitude because other scalar and pseudoscalar terms can be constructed from the available kinematic quantities in the decay. However, in each case one can show that such terms are already included in (9.3). For example, consider the pion four-momentum p_π, and let us write the quantities $\bar{u}_f \gamma^\alpha p_{\pi\alpha} u_i$ and $\bar{u}_f \gamma^\alpha p_{\pi\alpha} \gamma^5 u_i$. Now

$$p_{\pi\alpha} = p_i - p_f$$

where p_i and p_f are the four-momenta of the initial and final baryons. Thus,

$$\begin{aligned} \bar{u}_f \not{p}_\pi u_i &= \bar{u}_f(\not{p}_i - \not{p}_f)u_i \\ &= (m_i - m_f)\bar{u}_f u_i \end{aligned}$$

Such a term is already included in the A term of (9.3). Similarly

$$\begin{aligned} \bar{u}_f \not{p}_\pi \gamma^5 u_i &= -\,\bar{u}_f(\not{p}_f - \not{p}_i)\gamma^5 u_i \\ &= -\,(m_f + m_i)\bar{u}_f \gamma^5 u_i \end{aligned}$$

which is already included in the B term of (9.3).

Since the pion itself is pseudoscalar, the A term in (9.3) is pseudoscalar and corresponds to zero orbital angular momentum (s wave) for the final state.

On the other hand the B term is scalar, and corresponds to a final-state p wave. We now calculate the transition rate in terms of A and B in the rest frame of the initial hyperon. We have

$$u_i = \begin{pmatrix} \chi_i \\ 0 \end{pmatrix}$$

and

$$u_f = \sqrt{E_f + m_f} \begin{pmatrix} \chi_f \\ \dfrac{\boldsymbol{\sigma} \cdot \mathbf{p}_f}{E_f + m_f} \chi_f \end{pmatrix}$$

where E_f and m_f are the relativistic energy and rest mass, respectively, of the final baryon, and χ_i, χ_f are two-component Pauli spinors for initial and final baryon, respectively. Thus (9.3) becomes

$$\begin{aligned} \mathcal{M} &\propto \left(\chi_f^\dagger, -\chi_f^\dagger \frac{\boldsymbol{\sigma} \cdot \mathbf{p}_f}{E_f + m_f} \right) \begin{pmatrix} A & -B \\ -B & A \end{pmatrix} \begin{pmatrix} \chi_i \\ 0 \end{pmatrix} \\ &= A \chi_f^\dagger \chi_i + B \chi_f^\dagger \frac{\boldsymbol{\sigma} \cdot \mathbf{p}_f}{E_f + m_f} \chi_i \end{aligned} \tag{9.4}$$

Now defining

$$\hat{\mathbf{n}} = \frac{\mathbf{p}_f}{|\mathbf{p}_f|} \qquad s = A \qquad p = \frac{|\mathbf{p}_f|}{E_f + m_f} B$$

we have

$$\mathcal{M} \propto \chi_f^\dagger (s + p \boldsymbol{\sigma} \cdot \hat{\mathbf{n}}) \chi_i \tag{9.5}$$

The transition probability is therefore proportional to

$$\begin{aligned} |\chi_f^\dagger (s + p\boldsymbol{\sigma} \cdot \hat{\mathbf{n}}) \chi_i|^2 &= \operatorname{tr} [\chi_f \chi_f^\dagger (s + p\boldsymbol{\sigma} \cdot \hat{\mathbf{n}}) \chi_i \chi_i^\dagger (s^* + p^* \boldsymbol{\sigma} \cdot \hat{\mathbf{n}})] \\ &= \operatorname{tr} \left[\left(\frac{1 + \boldsymbol{\sigma} \cdot \hat{\boldsymbol{\eta}}_f}{2} \right) (s + p\boldsymbol{\sigma} \cdot \hat{\mathbf{n}}) \left(\frac{1 + \boldsymbol{\sigma} \cdot \hat{\boldsymbol{\eta}}_i}{2} \right) \right. \\ &\qquad \left. \times (s^* + p^* \boldsymbol{\sigma} \cdot \hat{\mathbf{n}}) \right] \end{aligned} \tag{9.6}$$

where $\hat{\boldsymbol{\eta}}_i$ and $\hat{\boldsymbol{\eta}}_f$ are unit vectors in the direction of initial and final baryon spins, respectively.

Now, using the following properties of σ matrices:

$$\begin{aligned} \operatorname{Tr} \sigma_i &= 0 \\ \operatorname{Tr} (\sigma_i \sigma_j) &= 2\delta_{ij} \\ \operatorname{Tr} (\sigma_i \sigma_j \sigma_k) &= 2i\varepsilon_{ijk} \end{aligned}$$

and

$$\operatorname{Tr} (\sigma_i \sigma_j \sigma_k \sigma_l) = 2\delta_{ij}\delta_{kl} - 2\delta_{ik}\delta_{jl} + 2\delta_{jk}\delta_{il}$$

we find that the transition rate is proportional to

$$R = 1 + \gamma\hat{\boldsymbol{\eta}}_f \cdot \hat{\boldsymbol{\eta}}_i + (1-\gamma)\hat{\boldsymbol{\eta}}_f \cdot \hat{\mathbf{n}}\hat{\boldsymbol{\eta}}_i \cdot \hat{\mathbf{n}} + \alpha(\hat{\boldsymbol{\eta}}_f \cdot \hat{\mathbf{n}} + \hat{\boldsymbol{\eta}}_i \cdot \hat{\mathbf{n}}) + \beta\hat{\mathbf{n}} \cdot (\hat{\boldsymbol{\eta}}_f \times \hat{\boldsymbol{\eta}}_i) \qquad (9.7)$$

where

$$\alpha = 2\,\mathrm{Re}\,\frac{sp^*}{|s|^2 + |p|^2} \qquad (9.8)$$

$$\beta = 2\,\mathrm{Im}\,\frac{sp^*}{|s|^2 + |p|^2} \qquad (9.9)$$

and

$$\gamma = \frac{|s|^2 - |p|^2}{|s|^2 + |p|^2} \qquad (9.10)$$

We note that α, β, and γ are not independent since

$$\alpha^2 + \beta^2 + \gamma^2 = 1$$

Let us consider the transition rate averaged over initial hyperon polarizations $\hat{\boldsymbol{\eta}}_i$ (decay of an unpolarized hyperon). Then (9.7) reduces to

$$R_{\mathrm{unpol}} \approx 1 + \alpha\hat{\boldsymbol{\eta}}_f \cdot \hat{\mathbf{n}} \qquad (9.11)$$

Thus, for example, in the decay of unpolarized Λ^0 at rest to $p\pi^-$, the protons are emitted with a longitudinal polarization α. Consider, on the other hand, the decay of a polarized hyperon, where we sum over final baryon polarizations $\hat{\boldsymbol{\eta}}_f$; then (9.7) reduces to

$$R \approx 1 + \alpha\hat{\boldsymbol{\eta}}_i \cdot \hat{\mathbf{n}} \qquad (9.12)$$

Thus, for example, in decay of polarized Λ^0 to $p\pi^-$, the protons (and pions) are emitted anisotropically, with asymmetry coefficient α.

It is easy to show that under time reversal, $\mathcal{M} \to \mathcal{M}' = \mathcal{M}^*$ if and only if $s = s^*$, $p = p^*$. Therefore R is time-reversal invariant if and only if β vanishes in (9.7) *for no baryon-meson final-state interaction.* Since there is in fact a strong final-state interaction, we have $s \neq s^*$, $p \neq p^*$ even for time-reversal invariance. Instead,

$$s = |s|e^{i\delta_s}$$

$$p = |p|e^{i\delta_p}$$

where δ_s and δ_p are the pion-baryon s- and p-wave strong-interaction phase shifts.[1] Thus, for T invariance,

$$\beta = \frac{2sp^*}{|s|^2 + |p|^2}\sin(\delta_s - \delta_p) \qquad (9.13)$$

[1] For simplicity we think here in terms of phase shifts δ_s and δ_p corresponding to a single final isospin I. This is the case, for example, for $\Lambda^0 \to p\pi^-$, provided the $|\Delta I| = \frac{1}{2}$ rule holds, since $I(\Lambda^0) = 0$.

Now, let us consider the problem of observing $\Lambda^0 \to p\pi^-$ decay, as an example. From Eq. (9.11), we see that the proton has a polarization parallel to its momentum in the Λ^0 rest frame. In the laboratory frame, however, the proton polarization has an appreciable transverse component, and this can be detected by scattering the proton on carbon since the proton-carbon interaction contains an appreciable spin-orbit term.

In an experiment performed by Overseth and Roth (Ove 67), $\Lambda^0 \to p\pi^-$ decay is observed in a thin-plate spark chamber, and the polarization of decay protons is determined from proton scattering in a carbon-plate spark chamber. The results are

$$\alpha = 0.645 \pm 0.017$$

$$\beta = -0.10 \pm 0.07 \qquad (9.14)$$

and

$$\gamma = 0.75 \pm 0.02$$

Thus we obtain

$$\Delta \equiv -\arctan\frac{\beta}{\alpha} = 9.0° \pm 5.5° \qquad (9.15)$$

and

$$\left|\frac{p}{s}\right| = 0.38 \pm 0.01 \qquad (9.16)$$

Result (9.15) is to be compared with the difference $\delta_s - \delta_p = 6.5° \pm 1.5°$ determined from pion-proton scattering experiments at low energies (Bar 60). Clearly these results are consistent with T invariance. Result (9.16) indicates the relative strength of *s*- and *p*-wave amplitudes in $\Lambda^0 \to p\pi^-$ decay. Table 9.1 summarizes similar data obtained for the other hyperon nonleptonic decays.

9.3 THE $|\Delta I| = \frac{1}{2}$ RULE

If the $|\Delta I| = \frac{1}{2}$ rule is valid, the effective weak hamiltonian for hyperon nonleptonic decays transforms like an isospinor under isotopic rotations. It is helpful in visualizing the effects of the $|\Delta I| = \frac{1}{2}$ rule to imagine that the initial hyperon absorbs a *spurion* (a fictitious particle with $I = \frac{1}{2}$, $I_3 = -\frac{1}{2}$) in making the transition to final baryon-plus-pion. With spurion included, the hypothetical transition process is isospin-invariant. Thus consider the decays of Λ^0:

	Λ^0 I, I_3	+ Spurion I, I_3	→ Nucleon I, I_3	+ Meson I, I_3
$\Lambda^0 \to p\pi^-$	0, 0	$(\frac{1}{2}, -\frac{1}{2})$	$(\frac{1}{2}, \frac{1}{2})$	$(1, -1)$
$\Lambda^0 \to n\pi^0$	0, 0	$(\frac{1}{2}, -\frac{1}{2})$	$(\frac{1}{2}, -\frac{1}{2})$	$(1, 0)$

The ratio of the amplitudes for $\Lambda^0 \to p\pi^-$ and $\Lambda^0 \to n\pi^0$ should then be proportional to a ratio of Clebsch–Gordan coefficients:

$$\frac{\mathscr{M}(\Lambda^0 \to p\pi^-)}{\mathscr{M}(\Lambda^0 \to n\pi^0)} = \frac{\langle \frac{1}{2}, \frac{1}{2}, 1, -1 | \frac{1}{2}, -\frac{1}{2} \rangle}{\langle \frac{1}{2}, -\frac{1}{2}, 1, 0 | \frac{1}{2}, -\frac{1}{2} \rangle} = -\sqrt{2}$$

Thus the branching ratio $R = \Gamma(\Lambda^0 \to p\pi^-)/\Gamma(\Lambda^0 \to n\pi^0)$ should be 2, a prediction which is confirmed by observation.

Next, consider the Σ nonleptonic decays

$$\Sigma_-{}^- : \quad \Sigma^- \to \pi^- n$$
$$\Sigma_0{}^+ : \quad \Sigma^+ \to \pi^0 p$$
$$\Sigma_+{}^+ : \quad \Sigma^+ \to \pi^+ n$$

Now we have the following:

Σ hyperon I, I_3	+	Spurion I, I_3	$\to$	Nucleon I, I_3	+	Meson I, I_3
$\Sigma_+{}^+$ 1, 1		$\frac{1}{2}, -\frac{1}{2}$		$\frac{1}{2}, -\frac{1}{2}$ n		1, 1 π^+
$\Sigma_0{}^+$ 1, 1		$\frac{1}{2}, -\frac{1}{2}$		$\frac{1}{2}, \frac{1}{2}$ p		1, 0 π^0
$\Sigma_-{}^-$ 1, -1		$\frac{1}{2}, -\frac{1}{2}$		$1, -\frac{1}{2}$ n		1, -1 π^-

On each side of the reaction equation we are combining an $I = 1$ object with an $I = \frac{1}{2}$ object, and we thus obtain for a total isospin $I = \frac{1}{2}$ or $\frac{3}{2}$. Writing out the Clebsch–Gordan coefficients, we find

$\Sigma_+{}^+ :$ Amplitude $\equiv A_+{}^+ = k \left\langle \dfrac{-\psi_{3,1} + \sqrt{2}\psi_{1,1}}{\sqrt{3}} \left| H_{\text{eff}} \right| \dfrac{-\psi_{3,1} + \sqrt{2}\psi_{1,1}}{\sqrt{3}} \right\rangle$

$\Sigma_0{}^+ :$ Amplitude $\equiv A_0{}^+ = k \left\langle \dfrac{\sqrt{2}\psi_{3,1} + \psi_{1,1}}{\sqrt{3}} \left| H_{\text{eff}} \right| \dfrac{-\psi_{3,1} + \sqrt{2}\psi_{1,1}}{\sqrt{3}} \right\rangle$

and

$\Sigma_-{}^- :$ Amplitude $\equiv A_-{}^- = k \langle \psi_{3,-3} | H_{\text{eff}} | \psi_{3,-3} \rangle$

where H_{eff} is isospin-invariant and k is a common constant.[1]   Thus we find

$$A_+{}^+ = \tfrac{1}{3}x + \tfrac{2}{3}y \qquad A_0{}^+ = -\frac{\sqrt{2}}{3}x + \frac{\sqrt{2}}{3}y$$

and

$$A_-{}^- = x$$

where x and y are two numbers.   Eliminating y we find

$$A_+{}^+ - A_-{}^- = \sqrt{2}\,A_0{}^+ \qquad (9.17)$$

[1] In these expressions $\psi_{3,1}$ refers to the $I = \frac{3}{2}$, $I_3 = \frac{1}{2}$ isospin state, etc.

Finally we consider the transitions $\Xi^- \to \Lambda\pi^-$ and $\Xi^0 \to \Lambda\pi^0$. This time we have

	Cascade hyperon I, I_3	+ Spurion I, I_3	→ Lambda I, I_3	+ Pion I, I_3
$\Xi^0 \to \Lambda\pi^0$	$\frac{1}{2}, \frac{1}{2}$	$\frac{1}{2}, -\frac{1}{2}$	$0, 0$	$1, 0$
$\Xi^- \to \Lambda\pi^-$	$\frac{1}{2}, -\frac{1}{2}$	$\frac{1}{2}, -\frac{1}{2}$	$0, 0$	$1, -1$

Thus,

$$\frac{\mathcal{M}(\Xi^0 \to \Lambda^0\pi^0)}{\mathcal{M}(\Xi^- \to \Lambda^0\pi^-)} = \frac{\langle \frac{1}{2}, \frac{1}{2}, \frac{1}{2}, -\frac{1}{2} | 1, 0 \rangle}{\langle \frac{1}{2}, -\frac{1}{2}, \frac{1}{2}, -\frac{1}{2} | 1, -1 \rangle} = -\frac{1}{\sqrt{2}} \tag{9.18}$$

To summarize, since each decay amplitude is really a linear combination of s- and p-wave amplitudes, the $|\Delta I| = \frac{1}{2}$ rule predicts that

$$s(\Lambda_-{}^0) = -\sqrt{2}\, s(\Lambda_0{}^0) \tag{9.19}$$

$$p(\Lambda_-{}^0) = -\sqrt{2}\, p(\Lambda_0{}^0) \tag{9.20}$$

$$s(\Sigma_+{}^+) - s(\Sigma_-{}^-) = \sqrt{2}\, s(\Sigma_0{}^+) \tag{9.21}$$

$$p(\Sigma_+{}^+) - p(\Sigma_-{}^-) = \sqrt{2}\, p(\Sigma_0{}^+) \tag{9.22}$$

and

$$s(\Xi^-) = -\sqrt{2}\, s(\Xi^0) \tag{9.23}$$

$$p(\Xi^-) = -\sqrt{2}\, p(\Xi^0) \tag{9.24}$$

These predictions are in good agreement with existing experimental results (Fil 69 and Table 9.1).

9.4 OCTET DOMINANCE; THE LEE–SUGAWARA RELATION

As we have mentioned earlier, the $|\Delta I| = \frac{1}{2}$ rule for nonleptonic decays of kaons and hyperons is somewhat difficult to understand in view of the fact that the relevant part of the weak lagrangian for these decays is proportional to $\mathfrak{J}^0 \mathfrak{J}_1{}^\dagger$ or $\mathfrak{J}_1 \mathfrak{J}_0{}^\dagger$. Since $\mathfrak{J}_0$ transforms like an isovector and $\mathfrak{J}_1$ like an isospinor, we naturally expect that the nonleptonic decay amplitudes satisfy the selection rules $|\Delta I| = \frac{1}{2}$ *and* $|\Delta I| = \frac{3}{2}$. Thus, we have to explain why it is that in all strange particle weak decays the $|\Delta I| = \frac{3}{2}$ component is missing or very small.

Two explanations for this puzzle readily suggest themselves. One is that the weak lagrangian really contains additional terms besides those we have already written down, and that these new terms represent products of *neutral* currents for each of which $\Delta Q = 0$. Consider the "ordinary" charged hadronic weak currents $\mathfrak{J}_0$ and $\mathfrak{J}_1$. We know that $\mathfrak{J}_0$ acts like an isovector and that $\mathfrak{J}_1$ acts like an isospinor:

$$\mathfrak{J}_0(I, m_I) = \mathfrak{J}_0(1, -1) \qquad \Delta S = 0 \qquad \Delta Q = -1$$

$$\mathfrak{J}_1(I, m_I) = \mathfrak{J}_1(\tfrac{1}{2}, -\tfrac{1}{2}) \qquad \Delta S = -1 \qquad \Delta Q = -1$$

Let us define new neutral hadronic currents:

$$\mathfrak{J}'_0(I, m_I) = \mathfrak{J}'_0(1, 0) \qquad \Delta S = 0 \qquad \Delta Q = 0$$

$$\mathfrak{J}'_1(I, m_I) = \mathfrak{J}'_1(\tfrac{1}{2}, +\tfrac{1}{2}) \qquad \Delta S = -1 \qquad \Delta Q = 0$$

[It follows that $\mathfrak{J}_1'^{\dagger}\,(I, m_I) = \mathfrak{J}_1'^{\dagger}\,(\tfrac{1}{2}, -\tfrac{1}{2})$.] In order to satisfy the $\Delta S = \Delta Q$ rule, which applies only to $|\Delta S| = 1$ leptonic and semileptonic decays of hadrons, we must assume that the neutral currents do not couple to the weak leptonic currents. However, to describe the nonleptonic strangeness-changing decays we might make the replacement

$$\mathfrak{J}_0(1, -1)\mathfrak{J}_1^{\dagger}(\tfrac{1}{2}, \tfrac{1}{2}) \rightarrow \mathfrak{J}_0(1, -1)\mathfrak{J}_1^{\dagger}(\tfrac{1}{2}, \tfrac{1}{2}) - \mathfrak{J}'_0(1, 0)\mathfrak{J}_1'^{\dagger}(\tfrac{1}{2}, -\tfrac{1}{2}) \qquad (9.25)$$

If the neutral currents are in some sense equal in strength to the charged currents, we can guarantee suppression of the $|\Delta I| = \frac{3}{2}$ amplitude by the very construction of (9.25). The $K^+ \rightarrow \pi^+\pi^0$ decay would then occur because of electromagnetic corrections. Of course, this scheme suffers from the obvious drawback that there is no experimental evidence for the existence of such neutral currents beyond the validity of the $|\Delta I| = \frac{1}{2}$ rule itself.[1]

Another possibility is that H_W, the effective weak hamiltonian for nonleptonic decays, *itself* transforms as an octet under SU_3 transformations. Actually $\mathfrak{J}_0$ and $\mathfrak{J}_1$ can each be expressed as a linear combination of the vector and axial vector octet currents j_i and g_i. Thus we expect that the product $\mathfrak{J}_0\,\mathfrak{J}_1^{\dagger}$ should contain not only an octet but also higher SU_3 representations, including a 27. This follows because of the decomposition

$$8 \times 8 = 27 + 10 + \overline{10} + 8_S + 8_A + 1 \qquad (9.26)$$

If H_W contained a component which transformed like a 27, we would expect $|\Delta I| = \frac{3}{2}$ transitions to be possible. On the other hand, if H_W transforms like an octet (through some dynamical mechanism, not understood, which inhibits the 27 and/or "enhances" the octet), then the $|\Delta I| = \frac{1}{2}$ rule is automatically guaranteed.

[1] But in many cases tests for the existence of neutral currents are merely tests for the existence of a neutral hadron-lepton coupling. See Sec. 10.6.

At this point we may ask the following questions: Suppose octet dominance is true; then

1 What octet component must the effective weak hamiltonian be proportional to?

2 What are the experimental consequences of the octet dominance assumption?

In the first place, we know that H_W as a whole must satisfy $\Delta Q = 0$ (charge conservation) even if the weak currents themselves are charged. Also it must satisfy $|\Delta S| = 1$. Of the eight octet components, only λ_3, λ_8, λ_6, and λ_7 correspond to $\Delta Q = 0$. Among these, only λ_6 and λ_7 correspond to $|\Delta S| = 1$. Now we know that the K^0 meson transforms like $\lambda_6 + i\lambda_7$ and that $\bar{K}^0$ transforms like $\lambda_6 - i\lambda_7$. Also, we know that $|\bar{K}^0\rangle = -\hat{C}\hat{P}|K^0\rangle$, where $|K^0\rangle$ and $|\bar{K}^0\rangle$ refer to states of K^0 and $\bar{K}^0$ mesons at rest. This follows (with a conventional choice of phase)[1] from the fact that K^0 and $\bar{K}^0$ are charge conjugates with odd intrinsic parity. As discussed in detail in subsequent chapters, neither K^0 nor $\bar{K}^0$ has a definite lifetime for weak decay, but with the assumption of CP invariance, the states

$$|K_s{}^0\rangle = \frac{1}{\sqrt{2}}[|K^0\rangle - |\bar{K}^0\rangle] \tag{9.27}$$

and

$$|K_L{}^0\rangle = \frac{1}{\sqrt{2}}[|K^0\rangle + |\bar{K}^0\rangle] \tag{9.28}$$

do have definite lifetimes. Obviously the CP eigenvalues of $|K_S{}^0\rangle$ and $|K_L{}^0\rangle$ are $+1$ and -1, respectively. Moreover, $K_S{}^0$ must transform like λ_7, and $K_L{}^0$ like λ_6, under SU_3 transformations. If we assume that H_W is CP-invariant, then it follows that H_W can transform as λ_6 *or* λ_7, but not as a linear combination of the two.

Let us assume to begin with that H_W transforms like λ_6. As in our previous discussion of the $|\Delta I| = \frac{1}{2}$ rule, we may examine the consequences of this assumption by pretending that the weak transition occurs by absorption of a spurion h:

$$h + \text{hyperon} \rightarrow \text{baryon} + \text{meson} \tag{9.29}$$

When the spurion is endowed with the quantum numbers of the effective hamiltonian, i.e., if we assume that h transforms like λ_6, then the *interaction* of (9.29) may be assumed SU_3-invariant. In this case the most general SU_3-invariant amplitude we can write to account for (9.29) must be a linear combination of the traces of products of the SU_3 matrices which contain creation and destruction operators for the particles of (9.29). (Here we disregard space and

[1] See Sec. 10.2.

spin indices.) The initial baryon destruction operators and final baryon creation operators are represented by the matrices B and $\bar{B}$, respectively:

$$B = \begin{pmatrix} \dfrac{\Sigma^0}{\sqrt{2}} + \dfrac{\Lambda^0}{\sqrt{6}} & \Sigma^+ & p \\ \Sigma^- & -\dfrac{\Sigma^0}{\sqrt{2}} + \dfrac{\Lambda^0}{\sqrt{6}} & n \\ \Xi^- & \Xi^0 & -\dfrac{2\Lambda^0}{\sqrt{6}} \end{pmatrix} \tag{9.30}$$

$$\bar{B} = \begin{pmatrix} \dfrac{\bar{\Sigma}^0}{\sqrt{2}} + \dfrac{\bar{\Lambda}^0}{\sqrt{6}} & \bar{\Sigma}^- & \bar{\Xi}^- \\ \bar{\Sigma}^+ & -\dfrac{\bar{\Sigma}^0}{\sqrt{2}} + \dfrac{\bar{\Lambda}^0}{\sqrt{6}} & \bar{\Xi}^0 \\ \bar{p} & \bar{n} & -\dfrac{2\bar{\Lambda}^0}{\sqrt{6}} \end{pmatrix} \tag{9.31}$$

while the final meson creation operators are represented by $\bar{M}$

$$\bar{M} = \begin{pmatrix} \dfrac{\bar{\pi}^0}{\sqrt{2}} + \dfrac{\bar{\eta}^0}{\sqrt{6}} & \bar{\pi}^- & \bar{K}^- \\ \bar{\pi}^+ & -\dfrac{\bar{\pi}^0}{\sqrt{2}} + \dfrac{\bar{\eta}^0}{\sqrt{6}} & \bar{K}^0 \\ \bar{K}^+ & K^0 & -\dfrac{2\bar{\eta}^0}{\sqrt{6}} \end{pmatrix} \tag{9.32}$$

and the spurion-annihilation operator h transforms like λ_6

$$\lambda_6 = \begin{pmatrix} 0 & 0 & 0 \\ 0 & 0 & 1 \\ 0 & 1 & 0 \end{pmatrix} \tag{9.33}$$

Taking into account that each of these matrices has zero trace, the amplitude must be

$$\mathcal{M} = \mathcal{M}_s + \mathcal{M}_p \tag{9.34}$$

where

$$\mathcal{M}_s = \sum_{i=1}^{9} s_i I_i$$

and

$$\mathcal{M}_p = \sum_{i=1}^{9} p_i I_i$$

are s- and p-wave amplitudes, respectively, and where

$$\begin{aligned} I_1 &= \operatorname{tr}(h\bar{B}\bar{M}B) & I_6 &= \operatorname{tr}(hB)\operatorname{tr}(\bar{B}\bar{M}) \\ I_2 &= \operatorname{tr}(hB\bar{M}\bar{B}) & I_7 &= \operatorname{tr}(h\bar{M}\bar{B}B) \\ I_3 &= \operatorname{tr}(h\bar{B}B\bar{M}) & I_8 &= \operatorname{tr}(h\bar{M}B\bar{B}) \\ I_4 &= \operatorname{tr}(hB\bar{B}\bar{M}) & I_9 &= \operatorname{tr}(h\bar{M})\operatorname{tr}(\bar{B}B) \\ I_5 &= \operatorname{tr}(h\bar{B})\operatorname{tr}(B\bar{M}) \end{aligned} \tag{9.35}$$

Now, the s-wave amplitude is pseudoscalar (odd under $\hat{P}$), while the p-wave amplitude is even under $\hat{P}$. Also, we require the amplitude as a whole to be CP-invariant. Therefore the s-wave amplitude must be odd under $\hat{C}$, while the p-wave amplitude is even under $\hat{C}$. However, under $\hat{C}$

$$\begin{aligned} B &\to \tilde{\bar{B}} \\ \bar{B} &\to \tilde{B} \\ \bar{M} &\to \tilde{\bar{M}} \\ h &\to h \end{aligned} \tag{9.36}$$

and

Hence we find that $I_1 \leftrightarrow I_1$, $I_2 \leftrightarrow I_2$, $I_9 \leftrightarrow I_9$, $I_3 \leftrightarrow I_7$, $I_4 \leftrightarrow I_8$, and $I_5 \leftrightarrow I_6$. Thus it follows that $s_1 = s_2 = s_9 = 0$, and

$$\mathcal{M}_s = s_3(I_3 - I_7) + s_4(I_4 - I_8) + s_5(I_5 - I_6) \tag{9.37}$$

It is now a simple, if tedious, matter to calculate the indicated traces. One obtains

$$\mathcal{M}_s = s_3(\bar{B}B\bar{M})_3{}^2 + s_4(B\bar{B}\bar{M})_3{}^2 + s_5(\bar{B}_3{}^2 + \bar{B}_3{}^2)\operatorname{tr}(B\bar{M}) \tag{9.38}$$

where the lower and upper indices indicate row and column, respectively. With the aid of (9.30) to (9.32), (9.38) becomes

$$\begin{aligned} \mathcal{M}_s = s_3\left[\frac{\bar{p}\bar{\pi}^-\Lambda^0}{\sqrt{6}} - \frac{\bar{p}\bar{\pi}^0\Sigma^+}{\sqrt{2}} + \bar{n}\bar{\pi}^-\Sigma^- - \frac{\bar{n}\bar{\pi}^0\Lambda^0}{\sqrt{12}} - \frac{2}{\sqrt{6}}\bar{\Lambda}^0\bar{\pi}^-\Xi^- + \frac{2}{\sqrt{12}}\bar{\Lambda}^0\bar{\pi}^0\Xi^0\right] \\ + s_4\left[\frac{\bar{\Lambda}^0\bar{\pi}^-}{\sqrt{6}}\Xi^- - \frac{\bar{\Lambda}^0\bar{\pi}^0\Xi^0}{\sqrt{12}} - \frac{2\bar{p}\bar{\pi}^-\Lambda^0}{\sqrt{6}} + \frac{2}{\sqrt{12}}\bar{n}\bar{\pi}^0\Lambda^0\right] \\ + s_5[\bar{\pi}^+\bar{n}\Sigma^+ + \bar{\pi}^-\bar{n}\Sigma^-] \end{aligned} \tag{9.39}$$

In other words, the s-wave amplitudes for the various hyperon nonleptonic decays are as follows:

$$\Lambda^0 \to p\pi^-(\Lambda_-{}^0): \qquad \frac{1}{\sqrt{6}}(s_3 - 2s_4) \tag{9.40}$$

$$\Lambda^0 \to n\pi^0(\Lambda_0{}^0): \qquad -\frac{1}{\sqrt{2}}\frac{1}{\sqrt{6}}(s_3 - 2s_4) \tag{9.41}$$

$$\Sigma^+ \to n\pi^+(\Sigma_+{}^+): \qquad s_5 \tag{9.42}$$

$$\Sigma^+ \to p\pi^0(\Sigma_0{}^+): \qquad -\frac{s_3}{\sqrt{2}} \tag{9.43}$$

$$\Sigma^- \to n\pi^-(\Sigma_-{}^-): \qquad s_5 + s_3 \tag{9.44}$$

$$\Xi^0 \to \Lambda^0\pi^0: \qquad \frac{1}{\sqrt{2}}\frac{1}{\sqrt{6}}(2s_3 - s_4) \tag{9.45}$$

$$\Xi^- \to \Lambda^0\pi^-: \qquad -\frac{1}{\sqrt{6}}(2s_3 - s_4) \tag{9.46}$$

Results (9.40) and (9.41) *by themselves* are merely a restatement of the $|\Delta I| = \frac{1}{2}$ rule for Λ^0 decays [see Eqs. (9.25) and (9.26)]. Similarly, (9.42) and (9.43) merely yield the result (9.27) again, while (9.45) and (9.46) are consistent with (9.29). However, (9.40) to (9.46) together also yield one further result, called the *Lee–Sugawara relation*[1]:

$$s(\Lambda_-{}^0) + 2s(\Xi^-) = \sqrt{3}\, s(\Sigma_0{}^+) \tag{9.47}$$

as may be verified easily. The relation (9.47) requires the assumption that the effective hamiltonian transforms like λ_6, which is a stronger condition than the $|\Delta I| = \frac{1}{2}$ rule.

No further restriction on the p-wave amplitudes beyond those imposed by the $|\Delta I| = \frac{1}{2}$ rule can be derived if the λ_6 assumption is made. On the other hand, if we assume that the weak hamiltonian transforms like λ_7, then the p-wave amplitudes satisfy the Lee–Sugawara relation and the s-wave amplitudes are not so restricted. In actual fact, the Lee–Sugawara relation appears to be quite well satisfied by both sets of amplitudes, according to the experimental results (Fil 69). (See Fig. 9.1.)

All of the preceding discussion has established certain ratios between s- and p-wave hyperon nonleptonic decay amplitudes, but nothing has been said about the formidable task of calculating the absolute magnitudes of these

[1] See Lee 64, Sug 64, Gel 64, and Oku 64.

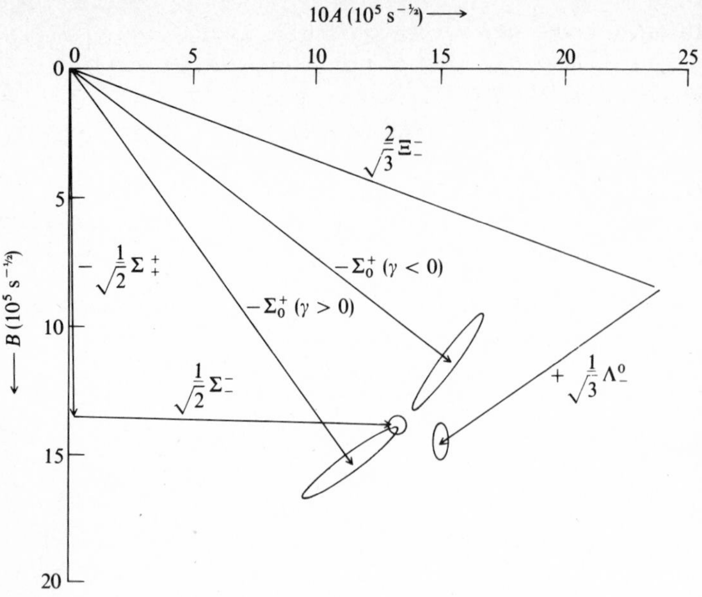

FIGURE 9.1
The $|\Delta I| = \frac{1}{2}$ rule and the Lee–Sugawara relation are tested by plotting the experimental values of A and B on this diagram. According to the $|\Delta I| = \frac{1}{2}$ rule, one should have $s(\Sigma_+{}^+ - s(\Sigma_-{}^-) = \sqrt{2}s(\Sigma_0{}^+)$ and $p(\Sigma_+{}^+) - p(\Sigma_-{}^-) = \sqrt{2}p(\Sigma_0{}^+)$. The Lee–Sugawara relation for s, p amplitudes predicts that $s(\Lambda_-{}^0) + 2s(\Xi_-{}^-) = \sqrt{3}\,s(\Sigma_0{}^+)$ and $p(\Lambda^0) + 2p(\Xi_-{}^-) = \sqrt{3}\,p(\Sigma_0{}^+)$. *(From H. Filthuth, Proc. Topical Conf. Weak Interactions, CERN, Geneva, 1969, p. 131.)*

transition amplitudes. Some genuine progress has been made along these lines, with the aid of current algebra. However, since the methods are hardly elementary, and because the results are still tentative and incomplete, no discussion of these matters is given here.[1]

REFERENCES

Bar 60 Barnes, S. W., H. Winnick, K. Miyake, and K. Kinsey: *Phys. Rev.*, **117**:238 (1960).

Fil 69 Filthuth, H.: Hyperon Decays, *Proc. Topical Conf. Weak Interactions, CERN, Geneva* (1969).

Gel 64 Gell–Mann, M.: *Phys. Rev. Letters*, **12**:155 (1964).

Lee 64 Lee, B. W.: *Phys. Rev. Letters*, **12**:83 (1964).

[1] See, for example, Lee 69 and Mar 68.

Lee 69 Lee, B. W.: Theory of Non-leptonic Hyperon Decays—A Survey, *Proc. Topical Conf. Weak Interactions, CERN, Geneva* (1969).

Mar 68 Marshak, R. E., Riazuddin, and C. P. Ryan: "Theory of Weak Interactions in Particle Physics," p. 579, Interscience, a division of Wiley, New York, 1968.

Oku 64 Okubo, S.: *Phys. Letters,* **8:**362 (1964).

Ove 67 Overseth, O. E., and R. F. Roth: *Phys. Rev. Letters,* **19:**391 (1967).

Sug 64 Sugawara, H.: *Progr. Theoret. Phys.* (*Kyoto*), **31:**213 (1964).

10

WEAK INTERACTIONS OF K MESONS, I

10.1 INTRODUCTION

The weak interactions of kaons are rich in variety and interest, and many features are of fundamental importance for the whole subject of weak interactions. All the K decays are summarized in Table 10.1. We note first the K_{l2} decays, which are analogous to the corresponding π_{l2} decays but proceed at comparatively slow rates governed by the decay constant $f_K = f_\pi \tan \Theta$, where Θ is the Cabibbo angle. (This was already discussed in Chap. 4.) Next, we have the K_{l3} and K_{l4} decays of $K^{\pm}$ and the long-lived neutral kaon. These are related to one another through CPT invariance and the $\Delta S = \Delta Q$ and $|\Delta I| = \frac{1}{2}$ rules. Additional restrictions on the amplitudes are obtained from Cabibbo's hypothesis and the current commutation relations, which lead to useful, if only approximate, *soft-pion* theorems connecting the amplitudes for K_{l2}, K_{l3}, and K_{l4} decays.

Experimental searches for the following decays of the long-lived neutral kaon,

$$K_L{}^0 \overset{?}{\rightarrow} \mu^+\mu^- \tag{10.1}$$

$$K_L{}^0 \overset{?}{\rightarrow} e^+e^- \tag{10.2}$$

Table 10.1 DECAY MODES OF *K* MESONS

Initial particle	Mass	J^P	I,	I_3	S	Mean life, s
K^+	493.82 ± 0.11	0^-	$\frac{1}{2}$	$\frac{1}{2}$	$+1$	$(1.235 \pm 0.004) \times 10^{-8}$
K^0	497.76 ± 0.16	0^-	$\frac{1}{2}$	$-\frac{1}{2}$	$+1$	
$\bar{K}^0$	497.76 ± 0.16	0^-	$\frac{1}{2}$	$\frac{1}{2}$	-1	
K^-	493.82 ± 0.11	0^-	$\frac{1}{2}$	$-\frac{1}{2}$	-1	$(1.235 \pm 0.004) \times 10^{-8}$
K_S^0‡						$(0.882 \pm 0.006) \times 10^{-10}$
K_L^0‡						$(5.181 \pm 0.091) \times 10^{-8}$

$$|K_S^0\rangle = \frac{1}{\sqrt{2(1+|\varepsilon|^2)}}\left[(1-\varepsilon)|K^0\rangle - (1-\varepsilon)|\bar{K}^0\rangle\right]$$

$$|K_L^0\rangle = \frac{1}{\sqrt{2(1+|\varepsilon|^2)}}\left[(1+\varepsilon)|K^0\rangle + (1-\varepsilon)|\bar{K}^0\rangle\right]$$

Decay mode		Branching ratio
$K^\pm \to \mu^\pm\nu$	$K_{l2}^\pm$	$(63.77 \pm 0.29)\%$
$e^\pm\nu$	$K_{l2}^\pm$	$(1.2 \pm 0.3) \times 10^{-5}$
$\pi^0\mu^\pm\nu$	$K_{l3}^\pm$	$(3.18 \pm 0.11)\%$
$\pi^0 e^\pm\nu$	$K_{l3}^\pm$	$(4.85 \pm 0.07)\%$
$\pi^+\pi^-\mu^\pm\nu$	$K_{l4}^\pm$	$(0.9 \pm 0.4) \times 10^{-5}$
$\pi^+\pi^- e^\pm\nu$	$K_{l4}^\pm$	$(3.3 \pm 0.3) \times 10^{-5}$
$\pi^\pm\pi^0$	$K_{\pi 2}^\pm$	$(20.93 \pm 0.30)\%$
$\pi^+\pi^-\pi^+$	$K_{\pi 3}^\pm$	$(5.57 \pm 0.04)\%$
$\pi^\pm\pi^0\pi^0$	$K_{\pi 3}^\pm$	$(1.70 \pm 0.05)\%$
$\pi^\pm\pi^0\gamma$	Miscellaneous	$(<1.9) \times 10^{-4}$
$\pi^\pm\pi^+\pi^-\gamma$	Miscellaneous	$(10 \pm 4) \times 10^{-5}$
$\pi^0 e^\pm\nu\gamma$	Miscellaneous	$(6 \pm 4) \times 10^{-4}$
$\pi^\pm e^+ e^-$	Miscellaneous	$(<0.4) \times 10^{-6}$
$\pi^\pm\mu^+\mu^-$	Miscellaneous	$(<2.4) \times 10^{-6}$
$\pi^\pm\gamma\gamma$	Miscellaneous	$(<1.1) \times 10^{-4}$
$K_S^0 \to \pi^+\pi^-$	$K_{S\pi 2}$	$(68.7 \pm 0.6)\%$
$\pi^0\pi^0$	$K_{S\pi 2}$	$(31.3 \pm 0.6)\%$
$\mu^+\mu^-$		$(<3.1) \times 10^{-7}$
e^+e^-		$(<2.2) \times 10^{-7}$
$\pi^+\pi^-\gamma$		$(3.3 \pm 1.2) \times 10^{-3}$
$K_L^0 \to \pi^\pm\mu^\mp\nu$§	K_{Ll3}^0	$(26.8 \pm 0.7)\%$
$\pi^\pm e^\mp\nu$§	K_{Ll3}^0	$(38.8 \pm 0.8)\%$
$\pi^0\pi^0\pi^0$	$K_{L\pi 3}^0$	$(21.5 \pm 0.7)\%$
$\pi^+\pi^-\pi^0$	$K_{L\pi 3}^0$	$(12.6 \pm 0.3)\%$
$\pi^+\pi^-$‖	$K_{L\pi 2}^0$	$(1.57 \pm 0.05) \times 10^{-3}$
$\pi^0\pi^0$‖	$K_{L\pi 2}^0$	$(1.21 \pm 0.29) \times 10^{-3}$
$\pi^+\pi^-\gamma$		$(<0.4) \times 10^{-3}$
$\gamma\gamma$		$(5.2 \pm 0.5) \times 10^{-4}$
$e^\pm\mu^\mp$#		$(<1.6) \times 10^{-9}$
$\mu^+\mu^-$††		$(<1.8) \times 10^{-9}$
e^+e^-††		$(<1.6) \times 10^{-9}$

‡ K^0, $\bar{K}^0$ do not have definite lifetimes for weak decay.
§ Because of CP violation the rates for $\pi^+ l^- \nu$ and $\pi^- l^+ \nu$ are not exactly equal.
‖ CP-violating.
Forbidden by lepton conservation.
†† Test for neutral currents.

provide sensitive upper limits for the existence of neutral weak currents (or, more precisely, an upper limit on the coupling of neutral weak hadronic and leptonic currents).

The nonleptonic decays

$$K \to 2\pi, 3\pi \qquad (10.3)$$

are important enough to comprise an entire subject by themselves. Here we have to deal first of all with the $K^0\bar{K}^0$ complex, with its peculiar and unique properties involving regeneration, and CP violation in the transitions

$$K_L{}^0 \to \pi^+\pi^-, \pi^0\pi^0 \qquad (10.4)$$

We also encounter in the decays (10.3) examples of the $|\Delta I| = \frac{1}{2}$ rule, already introduced in Chap. 7.

In the present chapter we shall begin with a summary of the general properties of kaons and then concentrate on the semileptonic K_{l3} and K_{l4} transitions. We shall also include a discussion of the *neutral current* problem. Chapter 11 is devoted to the study of nonleptonic decays of kaons and includes a lengthy and detailed discussion of CP violation.

10.2 GENERAL PROPERTIES OF KAONS

The kaons belong to the $J^P = 0^-$ meson octet illustrated in Fig. 7.1: (K^+, K^0) and $(\bar{K}^0, K^-)$ form two isodoublets in this octet with strangeness $S = +1$ and $S = -1$, respectively. K^+ and K^- are, in fact, charge conjugates, as are K^0 and $\bar{K}^0$. Let us define the states $|K^0\rangle$, $|\bar{K}^0\rangle$, representing K^0, $\bar{K}^0$ at rest. The $|K^0\rangle$, $|\bar{K}^0\rangle$ are by definition eigenstates of the hamiltonian of strong and electromagnetic interactions and simultaneously eigenstates of strangeness with eigenvalues ± 1, respectively. Since the strong and electromagnetic interactions conserve strangeness, the relative phase of the states $|K^0\rangle$ and $|\bar{K}^0\rangle$ is arbitrary, and we may choose it at will. Thus we may define

$$|\bar{K}^0\rangle \equiv -CP|K^0\rangle \qquad (10.5)$$

Neither K^0 nor $\bar{K}^0$ has a definite lifetime for weak decay, but one can form two independent linear superpositions of $|K^0\rangle$ and $|\bar{K}^0\rangle$, namely, the states $|K_L{}^0\rangle$ and $|K_S{}^0\rangle$. (L and S stand for *long* and *short*, respectively.) The particles $K_L{}^0$ and $K_S{}^0$ do have definite lifetimes, but they do not have definite strangeness or isospin. The short-lived $K_S{}^0$ decays in only two significant modes: $\pi^+\pi^-$ and $\pi^0\pi^0$, with total inverse lifetime $\gamma_S = 1.134 \times 10^{10}$ s^{-1}. The long-lived component $K_L{}^0$ decays in many modes, semileptonic and nonleptonic, and has the total inverse lifetime $\gamma_L = 1.93 \times 10^7$ s^{-1} (see Table 10.1).

Before the discovery of CP violation in 1964, it was thought that the states $|K_L{}^0\rangle$ and $|K_S{}^0\rangle$ could be expressed by the formulas

$$
\begin{aligned}
|K_S{}^0\rangle &= \frac{1}{\sqrt{2}}[|K^0\rangle - |\bar{K}^0\rangle] \\
|K_L{}^0\rangle &= \frac{1}{\sqrt{2}}[|K^0\rangle + |\bar{K}^0\rangle]
\end{aligned}
\tag{10.6}
$$

Since the final states $\pi^+\pi^-$ and $\pi^0\pi^0$ each have CP eigenvalue $+1$, the formulas (10.6), together with (10.5), "explained" why the decays $K_S{}^0 \to \pi^+\pi^-$ and $K_S{}^0 \to \pi^0\pi^0$ are allowed (CP $|K_S{}^0\rangle = +|K_S{}^0\rangle$) and why the decays $K_L{}^0 \to \pi^+\pi^-$, $\pi^0\pi^0$ are "forbidden" (CP $|K_L{}^0\rangle = -|K_L{}^0\rangle$) within the framework of CP invariance. Now, of course, we know that formulas (10.6) are incorrect and that they must be replaced by

$$|K_S{}^0\rangle = \frac{1}{\sqrt{2(1+|\varepsilon|^2)}}[(1+\varepsilon)|K^0\rangle - (1-\varepsilon)|\bar{K}^0\rangle] \tag{10.7a}$$

$$|K_L{}^0\rangle = \frac{1}{\sqrt{2(1+|\varepsilon|^2)}}[(1+\varepsilon)|K^0\rangle + (1-\varepsilon)|\bar{K}^0\rangle] \tag{10.7b}$$

(assuming that CPT invariance holds). The basis for (10.7a) and (10.7b) will be discussed in detail in Chap. 11. For the moment we note merely that the complex parameter ε characterizes CP violation and is found by experiment to be very small:

$$|\varepsilon| \approx 2 \times 10^{-3}$$

Thus CP violation is an extremely *small* effect, and in lowest approximation we may just neglect it in discussing many of the properties of the $K^0\bar{K}^0$ complex. Obviously, in the limit $\varepsilon \to 0$, formulas (10.7) reduce to (10.6).

For strong interactions it is appropriate to consider the neutral K complex from the point of view of the states $|K^0\rangle$ and $|\bar{K}^0\rangle$ rather than $|K_L{}^0\rangle$ and $|K_S{}^0\rangle$ since the former have definite strangeness. A typical strong production reaction is

$$\pi^- p \to K^0\Lambda^0 \tag{10.8}$$

where the total strangeness on each side of the equation is zero. Figure 10.1 represents the formation of a K^0 beam in this manner. Neglecting ε, the K^0 state at time $t = 0$ can be represented as

$$\psi(0) = K^0 = \frac{1}{\sqrt{2}}[K_S{}^0 + K_L{}^0] \tag{10.9}$$

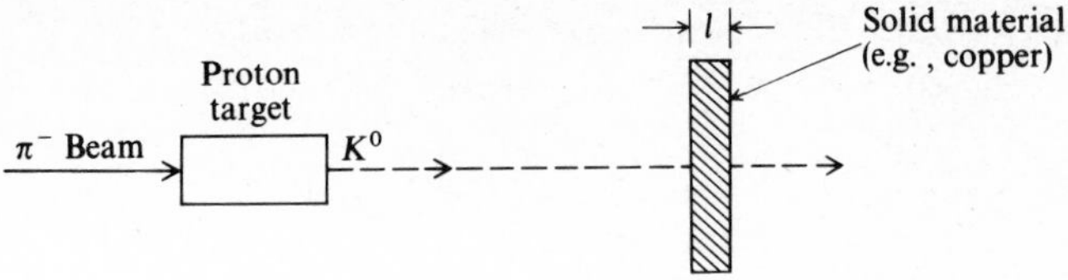

FIGURE 10.1

Downstream along the neutral kaon beam we have

$$\psi(t) = \frac{1}{\sqrt{2}}[K_S{}^0(t) + K_L{}^0(t)]$$

$$= \frac{1}{\sqrt{2}}[K_S{}^0(0)e^{-\lambda_S t} + K_L{}^0(0)e^{-\lambda_L t}]$$

where $$\lambda_S = \frac{\gamma_S}{2} + im_S \qquad \lambda_L = \frac{\gamma_L}{2} + im_L$$

and m_S and m_L are the masses of K_S and K_L, respectively. For times t such that

$$\frac{1}{\gamma_L} \gg t \gg \frac{1}{\gamma_S}$$

we have

$$\psi(t) \approx \frac{1}{\sqrt{2}} K_L{}^0(0)e^{-\lambda_L t} = \frac{1}{\sqrt{2}} K_L{}^0(0)e^{-\gamma_L t/2}e^{-im_L t}$$

Thus downstream from the source the entire beam is converted into the long-lived component $K_L{}^0$.

Next, consider the effect of a piece of matter of thickness l placed in the path of the beam (Fig. 10.1). Here one has collisions between the kaons in the beam and the matter nuclei. The wave function of a beam particle prior to collision is (neglecting the e^{-imt})

$$\psi = \frac{1}{\sqrt{2}} K_L{}^0 = \tfrac{1}{2}(K^0 + \overline{K}^0) \qquad (10.10)$$

The essential point now is that K^0 and $\overline{K}^0$ will scatter from the matter nuclei with very different amplitudes. A whole range of inelastic channels is open to $\overline{K}^0$, but for K^0 one has essentially elastic scattering only. This is because of strangeness conservation in strong interactions.

As a result of the scattering, then,

$$\psi = \tfrac{1}{2}(aK^0 + b\overline{K}^0) \qquad (10.11)$$

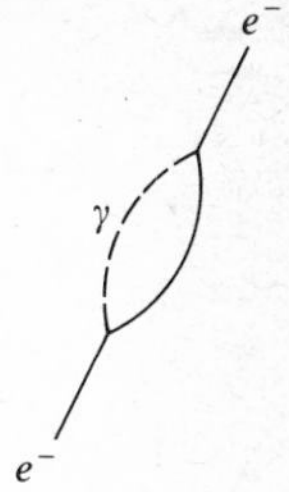

FIGURE 10.2

where a and b are two complex numbers, and $|a| > |b|$. Equation (10.11) can be written

$$\psi = \frac{1}{2}\left(\frac{a-b}{\sqrt{2}}\right)K_S{}^0 + \frac{1}{2}\left(\frac{b+a}{\sqrt{2}}\right)K_L{}^0 \qquad (10.12)$$

Thus the $K_S{}^0$ component dies away in the beam before the slab of matter but reappears (is "regenerated") owing to scattering with the matter nuclei.

Continuing to assume CP invariance, we can easily obtain a rough idea of why $K_S{}^0$ and $K_L{}^0$ should have different masses.[1] Remember that in quantum electrodynamics there is a contribution to the electron mass (it is infinite!) from self-energy diagrams, as shown in Fig. 10.2. Similarly, we should have a contribution to the mass of $K_S{}^0$ from the diagrams shown in Fig. 10.3, and a contribution to the $K_L{}^0$ mass from diagrams shown in Fig. 10.4. The $K_S{}^0$ and $K_L{}^0$ diagrams are different because of CP invariance (which we are assuming to be strictly true and which holds to a very good approximation).

If we had a theory of strong interactions somewhat similar to quantum electrodynamics, we might be able to calculate the contribution of each diagram to the mass of $K_S{}^0$ or $K_L{}^0$. Each contribution might be infinite, but we assume

[1] The mass difference has been determined experimentally by methods to be discussed in Chap. 11.

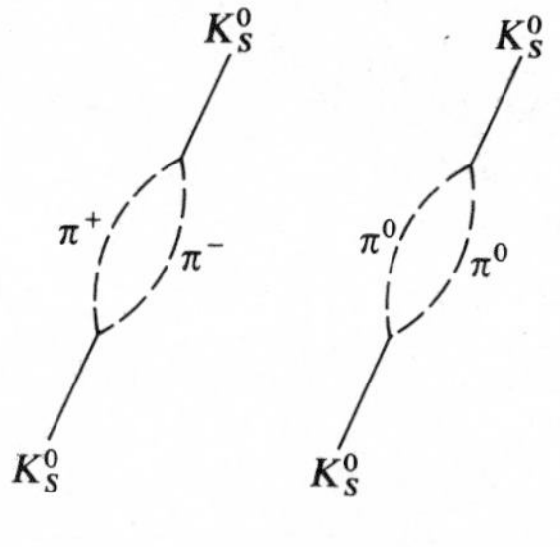

FIGURE 10.3

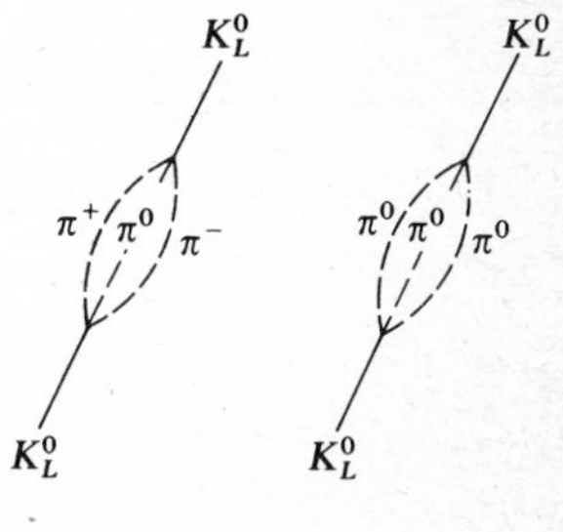

FIGURE 10.4

that the *difference* between the total $K_S{}^0$ and $K_L{}^0$ contributions is finite. Moreover, each diagram consists of two weak vertices, and each vertex contributes a factor G; and so we have

$$\Delta m = m_L - m_S = \alpha_1 G^2 \qquad (10.13)$$

where α_1 is some constant of proportionality. Since G^2 has dimension $(\text{mass})^{-4}$, α_1 must have dimension $(\text{mass})^5$. It remains to decide what the value of this unknown mass is. If we take

$$\alpha_1 \approx (m_\pi)^5 \qquad (10.14)$$

then

$$\alpha_1 \approx 6 \times 10^{-5}$$

and

$$\Delta m = \alpha_1 G^2 \approx 0.6 \times 10^{10}\ \text{s}^{-1}$$

If we choose

$$\alpha_1 \approx (m_K)^5$$

then

$$\Delta m \approx 0.3 \times 10^{13}\ \text{s}^{-1}$$

In actual fact the experimental value of the mass difference is

$$\Delta m = +0.44\gamma_S \approx 0.5 \times 10^{10}\ \text{s}^{-1} \qquad (10.15)$$

consistent with the choice $\alpha_1 \approx (m_\pi)^5$.

Of course, the estimate we have made is so rough it is practically worthless. But it serves to indicate at least that the mass difference arises through a second-order weak interaction rather than, for example, a first-order $|\Delta S| = 2$ transition between K^0 and $\bar{K}^0$. For, if it were the latter, we might expect the mass difference to be proportional to $G = 10^{-5}$ rather than $G^2 = 10^{-10}$. This would give a mass difference much larger than that actually observed:

$$\Delta m = G m_\pi{}^3 \approx 30\ \text{eV} \approx 10^{17}\ \text{s}^{-1}$$

Naturally this crude estimate gives no clue whatever to the sign of Δm, and, in fact, we have no reliable theory to explain this sign.

10.3 THE SEMILEPTONIC K_{l3} DECAYS

We now consider in some detail the decays

$$K^+ \to \pi^0 \mu^+ \nu \qquad K_L{}^0 \to \pi^- \mu^+ \nu$$

$$K^- \to \pi^0 \mu^- \bar{\nu} \qquad K_L{}^0 \to \pi^+ \mu^- \bar{\nu}$$

$$K^+ \to \pi^0 e^+ \nu \qquad K_L{}^0 \to \pi^- e^+ \nu$$

$$K^- \to \pi^0 e^- \bar{\nu} \qquad K_L{}^0 \to \pi^+ e^- \bar{\nu}$$

As we might expect, the amplitudes for these decays are not independent of one another but are related by CPT invariance, (approximate) CP invariance, the $\Delta S = \Delta Q$ rule, and the $|\Delta I| = \frac{1}{2}$ rule. Further restrictions on the amplitudes are also imposed by Cabibbo's hypothesis.

10.3.1 CPT Invariance

A weak decay (initial particle $a \to$ final particles b) is represented by the matrix element $\langle b|\mathcal{L}_W|a\rangle$ where $\mathcal{L}_W$ is the weak lagrangian. Let us consider a CPT transformation of this matrix element. Under C, particles are transformed to antiparticles and spins and three-momenta are left unchanged. Under P, three-momenta are reversed. Under T, which is antiunitary, initial and final states are interchanged and spins and momenta are reversed. Thus, the net effect of a CPT transformation is

$$\langle b|\mathcal{L}|a\rangle \xrightarrow{CPT} \langle \bar{a}'|\mathcal{L}|\bar{b}'\rangle = \langle \bar{b}'|\mathcal{L}|\bar{a}'\rangle^*$$

where the bar means antiparticle and the prime signifies that spins are reversed. Since the weak interaction is believed to be CPT-invariant, we conclude that

$$\langle b|\mathcal{L}|a\rangle = \langle \bar{b}'|\mathcal{L}|\bar{a}'\rangle^* \qquad (10.16)$$

Now we consider the amplitudes for the various K_{l3} decays. In particular, we denote the hadronic amplitude for $K^+ \to \pi^0\mu^+\nu$ or $\pi^0 e^+\nu$ by A_+:

$$A_+ \equiv \langle \pi^0|\mathcal{V}_1{}^\lambda|K^+\rangle \qquad (10.17)$$

Then from (10.16) we have

$$A_- \equiv \langle \pi^0|\mathcal{V}_1{}^{\lambda\dagger}|K^-\rangle = A_+{}^* \qquad (10.18)$$

In similar fashion we define f and g by the following equations:

$$f \equiv \langle \pi^-|\mathcal{V}_1{}^\lambda|K^0\rangle \qquad \Delta S = \Delta Q \qquad (10.19)$$

and

$$g \equiv \langle \pi^+|\mathcal{V}_1{}^{\lambda\dagger}|K^0\rangle \qquad \Delta S = -\Delta Q \qquad (10.20)$$

Then from (10.16) we have

$$\langle \pi^+|\mathcal{V}_1{}^{\lambda\dagger}|\bar{K}^0\rangle = f^* \qquad \Delta S = \Delta Q \qquad (10.21)$$

and

$$\langle \pi^-|\mathcal{V}_1{}^{\lambda}|\bar{K}^0\rangle = g^* \qquad \Delta S = -\Delta Q \qquad (10.22)$$

10.3.2 The $\Delta S = \Delta Q$ Rule

The transitions $K^+ \to \pi^0 l^+\nu$ and $K^- \to \pi^0 l^-\nu$ automatically satisfy the $\Delta S = \Delta Q$ rule. However, $K_L{}^0$ is a superposition of K^0 and $\bar{K}^0$ states; therefore, in transitions $K_L{}^0 \to \pi^\pm l^\mp \nu$ one can have contributions from amplitudes f or f^* ($\Delta S = \Delta Q$) and g or g^* ($\Delta S = -\Delta Q$) a priori. We shall now see how the $\Delta S = \Delta Q$ rule can be tested.

Let us first assume that CP invariance is valid. In this case, we can use Eqs. (10.6) to relate $K_S{}^0$, $K_L{}^0$ to K^0, $\bar{K}^0$; moreover, since CP invariance and CPT invariance imply T invariance, the amplitudes f and g are (relatively) real. We consider a neutral K beam formed in the state $|K^0\rangle$ at time $t = 0$. Then, at a later time t,

$$|K^0(t)\rangle = \frac{1}{\sqrt{2}}\{|K_S{}^0(t)\rangle + |K_L{}^0(t)\rangle\}$$

$$= \frac{1}{\sqrt{2}}\{|K_S{}^0(0)\rangle e^{-\lambda_S t} + |K_L{}^0(0)\rangle e^{-\lambda_L t}\}$$

$$= \frac{1}{2}\{|K^0(0) - \bar{K}^0(0)\rangle e^{-\lambda_S t} + |K^0(0) + \bar{K}^0(0)\rangle e^{-\lambda_L t}\}$$

We take the matrix elements corresponding to the transitions

$$K^0 \to \pi^- e^+ \nu_e \qquad \text{and} \qquad K^0 \to \pi^+ e^- \bar{\nu}_e$$

From (10.19) to (10.22) we thus have

$$\langle \pi^- | \mathcal{V}_1{}^\lambda | K^0(t)\rangle = \tfrac{1}{2}\{(f - g^*)e^{-\lambda_S t} + (f + g^*)e^{-\lambda_L t}\}$$

$$\langle \pi^+ | \mathcal{V}_1{}^{\lambda\dagger} | K^0(t)\rangle = \tfrac{1}{2}\{(g - f^*)e^{-\lambda_S t} + (g + f^*)e^{-\lambda_L t}\}$$

Therefore, the rate of production of $e^\pm$ at time t is proportional to

$$R^\pm(t) = (1 + x)^2 e^{-\gamma_S t} + (1 - x)^2 e^{-\gamma_L t} \pm 2(1 - x^2)e^{-((\gamma_S + \gamma_L)/2)t} \cos(m_L - m_S)t$$

$$\text{(CP invariance assumed)} \qquad (10.23)$$

where $x = -g/f$.

More generally, if we do not assume CP invariance, then x can be complex and, in addition, we have to use formulas (10.7) for $K_L{}^0$ and $K_S{}^0$. However, for the purposes of testing the $\Delta S = \Delta Q$ rule, it is sufficient to neglect ε, in which case Eqs. (10.6) are adequate and we find

$$R^\pm(t) = |1 + x|^2 e^{-\gamma_S t} + |1 - x|^2 e^{-\gamma_L t}$$
$$\pm 2(1 - |x|^2)e^{-(\gamma_S + \gamma_L)t/2} \cos \Delta m t$$
$$- 4 \,\mathrm{Im}(x) e^{-(\gamma_S + \gamma_L)t/2} \sin \Delta m t \qquad \text{(CP invariance not assumed)} \qquad (10.24)$$

where $x = -g^*/f$.

If we assume x is real (CP invariance), then the combined results of recent experiments on e^+, e^- time distributions give (Lit 69)

$$x = 0.037^{+0.025}_{-0.033} \qquad (10.25)$$

If we allow x to be complex (CP violation), then the world averages of different experiments are (Rub 69)

$$\begin{aligned} \text{Im}(x) &= -0.13 \pm 0.043 \\ \text{Re}(x) &= 0.14 \pm 0.05 \end{aligned} \tag{10.26}$$

in apparent violation of $\Delta S = \Delta Q$. However, no *single* experiment gives a statistically significant nonzero result for x. Our tentative conclusion is that there is no clear evidence for violation of the $\Delta S = \Delta Q$ rule in $K_{l3}{}^0$ decay, and we shall generally assume that the rule is valid in what follows.

10.3.3 The $|\Delta I| = \frac{1}{2}$ Rule

The $|\Delta I| = \frac{1}{2}$ rule establishes further connections between the amplitudes for K_{l3} decays. It states that the operators $\mathcal{V}_1{}^\lambda$ and $\mathcal{V}_1{}^{\lambda\dagger}$ transform like components of an isospinor. Thus we can write the hadronic amplitudes for the various K_{l3} decays as Clebsch–Gordan coefficients multiplied by a common numerical factor: the reduced matrix element $\mathcal{M}^0$. If $|\Delta I| = \frac{1}{2}$, we thus have

$$\langle \pi^0 | \mathcal{V}_1{}^\lambda | K^+ \rangle = A_+ = \langle \tfrac{1}{2}, \tfrac{1}{2}; \tfrac{1}{2}, -\tfrac{1}{2} | 1, 0 \rangle \, \mathcal{M}_0 = \frac{1}{\sqrt{2}} \mathcal{M}_0 \tag{10.27}$$

$$\langle \pi^0 | \mathcal{V}_1{}^{\lambda\dagger} | K^- \rangle = A_- = \langle \tfrac{1}{2}, -\tfrac{1}{2}; \tfrac{1}{2}, +\tfrac{1}{2} | 1, 0 \rangle \, \mathcal{M}_0 = \frac{1}{\sqrt{2}} \mathcal{M}_0 \tag{10.28}$$

$$\langle \pi^- | \mathcal{V}_1{}^\lambda | K^0 \rangle = f = \langle \tfrac{1}{2}, -\tfrac{1}{2}; \tfrac{1}{2}, -\tfrac{1}{2} | 1, -1 \rangle \, \mathcal{M}_0 = \mathcal{M}_0 \tag{10.29}$$

$$\langle \pi^+ | \mathcal{V}_1{}^{\lambda\dagger} | K^0 \rangle = g = \langle \tfrac{1}{2}, -\tfrac{1}{2}; \tfrac{1}{2}, +\tfrac{1}{2} | 1, +1 \rangle \, \mathcal{M}_0 = 0 \tag{10.30}$$

The $|\Delta I| = \frac{1}{2}$ rule implies the $\Delta S = \Delta Q$ rule, which requires that $g = 0$, as we see from Eq. (10.30). Furthermore, we obtain from (10.27) to (10.29)

$$A_+ = A_- = \frac{1}{\sqrt{2}} f \tag{10.31}$$

Now, neglecting ε, we obtain

$$\langle \pi^- | \mathcal{V}_1{}^\lambda | K_L{}^0 \rangle = \frac{1}{\sqrt{2}} \langle \pi^- | \mathcal{V}_1{}^\lambda | K^0 \rangle = \frac{1}{\sqrt{2}} f = A_+$$

and
$$\langle \pi^+ | \mathcal{V}_1{}^{\lambda\dagger} | K_L{}^0 \rangle = \frac{1}{\sqrt{2}} \langle \pi^+ | \mathcal{V}_1{}^{\lambda\dagger} | \bar{K}_0 \rangle = \frac{1}{\sqrt{2}} f^*$$

Thus, the rates for $K^+ \to \pi^0 \mu^+ \nu$, $K_L{}^0 \to \pi^+ \mu^- \bar{\nu}$, and $K_L{}^0 \to \pi^- \mu^+ \bar{\nu}$ should be the same (ignoring ε and after radiative corrections and corrections for differing phase-space factors have been applied); the same is true for $K^+ \to \pi^0 e^+ \nu$,

$K_L{}^0 \to \pi^+ e^- \bar{\nu}$, and $K_L{}^0 \to \pi^- e^+ \bar{\nu}$. Furthermore, the spectra and polarizations for these decays should be the same (although, of course, we have opposite polarization for l^+ and l^-).

The experimental K_{l3} decay rates are in very good agreement with the $|\Delta I| = \frac{1}{2}$ rule (Cro 68). Specifically, suppose we let α_{11}, α_{31}, and α_{33} be decay amplitudes corresponding to $(\Delta I, \Delta I_3) = (\frac{1}{2}, \frac{1}{2})$, $(\frac{3}{2}, \frac{1}{2})$, and $(\frac{3}{2}, \frac{3}{2})$, respectively. Then the experimental decay rates enable us to establish that

$$\text{Re}\left[\frac{\alpha_{31}}{\alpha_{11}} + \sqrt{3}\,\frac{\alpha_{33}}{\alpha_{11}}\right] = (3.4 \pm 1.3) \times 10^{-3} \qquad (10.32)$$

which we shall assume is consistent with $\alpha_{31} = \alpha_{33} = 0$.

10.3.4 Form of the K_{l3} Amplitude; Cabibbo's Hypothesis

Let us now establish an explicit form for the K_{l3} amplitude (for example, $K^+ \to \pi^0 l^+ \nu$) and compare it with the corresponding quantity for pion beta decay $\pi^+ \to \pi^0 e^+ \nu$. For K^+ decay we have

$$\mathcal{M}(K^+) = \frac{G}{\sqrt{2}} \langle \pi^0 | \mathcal{V}_1{}^\lambda | K^+ \rangle L_\lambda$$

$$= \frac{G}{\sqrt{2}} \sin \Theta [f_+(q^2)(p_K + p_\pi)^\lambda + f_-(q^2)(p_K - p_\pi)^\lambda] L_\lambda \qquad (10.33)$$

where $f_+(q^2)$ and $f_-(q^2)$ are form factors to be determined. For $\pi^+ \to \pi^0 e^+ \nu$ we have from Sec. 4.6

$$\mathcal{M}(\pi^+) = \frac{G}{\sqrt{2}} \langle \pi^0 | \mathcal{V}_0{}^\lambda | \pi^+ \rangle L_\lambda$$

$$= \frac{G}{\sqrt{2}} \cos \Theta \sqrt{2}\, [F_1(q^2)(p_{\pi^+} + p_{\pi^0})^\lambda + F_2(q^2)(p_{\pi^+} - p_{\pi^0})^\lambda] L_\lambda \qquad (10.34)$$

As will be recalled from Chap. 4 [Eqs. (4.56) to (4.59)], the conserved vector current hypothesis leads to the conclusion that $F_2(q^2) = 0$ and $F_1(q^2) \to 1$ as $q^2 \to 0$. Moreover, the factor $\sqrt{2}$ in the numerator of (10.34) arises from the fact that $\mathcal{V}_0{}^\lambda$ acts like an isospin-lowering operator between two states of isospin unity.

Now, according to Cabibbo's hypothesis, the operators $\mathcal{V}_1{}^\lambda$ and $\mathcal{V}_0{}^\lambda$ are related by the equations

$$\mathcal{V}_0{}^\lambda = (j_1 - ij_2) \cos \Theta$$

$$\mathcal{V}_1{}^\lambda = (j_4 - ij_5) \sin \Theta$$

where the j's are vector current operators belonging to the *same* octet. It follows that in the limit of perfect SU_3 symmetry, $\mathcal{V}_1{}^\lambda$ should also be a conserved current, and therefore we expect that in Eq. (10.33) the following equations hold:

$$\text{Perfect } SU_3 \text{ Symmetry} \quad \begin{cases} f_-(q^2) = 0 \\ f_+(q^2) = 1 \quad \text{for } q^2 \to 0 \end{cases} \tag{10.35}$$

Because of departures from perfect SU_3 symmetry, relations (10.35) are not expected to be exactly true. However, according to the Ademollo–Gatto theorem, the change induced in $f_+(q^2)$ is only of second order in the symmetry-breaking parameter.

10.3.5 The Ademollo–Gatto Theorem

We now outline a proof of the Ademollo–Gatto theorem (Ade 64) as it applies to K_{l3} decays. Consider the charges $F_i(t)$ corresponding to the vector octet of current operators $j_i(x, t)$:

$$F_i(t) = \int j_i{}^0(\mathbf{x}, t)\, d^3x \tag{10.36}$$

These charges satisfy the basic SU_3 commutation relations

$$[(F_4 + iF_5), (F_4 - iF_5)] = F_3 + \tfrac{3}{2} Y \tag{10.37}$$

as can be verified from Eqs. (7.23) to (7.25) and Table 7.1. We take the matrix element between states $|\pi^+(\mathbf{p}')\rangle$ and $|\pi^+(\mathbf{p})\rangle$ of both sides of (10.37)

$$\langle\pi^+(\mathbf{p}')|\,[F_4 + iF_5, F_4 - iF_5]\,|\pi^+(\mathbf{p})\rangle = \langle\pi^+(\mathbf{p}')|\,F_3 + \tfrac{3}{2}Y\,|\pi^+(\mathbf{p})\rangle = \langle\pi^+(\mathbf{p}')|\pi^+(\mathbf{p})\rangle \tag{10.38}$$

and we normalize the π states so that the right-hand side of (10.38) is

$$\langle\pi^+(\mathbf{p}')|\pi^+(\mathbf{p})\rangle = 2p^0(2\pi)^3\delta^3(\mathbf{p}' - \mathbf{p}) \tag{10.39}$$

where $p^0 = \sqrt{\mathbf{p}^2 + m_\pi{}^2}$ is the pion energy. As for the left-hand side of (10.38), we insert $(1/2p_0) \sum |\alpha\rangle\langle\alpha| = 1$ and write

$$\frac{1}{2p_0} \sum_\alpha \langle\pi^+(\mathbf{p}')|\,F_4 + iF_5\,|\alpha\rangle\langle\alpha|\,F_4 - iF_5\,|\pi^+(\mathbf{p})\rangle - \sum_\alpha \langle\pi^+(\mathbf{p}')|\,F_4 - iF_5\,|\alpha\rangle\langle\alpha|\,F_4 + iF_5\,|\pi^+(\mathbf{p})\rangle$$

with the normalization of (10.39). As in the discussion of the Adler–Weisberger relation, we break up the sum over α states into two portions. This time, we divide the α states into those corresponding to a $\overline{K}^0$ meson in various momentum states k and, on the other hand, all other states α. In the latter category are particle states not belonging to the same SU_3 multiplet as the pion. The $\bar{K}^0$, of course, *does* belong to the 0^- octet along with the pion. Then

$$\sum_{\alpha} = \sum_{\alpha=\bar{K}^0} + \sum_{\alpha\neq\bar{K}_0}$$
$$\frac{1}{2p_0}\sum_{\alpha=\bar{K}_0} = \int \frac{d^3\mathbf{k}}{2p^0(2\pi)^3} \tag{10.40}$$

Therefore,

$$\begin{aligned}
\frac{1}{2p_0}\sum_{\alpha=\bar{K}_0} &= \frac{1}{(2\pi)^3}\int \frac{d^3k}{2p^0} \langle \pi^+(\mathbf{p}')|F_4 + iF_5|\bar{K}^0(\mathbf{k})\rangle\langle \bar{K}^0(\mathbf{k})|F_4 - iF_5|\pi^+(\mathbf{p})\rangle \\
&= \frac{1}{(2\pi)^3}\int \frac{d^3k}{2p^0}\int d^3\mathbf{x}\, d^3\mathbf{x}' \langle \pi^+(\mathbf{p}')|j_4{}^0(\mathbf{x},t) + ij_5'(\mathbf{x},t)|\bar{K}^0(\mathbf{k})\rangle \\
&\qquad \langle \bar{K}^0(\mathbf{k})|j_4{}^0(\mathbf{x}',t) - ij_5{}^0(\mathbf{x}',t)|\pi^+(\mathbf{p})\rangle \\
&= \frac{1}{(2\pi)^3}\int \frac{d^3k}{2p^0}\int d^3\mathbf{x}\, d^3\mathbf{x}'[f_+(q'^2)(k^0 + p'^0) + f_-(q'^2)(k^0 - p'^0)]e^{i(p'^0-k^0)t} \\
&\qquad \times [f_+(q^2)(k^0 + p^0) + f_-(q^2)(k^0 - p^0)]e^{i(k^0-p^0)t}e^{i(\mathbf{k}-\mathbf{p}')\cdot\mathbf{x}}e^{i(\mathbf{p}-\mathbf{k})\cdot\mathbf{x}}
\end{aligned} \tag{10.41}$$

where $q'^2 = (k - p')^2$ and $q^2 = (k - p)^2$.

Now, performing the integrations over $\mathbf{x}$, $\mathbf{x}'$ in (10.41), we obtain

$$\begin{aligned}
\frac{1}{2p_0}\sum_{\alpha=\bar{K}^0} = \frac{1}{(2\pi)^3}\int \Bigg\{&\frac{d^3k}{2p^0}(2\pi)^6\,\delta^3(\mathbf{k}-\mathbf{p}')\,\delta^3(\mathbf{p}-\mathbf{k}) \\
&\times [f_+(q^{02})(k^0 + p'^0) + f_-(q^{02})(k^0 - p'^0)] \\
&\times [f_+(q^{02})(k^0 + p^0) + f_-(q^{02})(k^0 - p^0)]\Bigg\} e^{i(p'^0-p^0)t}
\end{aligned}$$

or

$$\frac{1}{2p_0}\sum_{\alpha=\bar{K}^0} = \frac{(2\pi)^3}{2p^0}\,\delta^3(\mathbf{p}'-\mathbf{p})[f_+(q^{02})(k^0 + p^0) + f_-(q^{02})(k^0 - p^0)]^2 \tag{10.42}$$

Since $k^0 = \sqrt{\mathbf{p}^2 + m_K{}^2}$, $p^0 = \sqrt{\mathbf{p}^2 + m_\pi{}^2}$, and $q^0 = k^0 - p^0$, we obtain in the limit of large p^0

$$\lim \frac{1}{2p_0}\sum_{\alpha=\bar{K}^0} = (2\pi)^3 2p^0\delta^3(\mathbf{p}'-\mathbf{p})f_+(0) \tag{10.43}$$

Now, comparing (10.43) with (10.41), (10.39), and (10.38), we see that $f_+(0) \approx 1$, provided $\sum_{\alpha\neq K_0}$ can in fact be neglected to a good approximation. Thus, we turn our attention to the total hamiltonian H,

$$H = H_0 + \lambda H_m$$

and separate it into two portions: an SU_3-invariant part H_0 and an SU_3-breaking part H_m. The parameter λ is small because otherwise the hadron mass spectrum would not exhibit any recognizable SU_3 symmetry.

Consider the matrix element

$$\langle\pi^+|F_4 + iF_5|\alpha\rangle = \frac{\langle\pi^+|[H, F_4 + iF_5]|\alpha\rangle}{E_\pi - E_\alpha} = \lambda\frac{\langle\pi^+|[H_m, F_4 + iF_5]|\alpha\rangle}{E_\pi - E_\alpha} \tag{10.44}$$

When we calculate sums over products of the form

$$\langle\pi^+|F_4 + iF_5|\alpha\rangle\langle\alpha|F_4 - iF_5|\pi^+\rangle \tag{10.45}$$

only those terms survive for which $p_\alpha = p_\pi$; thus $E_\pi - E_\alpha = \mathcal{O}(m_\pi - m_\alpha)$ for small momentum. If α represents a particle belonging to the same SU_3 multiplet (0^- octet) as π, then $E_\pi - E_\alpha = \mathcal{O}(\lambda)$. In this case the product (10.45) is of order unity from (10.44). However, if α belongs to a different SU_3 multiplet than π, then $E_\pi - E_\alpha$ is of order unity and the product (10.45) is of order λ^2 from (10.44). It follows that

$$\frac{\sum_{\alpha \neq K^0}}{\sum_{\alpha = K^0}} = \mathcal{O}(\lambda^2)$$

Thus to first order in λ we can ignore $\sum_{\alpha \neq K^0}$ and indeed we have the result

$$f_+(0) \approx 1 \tag{10.46}$$

Although the conclusion we have arrived at seems reasonable, we must enter the qualification that the proof is based on the assumption that $E_\pi - E_\alpha$ is small if α is in the same octet as π^-. However, the mass difference between K and π is not very small. Thus it is hard to interpret the foregoing theorem unambiguously.

10.3.6 A Detailed Discussion of K_{l3} Decay

Let us consider the amplitude for K_{l3} decay once more. We have[1]

$$\mathcal{M} = \frac{G}{\sqrt{2}} \sin\Theta[f_+(q^2)(p_K + p_\pi)^\lambda + f_-(q^2)(p_K - p_\pi)^\lambda]\bar{u}_l\gamma_\lambda(1 + \gamma^5)v_\nu \tag{10.47}$$

Since $p_K = p_\pi + p_l + p_{\bar\nu}$, this becomes

$$\mathcal{M} = \frac{G}{\sqrt{2}} \sin\Theta[f_+(q^2)(p_K + p_\pi)^\lambda\bar{u}_l\gamma_\lambda(1 + \gamma^5)v_{\bar\nu} + f_-(q^2)\bar{u}_l(\not{p}_l + \not{p}_{\bar\nu})(1 + \gamma^5)v_{\bar\nu}]$$

[1] The amplitude of (10.47) actually corresponds to $K^- \to \pi^0 l^- \bar\nu$.

or, since

$$\bar{u}_l \not{p}_l = m_l \bar{u}_l$$

and

$$\not{p}_{\bar{\nu}}(1 + \gamma^5)v_{\bar{\nu}} = (1 - \gamma^5)\not{p}_{\bar{\nu}} v_{\bar{\nu}} = 0$$

we obtain

$$\mathcal{M} = \frac{G}{\sqrt{2}} \sin \Theta[f_+(q^2)(p_K + p_\pi)^\lambda \bar{u}_l \gamma_\lambda(1 + \gamma^5)v_{\bar{\nu}} + f_-(q^2)m_l \bar{u}_l(1 + \gamma^5)v_{\bar{\nu}} \tag{10.48}$$

Defining $\xi = f_-(q^2)/f_+(q^2)$, we arrive at

$$\mathcal{M} = \frac{G}{\sqrt{2}} \sin \Theta f_+(q^2)[(p_K + p_\pi)^\lambda \bar{u}_l \gamma_\lambda(1 + \gamma^5)v_{\bar{\nu}} + m_l \xi \bar{u}_l(1 + \gamma^5)v_{\bar{\nu}}] \tag{10.49}$$

$$= \frac{G}{\sqrt{2}} \sin \Theta f_+(q^2)[2p_K{}^\lambda \bar{u}_l \gamma_\lambda(1 + \gamma^5)v_{\bar{\nu}} + (\xi - 1)m_l \bar{u}_l(1 + \gamma^5)v_{\bar{\nu}}] \tag{10.50}$$

In K_{e3} decay, $m_l = m_e \ll m_K$, and so we can safely neglect the term in $(\xi - 1)$ of (10.50). Thus for K_{e3} decay,

$$\mathcal{M}(K_{e3}) = \frac{2G}{\sqrt{2}} \sin \Theta f_+(q^2) p_K{}^\lambda \bar{u}_e \gamma_\lambda(1 + \gamma^5)v_{\bar{\nu}} \tag{10.51}$$

The K_{e3} transition probability is

$$dW = \frac{|\mathcal{M}|^2 \, d^3\mathbf{p}_\pi \, d^3\mathbf{p}_e \, d^3\mathbf{p}_{\bar{\nu}} \, \delta^4(p_K - p_\pi - p_e - p_\nu)}{(2\pi)^5 2E_e 2E_{\bar{\nu}} 2E_\pi 2E_K} \tag{10.52}$$

We next calculate $|\mathcal{M}|^2$:

$$|\mathcal{M}|^2 = 2G^2 \sin^2 \Theta f_+{}^2 p_K{}^\alpha p_K{}^\beta \operatorname{tr}[\Lambda_e \gamma_\alpha(1 + \gamma^5)\Lambda_{\bar{\nu}} \gamma_\beta(1 + \gamma^5)] \tag{10.53}$$

We are interested in a sum over final lepton spin states, and so we write

$$\Lambda_e = \not{p}_e + m_e$$

and

$$\Lambda_{\bar{\nu}} = \not{p}_{\bar{\nu}}$$

Thus the trace in (10.53) becomes

$$\operatorname{tr}[(\not{p}_e + m)\gamma_\alpha(1 + \gamma^5)\not{p}_{\bar{\nu}} \gamma_\beta(1 + \gamma^5)] = 2\operatorname{tr}[\not{p}_e \gamma_\alpha \not{p}_\nu \gamma_\beta(1 + \gamma^5)]$$
$$= 8p_e{}^\rho p_{\bar{\nu}}{}^\sigma \chi_{\rho\alpha\sigma\beta}$$

and

$$8p_K{}^\alpha p_K{}^\beta p_e{}^\rho p_{\bar{\nu}}{}^\sigma \chi_{\rho\alpha\sigma\beta} = 8p_K{}^\alpha p_K{}^\beta p_e{}^\rho p_{\bar{\nu}}{}^\sigma \times [g_{\rho\alpha} g_{\sigma\beta} + g_{\alpha\sigma} g_{\rho\beta} - g_{\rho\sigma} g_{\alpha\beta} - i\varepsilon_{\rho\alpha\sigma\beta}]$$
$$= 8[p_K \cdot p_e \, p_K \cdot p_{\bar{\nu}} + p_K \cdot p_{\bar{\nu}} \, p_K \cdot p_e - p_K{}^2 \, p_e \cdot p_{\bar{\nu}}]$$
$$= 8[2p_K \cdot p_e \, p_K \cdot p_{\bar{\nu}} - p_K{}^2 \, p_e \cdot p_{\bar{\nu}}]$$

and

$$|\overline{\mathcal{M}}|^2 = 16G^2 \sin^2 \Theta f_+{}^2[2p_K \cdot p_e \, p_K \cdot p_{\bar{\nu}} - p_K{}^2 p_e \cdot p_{\bar{\nu}}] \tag{10.54}$$

Here the bar over $|\mathcal{M}|^2$ denotes a sum over e, $\bar{\nu}$ spins.

Using (10.54) and (10.52), we shall now discuss

1 The pion energy spectrum
2 The total decay rate
3 The electron energy spectrum at fixed pion energy

10.3.6.1 The pion energy spectrum Here we have to integrate (10.53) over lepton momenta:

$$dW = \frac{G^2 \sin^2 \Theta f_+{}^2(q^2)\, d^3p_\pi(2p_K{}^\alpha p_K{}^\beta - p_K{}^2 g^{\alpha\beta})I_{\alpha\beta}}{(2\pi)^5 E_K E_\pi} \tag{10.55}$$

where
$$I_{\alpha\beta} = \iint \frac{d^5\mathbf{p}_{\bar{\nu}}\, d^3\,\mathbf{p}_e}{E_e E_{\bar{\nu}}}\, p_{e\alpha}\, p_{\bar{\nu}\beta}\, \delta^4(p_e + p_{\bar{\nu}} - q) \tag{10.56}$$

This integral was encountered in Chap. 2 in the discussion of muon decay. The value of $I_{\alpha\beta}$ was found to be

$$I_{\alpha\beta} = \frac{\pi}{6}(q^2 g_{\alpha\beta} + 2q_\alpha q_\beta) \tag{10.57}$$

Inserting (10.57) in (10.55) and integrating over all directions of pion emission, we find

$$dW = \frac{4\pi^2 G^2 \sin^2 \Theta f_+{}^2 |\mathbf{p}_\pi|^2\, d\mathbf{p}_\pi}{6(2\pi)^5 E_K E_\pi}[2p_K{}^\alpha p_K{}^\beta - p_K{}^2 g^{\alpha\beta}][q^2 g_{\alpha\beta} + 2q_\alpha q_\beta] \tag{10.58}$$

Multiplying the factors in brackets we obtain

$$\begin{aligned}(2p_K{}^\alpha\, p_K{}^\beta - p_K{}^2 g^{\alpha\beta})(q^2 g_{\alpha\beta} + 2q_\alpha q_\beta) &= 4[(p_K\cdot q)^2 - p_K{}^2 q^2] \\ &= 4m_K{}^2|\mathbf{p}_\pi|^2\end{aligned} \tag{10.59}$$

in the rest frame of *K*. Inserting (10.59) in (10.58), we finally have

$$dW = \frac{m_K G^2 f_+{}^2 \sin^2 \Theta |\mathbf{p}_\pi|^3\, dE_\pi}{12\pi^3} \tag{10.60}$$

Formula (10.60) may be compared with experimental observations of the energy spectrum to determine the q^2 dependence of f_+. One expects that f_+ will have the following approximate form:

$$f_+(q^2) \approx f_+(0)\left[1 + \lambda\left(\frac{q}{m_\pi}\right)^2\right] \tag{10.61}$$

the variation with q^2 arising because of virtual strong interactions. An average of recent experiments yields the values (Rub 69)

$$\begin{aligned}\lambda &= +0.029 \pm 0.010 \qquad \text{for } K^+ \to \pi^0 e^+ \nu \\ \lambda &= +0.019 \pm 0.008 \qquad \text{for } K_L{}^0 \to \pi^\pm e^\mp \nu\end{aligned} \tag{10.62}$$

In other words, the form factor f_+ varies quite slowly with q^2.

10.3.6.2 The total decay rate Integrating (10.60) over the pion energy with a definite assumed value for λ [Eq. (10.62)], we obtain an expression for the total decay rate in terms of $f_+{}^2(0)\sin^2\Theta$ which can be compared with the known total decay rate and also with the experimental and theoretical rates for $\pi_{e3}{}^+$ decay. If we assume $f_+(0) = 1$, which is reasonable because of the Ademollo–Gatto theorem as discussed above, we obtain a new value of the Cabibbo angle:

$$\Theta_V{}^M = 0.235 \pm 0.019 \qquad (10.63)$$

Here, the superscript M refers to *meson* and the subscript V is inserted because the K_{e3} decay is pure vector. This result, which includes radiative corrections (see, e.g., Gin 66, 67, 68), should be compared with the value of Θ obtained from the pure axial decays $K_{\mu 2}$ and $\pi_{\mu 2}$ discussed in Chap. 4

$$\Theta_A{}^M = 0.2688 \pm 0.0008 \qquad (10.64)$$

and the values of Θ obtained from the baryon semileptonic decays discussed in Chap. 8, which are well accounted for by the same Cabibbo angle for vector and axial vector amplitudes:

$$\Theta_V{}^B = \Theta_A{}^B = 0.235 \pm 0.006 \qquad (10.65)$$

We should also consider the value of Θ obtained from the $0^+ \to 0^+$ nuclear beta decays:

$$\Theta^\beta = 0.188 \pm 0.007$$

The discrepancies between Cabibbo angles as determined by these various means presumably arises from SU_3 symmetry breaking, but unfortunately we still lack the means to give fully convincing explanations for these small differences or indeed to explain the Cabibbo angle itself in terms of more fundamental concepts.[1]

10.3.6.3 The electron energy spectrum at fixed pion energy We have assumed that the hadronic weak current $\mathfrak{J}_1$ is a combination of vector plus axial vector and then that only the vector portion is operative in K_{e3} decay. A more general formulation, still consistent with proper Lorentz invariance, could

[1] An interesting speculation is offered by Fischbach et al. (Fis 71a, 71b) who propose that the pseudoscalar mesons are described by a Kemmer equation (see, e.g., Kem 39) rather than the well-known Klein–Gordon equation. It is shown that the Kemmer formulation provides an explanation for the large negative value of ξ (see following section) and it also yields a new value for $\Theta_V{}^M$, viz.,

$$\Theta_V{}^M(\text{Kemmer}) = 0.192 \pm 0.016$$

in agreement with that value obtained for the $0^+ \to 0^+$ nuclear beta decays.

include scalar and tensor as well as vector couplings. The amplitude would then take the form

$$\mathcal{M} \to 2m_K f_S \bar{u}_e(1+\gamma^5)v_{\nu e} + f_+(p_K+p_\pi)^\alpha \bar{u}_e \gamma_\alpha(1+\gamma^5)v_{\nu e} + \frac{2}{m_K} f_T p_K{}^\alpha p_K{}^\beta \bar{u}_e \sigma_{\alpha\beta}(1+\gamma^5)v_{\nu e} \tag{10.66}$$

Now the form factors f_+, f_S, and f_T depend only on the momentum transfer q^2. Moreover,

$$q = p_K - p_\pi = p_l + p_\nu$$

Thus

$$q^2 = (p_K - p_\pi)^2 = m_K{}^2 + m_\pi{}^2 - 2p_K \cdot p_\pi$$

In the K rest frame this becomes

$$q^2 = m_K{}^2 + m_\pi{}^2 - 2m_K E_\pi \tag{10.67}$$

so that q^2 depends only on the pion energy. Therefore a measurement of the electron energy spectrum at fixed pion energy is the same as a measurement at fixed q^2; the result hence cannot depend on the variation of f_+, f_S, or f_T with q^2.

To calculate the electron spectrum at fixed pion energy we must integrate the transition probability over neutrino momentum. Assuming the vector coupling, we obtain from (10.52) and (10.54)

$$dW = \frac{G^2 \sin^2 \Theta}{(2\pi)^5 E_K E_\pi E_e} f_+{}^2 \, d^3p_\pi \, d^3p_e \, p_{e\alpha}(2p_{K\alpha}p_K{}^\beta - p_K{}^2 g^{\alpha\beta}) \times \int \frac{d^3p_{\bar{\nu}}}{E_{\bar{\nu}}} p_{\bar{\nu}\beta} \, \delta(E_K - E_\pi - E_e - E_{\bar{\nu}})\delta^3(\mathbf{p}_K - \mathbf{p}_\pi - \mathbf{p}_e - \mathbf{p}_{\bar{\nu}}) \tag{10.68}$$

Since

$$\int d^3\mathbf{p}_{\bar{\nu}} \delta^3(\mathbf{p}_K - \mathbf{p}_\pi - \mathbf{p}_e - \mathbf{p}_{\bar{\nu}}) = 1$$

we obtain in the K rest frame

$$\begin{aligned} dW &= \frac{G^2 \sin^2 \Theta}{(2\pi)^5 m_K E_\pi E_e E_{\bar{\nu}}} f_+{}^2 \, d^3\mathbf{p}_\pi \, d^3\mathbf{p}_e \, \delta(m_K - E_\pi - E_e - E_{\bar{\nu}}) \\ &\qquad \times [2E_e E_{\bar{\nu}} m_K{}^2 - m_K{}^2(E_e E_{\bar{\nu}} - \mathbf{p}_e \cdot \mathbf{p}_{\bar{\nu}})] \\ &= \frac{G^2 \sin^2 \Theta m_K f_+{}^2 \, d^3\mathbf{p}_\pi \, d^3\mathbf{p}_e}{(2\pi)^5 E_\pi E_e E_{\bar{\nu}}} [E_e E_{\bar{\nu}} + \mathbf{p}_e \cdot \mathbf{p}_{\bar{\nu}}] \, \delta(E_e + E_{\bar{\nu}} + E_\pi - m_K) \end{aligned} \tag{10.69}$$

Now $$\mathbf{p}_e + \mathbf{p}_{\bar{\nu}} = -\mathbf{p}_\pi$$

Thus $$|\mathbf{p}_e|^2 + |\mathbf{p}_{\bar{\nu}}|^2 + 2\mathbf{p}_e \cdot \mathbf{p}_\nu = |\mathbf{p}_\pi|^2$$

or $$\mathbf{p}_e \cdot \mathbf{p}_{\bar{\nu}} = \tfrac{1}{2}(|\mathbf{p}_\pi|^2 - |\mathbf{p}_e|^2 - |\mathbf{p}_\nu|^2)$$

Then, neglecting the electron mass, we have

$$\mathbf{p}_e \cdot \mathbf{p}_{\bar{\nu}} = \frac{|\mathbf{p}_\pi|^2 - E_e^2 - E_{\bar{\nu}}^2}{2}$$

and so $$E_e E_{\bar{\nu}} + \mathbf{p}_e \cdot \mathbf{p}_{\bar{\nu}} = \tfrac{1}{2}[|\mathbf{p}_\pi|^2 - (E_{\bar{\nu}} - E_e)^2]$$

However, $$E_{\bar{\nu}} = m_K - E_\pi - E_e$$

hence $$E_e E_{\bar{\nu}} + \mathbf{p}_e \cdot \mathbf{p}_\nu = \tfrac{1}{2}[|\mathbf{p}_\pi|^2 - (m_K - E_\pi - 2E_e)^2] \tag{10.70}$$

It remains to integrate over electron and pion solid angles. We have

$$\frac{d^3\mathbf{p}_e\, d^3\mathbf{p}_\pi}{E_\pi E_e E_{\bar{\nu}}} = \frac{E_e\, dE_e |\mathbf{p}_\pi| d\mathbf{p}_\pi}{E_{\bar{\nu}}}\, d\Omega_e\, d\Omega_{e\pi}$$

Thus $$\frac{E_e\, dE_e |\mathbf{p}_\pi|\, dE_\pi}{E_{\bar{\nu}}} \int d\Omega_e\, d\Omega_\pi$$

$$\times\, \delta(E_e + E_\pi + \sqrt{|\mathbf{p}_e|^2 + |\mathbf{p}_\pi|^2 + 2|\mathbf{p}_e|\,|\mathbf{p}_\pi| \cos\theta_{e\pi}} - m_K)$$

$$= \frac{8\pi^2 E_e\, dE_e |\mathbf{p}_\pi|\, dE_\pi}{E_{\bar{\nu}}} \int_0^\pi \sin\theta_{e\pi}\, d\theta_{e\pi}$$

$$\times\, \delta(E_e + E + \sqrt{|\mathbf{p}_e|^2 + |\mathbf{p}_\pi|^2 + 2|\mathbf{p}_e|\,|\mathbf{p}_\pi| \cos\theta_{e\pi}} - m_K)$$

$$= \frac{8\pi^2 E_e\, dE_e |\mathbf{p}_\pi|\, dE_\pi}{E_{\bar{\nu}}} \frac{\sin\theta_{e\pi}}{2|\mathbf{p}_e|\,|\mathbf{p}_\pi| \sin\theta_{e\pi}/E_{\bar{\nu}}}$$

Therefore $$\int \frac{d^3\mathbf{p}_e\, d^3\mathbf{p}_\pi}{E_\pi E_e E_{\bar{\nu}}}\, \delta(E_e + E_{\bar{\nu}} + E_\pi - m_K) = 8\pi^2\, dE_e\, dE_\pi \tag{10.71}$$

Now, substituting (10.71) and (10.70) into (10.69), we obtain

$$dW = \frac{G^2 \sin^2\Theta m_K f_+^2}{8\pi^3}\, [|\mathbf{p}_\pi|^2 - (m_K - E_\pi - 2E_e)^2]\, dE_\pi\, dE_e \tag{10.72}$$

where $$\frac{m_K - E_\pi - |\mathbf{p}_\pi|}{2} \le E_e \le \frac{m_K - E_\pi + |\mathbf{p}_\pi|}{2}$$

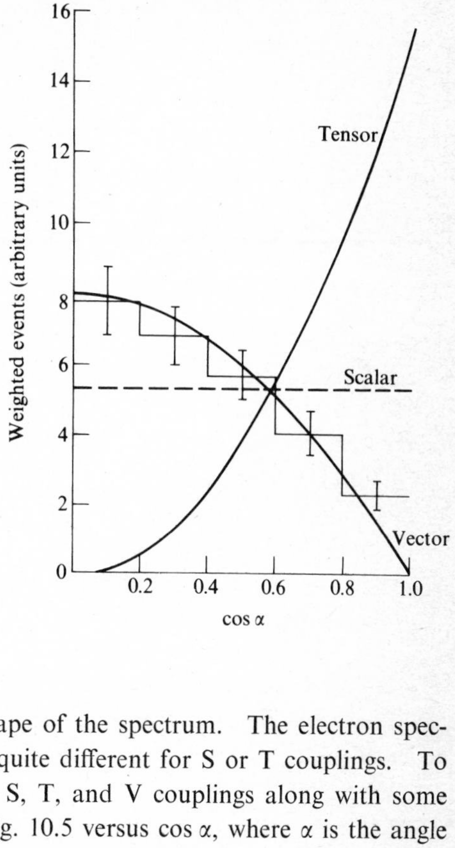

FIGURE 10.5
Measurements of the electron spectrum at fixed pion energy plotted as a function of cos α. Theoretical spectra for S, V, and T couplings are also shown. [*From G. E. Kalmus et al., Univ. of Calif. Lawrence Rad. Lab. Rept., UCRL-11553, 1964.*]

Equation (10.72) reveals the shape of the spectrum. The electron spectrum at fixed pion energy would be quite different for S or T couplings. To show this, the calculated spectra for S, T, and V couplings along with some experimental results are plotted in Fig. 10.5 versus cos α, where α is the angle between π and e in the $e\nu$ CM frame. (We can easily show that for fixed pion energy the electron energy depends uniquely on cos α:

$$E_e = a + b \cos \alpha$$

where a and b are constants.) As shown in Fig. 10.5, the experimental results confirm very well the assumption of vector coupling. Quantitatively we may place an upper limit on the ratios $|f_S/f_+|$ and $|f_T/f_+|$ as defined in Eq. (10.66) from the results of measurements of the Dalitz plot for K_{l3} decay as well as from the electron spectrum at fixed pion energy. A combination of world data gives (Rub 69)

$$\left|\frac{f_S}{f_+}\right| < 0.2 \qquad \left|\frac{f_T}{f_+}\right| < 0.58 \qquad (10.73)$$

10.3.7 $K_{\mu 3}$ Decay

We return to the general form (10.49) for the $K_{\mu 3}$ amplitude:

$$\mathcal{M} = \frac{G}{\sqrt{2}} \sin \Theta f_+(q^2)[(p_K + p_\pi)^\lambda \bar{u}_\mu \gamma_\lambda (1 + \gamma^5) v_{\bar{\nu}} + m_\mu \xi \bar{u}_\mu (1 + \gamma^5) v_{\bar{\nu}}] \qquad (10.74)$$

In the present case it is no longer legitimate to neglect the lepton mass (now m_μ) in comparison to the kaon mass. Consequently we have to take into account the *induced scalar* term in $m_\mu \xi$. The q^2 dependence of ξ is generally expressed as follows:

$$\xi(q^2) = \xi(0) + \frac{\Lambda q^2}{m_\pi^2} \qquad (10.75)$$

where $\xi(0)$ and Λ are constants to be determined by experiment. (If SU_3 symmetry were perfect, ξ would be zero.)

Unfortunately, various experimental determinations of ξ yield conflicting results; thus $\xi(0)$ and Λ are still highly uncertain. In principle, measurements of muon polarization offer an effective means of determining ξ. Qualitatively, if ξ were zero the decay amplitude would be pure vector and the μ^+ would be right-handed in $K^+ \to \pi^0 \mu^+ \nu$ or $K_L^0 \to \pi^- \mu^+ \nu$ decays. With nonzero ξ we also have an effective scalar amplitude which alone would produce left-handed μ^+. The combination of V and S amplitudes leads to a complicated interference in the polarization, which depends on ξ. Quantitatively one finds the positive-muon polarization vector **P** to have unit magnitude (Cab 64a, 64b, 65) and the following angular dependence:

$$\mathbf{P} = \frac{\mathbf{A}}{|\mathbf{A}|}$$

$$\mathbf{A} = a_1(\xi)\mathbf{p}_\mu - a_2(\xi)\left\{\mathbf{p}_\mu\left[\frac{(m_K - E_\pi)}{m_\mu} + \frac{(\mathbf{p}_\pi \cdot \mathbf{p}_\mu)(E_\mu - m_\mu)}{m_\mu |\mathbf{p}_\mu|^2}\right] + \mathbf{p}_\pi\right\}$$
$$+ m_K \operatorname{Im} \xi(q^2)(\mathbf{p}_\pi \times \mathbf{p}_\mu) \qquad (10.76)$$

where
$$a_1(\xi) = \frac{2m_K^2}{m_\mu}[E_\nu + \operatorname{Re} b(q^2)(E_\pi^* - E_\pi)]$$

$$a_2(\xi) = m_K^2 + 2 \operatorname{Re} b(q^2) m_K E_\mu + |b(q^2)|^2 m_\mu^2$$

$$b(q^2) = \tfrac{1}{2}[\xi(q^2) - 1]$$

and
$$E_\pi^* = \frac{m_K^2 + m_\pi^2 - m_\mu^2}{2m_K}$$

If time-reversal invariance holds, ξ is real; and the polarization vector lies entirely in the $\mathbf{p}_\pi$, $\mathbf{p}_\mu$ plane and is defined completely by the angle it makes with the muon momentum.

Precise experiments fail to detect a transverse polarization (You 67)

$$P_\perp = (0.25 \pm 1.25) \times 10^{-2} \qquad (10.77)$$

thus leading to the conclusion that

$$\text{Im}\,\xi = -(1.25 \pm 6.6) \times 10^{-2} \qquad (10.78)$$

in agreement with T invariance. The results of a number of measurements of the longitudinal polarization are listed in Table 10.2. These yield values of ξ in moderately good agreement.

Another method for the determination of ξ consists in analysis of the density of $K_{\mu 3}$ events in the Dalitz plot, which is sensitive to ξ mainly in the region of low pion energy (where the Dalitz plot is thinly populated). Unfortunately the experimental results (Bet 69, Bas 68) are not in agreement, and no definite conclusions about ξ are forthcoming.

Finally we mention measurements of the branching ratio

$$R = \frac{K \to \pi\mu\nu}{K \to \pi e \nu}$$

which in principle yields ξ, provided $e\mu$ universality is valid (in which case f_+ for $K_{\mu 3}$ and K_{e3} decays is the same). One finds that

$$R = 0.649 + 0.127\xi + 0.00193\xi^2 \qquad (10.79)$$

Table 10.2 VALUES OF ξ AS DETERMINED FROM LONGITUDINAL MUON POLARIZATION

Decay	ξ	Reference
$K_L{}^0 \to \pi^-\mu^+\nu$	-1.2 ± 0.5	Aue 66
$K_L{}^0 \to \pi^-\mu^+\nu$	-1.1 ± 0.5	Abr 66
$K_L{}^0 \to \pi^-\mu^+\nu$	$-1.81\,{}^{+0.5}_{-0.26}$	Hel 68
$K^+ \to \pi^0\mu^+\nu$	-1.0 ± 0.3	Bet 68†
$K^+ \to \pi^0\mu^+\nu$	-0.95 ± 0.3	Cut 68†

† In these measurements ξ was determined at several values of $q^2/m_\pi{}^2$ $(1 < q^2/m_\pi{}^2 < 7)$ with the conclusion that $\xi \approx$ constant.

If we assume that $\xi = -0.98 \pm 0.20$ for $K^+ \to \pi^0 \mu^+ \nu$ decay and $\xi = -1.45 \pm 0.26$ for $K_L{}^0 \to \pi^- \mu^+ \nu$ decay, from the polarization measurements of Table 10.2, we predict from (10.79) that

$$R = 0.53 \pm 0.02 \quad \text{for } K^+$$

$$R = 0.50 \pm 0.025 \quad \text{for } K_L{}^0 \qquad (10.80)$$

The experimental measurements of R (Sti 68) yield values

$$R \approx 0.7 \qquad (10.81)$$

with the exception of one result (Aac 68) which is in agreement with (10.80). The discrepancy between the bulk of the R measurements (10.81) and the values of (10.80) is difficult to understand.

10.4 THE SOFT-PION THEOREM

We now establish connections between the form factors for K_{l3} and K_{l4} decays (the latter to be discussed below) by means of the so-called *soft-pion* relation of current algebra.[1] Here we consider a matrix element of the following form:

$$M = \langle \beta, \pi^i | Q | \alpha \rangle \qquad (10.82)$$

where Q is some local operator, $|\alpha\rangle$ is an initial state, and $|\beta, \pi^i\rangle$ is a final state consisting of zero or more particles β and, in addition, a pion. Index i refers to the pion charge state

$$i = 1 + i2: \quad \pi^-$$

$$i = 1 - i2: \quad \pi^+$$

$$i = 3 \quad : \quad \pi^0$$

For example, in $K_{l3}{}^+$ decay, α represents K^+, β represents the vacuum state, and Q represents $\mathcal{V}_1{}^\lambda$; M is then the matrix element of the hadronic portion of the weak current.

We may demonstrate the following useful theorem:

$$\lim_{q \to 0} \frac{f_\pi}{\sqrt{2}} \langle \beta, \pi^i | Q | \alpha \rangle = i \cos \Theta \langle \beta | [F_i{}^5, Q] | \alpha \rangle$$

[1] It is possible to establish a relationship between f_+, f_-, f_π, and f_K, using the soft-pion technique. Here f_π and f_K are the decay constants for π_{l2} and K_{l2} decays, respectively, and $f_K/f_\pi = \tan \Theta_A{}^M$. Also, f_+ and f_- are the K_{l3} form factors, which must be evaluated in the unphysical limit of zero pion mass. Recent theoretical attempts to calculate the K_{l3} form factors are summarized in a review by Callan (Cal 69).

where q is the pion four-momentum. Without giving much detail, we now merely outline the proof since it depends on some knowledge of formal quantum field theory. First we write

$$M = \langle \beta, \pi^i | Q | \alpha \rangle$$
$$= \langle \beta | a_i(q) Q | \alpha \rangle$$

where $a_i(q)$ is the annihilation operator for the pion field corresponding to a pion of four-momentum q. Thus

$$M = \langle \beta | Q a_i(q) | \alpha \rangle + \langle \beta | [a_i(q), Q] | \alpha \rangle \qquad (10.83)$$

The first term on the right-hand side of (10.83) is zero since the state $|\alpha\rangle$, by assumption, contains no pions. To examine the second term, we write the pion field as

$$\phi_i(\mathbf{x}, t) \equiv \int d^3\mathbf{q} [a_i(q) e^{-iq\cdot x} + a_i^\dagger(q) e^{iq\cdot x}]$$

where $$a_i(q) = i \int d^3\mathbf{x}\, e^{iq\cdot x} \overleftrightarrow{\partial_0} \phi_i(x)$$

It is then possible to show that M can be expressed as a *retarded commutator:*

$$M = +i \int d^4x\, e^{-iq\cdot x} (\Box^2 + m_\pi^2) \{ \langle \beta | [\phi_i(x), Q] | \alpha \rangle \Theta(x_0) \} \qquad (10.84)$$

where
$$\Theta(x_0) = 1 \qquad x_0 > 0$$
$$\Theta(x_0) = 0 \qquad x_0 < 0$$

and
$$\Box^2 = \frac{\partial^2}{\partial t^2} - \nabla^2$$

Now we can integrate M by parts, so that (10.84) becomes

$$M = +i \int d^4x\, e^{-iq\cdot x} (-q^2 + m_\pi^2) \langle \beta | [\phi_i(x), Q] | \alpha \rangle \Theta(x_0) \qquad (10.85)$$

Let us also introduce a new quantity T^μ, defined by the following equation:

$$T^\mu = \int d^4x\, e^{-iq\cdot x} \langle \beta | [g_i^\mu(x), Q] | \alpha \rangle \Theta(x_0) \qquad (10.86)$$

where the $g_i^\mu(x)$ are the members of the axial vector current octet. From (10.86) we have by partial integration

$$q_\mu T^\mu = \int d^4x \left(i \frac{\partial}{\partial x^\mu} e^{-iq\cdot x} \right) \langle \beta | [g_i^\mu(x), Q] | \alpha \rangle \Theta(x_0)$$
$$= -i \int d^4x e^{-iq\cdot x} \Theta(x_0) \left\langle \beta \left| \left[\frac{\partial}{\partial x^\mu} g_i^\mu(x), Q \right] \right| \alpha \right\rangle$$
$$-i \int d^4x e^{-iq\cdot x} \langle \beta | [g_1^\mu(x), Q] | \alpha \rangle \frac{\partial}{\partial x^\mu} \Theta(x_0) \qquad (10.87)$$

Since
$$\frac{\partial}{\partial x^\mu}\Theta(x_0) = \frac{\partial}{\partial x^0}\Theta(x_0) = \delta(x_0)$$

the second term on the right-hand side of (10.87) becomes

$$-i\int d^4x e^{-iq\cdot x}\langle\beta|[g_i{}^0(x), Q]|\alpha\rangle\delta(x_0) = -i\int d^3x\, e^{iq\cdot x}\langle\beta|[g_i{}^0(\mathbf{x}, 0), Q]|\alpha\rangle \tag{10.88}$$

Now let us take the limit of $q_\mu T^\mu$ as $q \to 0$. In this case, if T^μ is nonsingular at $q = 0$, the left-hand side of (10.87) becomes zero and we have from (10.88)

$$i\int d^4x\Theta(x_0)\left\langle\beta\left|\left[\frac{\partial}{\partial x^\mu}g_i{}^\mu(x), Q\right]\right|\alpha\right\rangle = -i\int d^3x\langle\beta|[g_i{}^0(\mathbf{x}, 0), Q]|\alpha\rangle$$
$$= -i\langle\beta|[F_i{}^5, Q]|\alpha\rangle \tag{10.89}$$

Now consider $\partial/\partial x^\mu g_i{}^\mu(x)$. We have

$$A_0{}^\mu(x) = \cos\Theta[g_1{}^\mu(x) - ig_2{}^\mu(x)] \tag{10.90a}$$

and
$$A_0{}^{\mu\dagger}(x) = \cos\Theta[g_1{}^\mu(x) + ig_2{}^\mu(x)] \tag{10.90b}$$

From the PCAC relation (8.9) we have

$$\frac{\partial}{\partial x^\mu}A_0{}^{\mu\dagger}(x) = a(\phi_1 + i\phi_2) \tag{10.91}$$

Thus
$$\cos\Theta\frac{\partial}{\partial x^\mu}(g_1{}^\mu \pm ig_2{}^\mu) = a(\phi_1 \pm i\phi_2)$$

and we assume this is also true for index $i = 3$, or

$$\cos\Theta\frac{\partial}{\partial x^\mu}g_i{}^\mu(x) = a\phi_i(x) \tag{10.92}$$

in general. Now, substituting (10.92) in the left-hand side of (10.89), we obtain

$$a\int d^4x\Theta(x_0)\langle\beta|[\phi_i(x), Q]|\alpha\rangle = -\langle\beta|[F_i{}^5, Q]|\alpha\rangle\cos\Theta \tag{10.93}$$

However, in the limit as $q \to 0$, Eq. (10.85) becomes

$$\lim_{q\to 0} M = +im_\pi{}^2\int d^4x\,\Theta(x_0)\langle\beta|[\phi_i(x), Q]|\alpha\rangle \tag{10.94}$$

Comparing (10.93) and (10.94) and recalling that

$$a = \frac{m_\pi{}^2 f_\pi}{\sqrt{2}}$$

we finally obtain

$$\lim_{q\to 0} \frac{f_\pi}{\sqrt{2}} \langle \beta, \pi^i | Q | \alpha \rangle = i \cos \Theta \langle \beta | [F_i^5, Q] | \alpha \rangle \qquad (10.95)$$

Equation (10.95) is useful because a matrix element between two states α and $\beta + \pi$ is reduced, in the limit of vanishing pion energy-momentum, to the matrix element of $[F_i^5, Q]$ between α and β. By successive applications of the theorem, we can reduce the number of pions appearing in the final state from n to $n-1$, to $n-2$, to ..., to zero. Of course, the theorem is useful only if the unphysical limit $q \to 0$ (the soft-pion limit) is a reasonably accurate extrapolation. The result (10.95) will be applied to the K_{l4} decays, which we now consider.

10.5 K_{l4} DECAYS

Consider the weak transitions

$$K^+ \to \pi^+\pi^- e^+ \nu \qquad (K_{e4}(e^+)) \qquad (10.96)$$

$$K^+ \to \pi^+\pi^+ e^- \bar{\nu} \qquad (K_{e4}(e^-)) \qquad (10.97)$$

The former satisfies $\Delta S = \Delta Q$; its rate has been established at $(2.60 \pm 0.30) \times 10^3\ s^{-1}$ on the basis of several hundred observed events (Ely 68). The decay $K_{e4}(e^-)$ is forbidden ($\Delta S = -\Delta Q$) and has never been observed; an upper limit of 7×10^{-7} has been established on the branching ratio.

Earlier [Eqs. (4.25) and (4.26)] we wrote down possible general forms for the vector and axial vector hadronic amplitudes for a transition

$$\text{Mesons} \to 2 \text{ mesons} + \text{lepton pair}$$

These are applicable to K_{l4} decay, but it is more convenient to change variables and write

$$\mathcal{M} = \langle \pi^+\pi^- | \mathcal{V}_1^\lambda + \mathcal{A}_1^\lambda | K^+ \rangle$$

with

$$\langle \pi^+\pi^- | \mathcal{A}_1^\lambda | K^+ \rangle = \sin\Theta[a_1(p_+^\lambda + p_-^\lambda) + a_2(p_+^\lambda - p_-^\lambda) + a_3(p_K^\lambda - p_+^\lambda - p_-^\lambda)] \qquad (10.98a)$$

and

$$\langle \pi^+\pi^- | \mathcal{V}_1^\lambda | K^+ \rangle = \sin\Theta[a_4 \varepsilon^{\lambda\alpha\beta\gamma} p_{K\alpha} p_{+\beta} p_{-\gamma} \qquad (10.98b)$$

where p_+, p_-, and p_K are the four-momenta of π^+, π^-, and K^+, while a_1, a_2, a_3, and a_4 are form factors depending on the kaon and pion masses and the other kinematic invariants which can be constructed from p_K, p_+, and p_-.

We apply Eq. (10.95) to K_{l4} decay first by taking the limit as the energy and momentum of $\pi^+ \to 0$. Then

$$\lim_{p_+ \to 0} \frac{f_\pi}{\sqrt{2}} \langle \pi^+ \pi^- | \mathcal{V}_1{}^\lambda + \mathcal{A}_1{}^\lambda | K^+ \rangle = i \langle \pi^- | [F_-{}^5, \mathcal{V}_1{}^\lambda + \mathcal{A}_1{}^\lambda] | K^+ \rangle$$

However, $\quad [F_-{}^5, \mathcal{V}_1{}^\lambda + \mathcal{A}_1{}^\lambda] = [F_-{}^5, \sin\Theta(j_4{}^\lambda - ij_5{}^\lambda + g_4{}^\lambda - ig_5{}^\lambda)] = 0$

[This can be seen from Eqs. (7.87) and (7.88) and Table 7.1.] Thus, from (10.98), in the limit as $p_+{}^\lambda \to 0$, we have

$$a_1 p_-{}^\lambda - a_2 p_-{}^\lambda + a_3(p_K{}^\lambda - p_-{}^\lambda) = 0$$

or

$$a_1 = a_2 \qquad (10.99)$$

and

$$a_3 = 0 \qquad (10.100)$$

The experimental data are in agreement with (10.99). Relation (10.100) has not been tested adequately.

Next we take the limit as $p_- \to 0$:

$$\lim_{p_- \to 0} \frac{f_\pi}{\sqrt{2}} \langle \pi^+ \pi^- | \mathcal{V}_1{}^\lambda + \mathcal{A}_1{}^\lambda | K^+ \rangle$$

$$= i \langle \pi^+ | [F_+{}^5, \sin\Theta(j_4 - ij_5 + g_4 - ig_5)] | K^+ \rangle \qquad (10.101)$$

The left-hand side of (10.101) is

$$\frac{f_\pi}{\sqrt{2}} \sin\Theta[a_1 p_+{}^\lambda + a_2 p_+{}^\lambda + a_3(p_K{}^\lambda - p_+{}^\lambda)]$$

while the right-hand side is

$$i \sin\Theta \langle \pi^+ | [F_+{}^5, j_4{}^\lambda - ij_5{}^\lambda + g_4{}^\lambda - ig_5{}^\lambda] | K^+ \rangle$$
$$= i \sin\Theta \langle \pi^+ | j_6{}^\lambda - ij_7{}^\lambda + g_6{}^\lambda - ig_7{}^\lambda | K^+ \rangle$$
$$= i\sqrt{2} \sin\Theta \langle \pi^0 | j_4{}^\lambda - ij_5{}^\lambda | K^+ \rangle = i\sqrt{2} \sin\theta [f_+(p_K{}^\lambda + p_\pi{}^\lambda) + f_-(p_K{}^\lambda - p_\pi{}^\lambda)]$$

Thus (10.103) yields in the limit as $p_- \to 0$

$$\left| \frac{f_\pi}{2} a_3 \right| = (f_+ + f_-) \qquad (10.102)$$

and

$$\left| \frac{f_\pi}{4} (a_1 + a_2) \right| = f_+ \qquad (10.103)$$

Note that (10.102) and (10.103) give very different conditions on a_3 as $p_+ \to 0$ or $p_- \to 0$, but these are not necessarily in contradiction.

For the purpose of comparison with experiment, we neglect the term in a_3 (10.100) and also the vector matrix element proportional to a_4, and set $a_1 = a_2$ [Eq. (10.99)]. Equation (10.103) then predicts that

$$a_1 = a_2 = \left| \frac{2f_+}{f_\pi} \right| \tag{10.104}$$

A straightforward calculation then yields the prediction for the total decay rate

$$\Gamma(K^+ \to \pi^+ \pi^- e^+ \nu_e) = 2.5 \times 10^3 \text{ s}^{-1} \tag{10.105}$$

which is in good agreement with the experimental rate.

10.6 DO NEUTRAL WEAK CURRENTS EXIST?

Until now we have assumed that the weak currents are *charged*, i.e., that for $j_e, j_\mu, \mathfrak{J}_0$, and $\mathfrak{J}_1$, $\Delta Q = -1$. What evidence is there for or against the existence of neutral currents for which $\Delta Q = 0$?‡ In purely leptonic weak interactions the transitions

$$\mu^\pm \to e^\pm \gamma$$

and

$$\mu^- \to e^+ e^- e^+$$

are forbidden by the selection rule *no neutral currents* and also by lepton conservation. Neutral currents conceivably could manifest themselves in the weak coupling of hadrons to leptons through such decays as

$$K^+ \to \pi^+ \pi^0 e^+ e^-$$

$$K^+ \to \pi^+ \nu \bar{\nu}$$

$$K^+ \to \pi^+ \mu^+ \mu^- \qquad \text{or} \qquad \pi^+ e^+ e^-$$

and

$$K_L{}^0 \to \mu^+ \mu^- \tag{10.106}$$

$$K_L{}^0 \to e^+ e^- \tag{10.107}$$

As it happens, an experimental search for the last-two-named decays provides the smallest upper limits for the neutral-current hadron-lepton weak coupling (Fri 71). As recorded in Table 10.1, the branching ratios for $K_L{}^0 \to \mu^+ \mu^-$ and $e^+ e^-$ are each less than 1.6×10^{-9}.

Actually, this result is somewhat surprising in view of the fact that $K_L{}^0 \to \mu^+ \mu^-$ ought to proceed at a small but finite rate because of the existence

‡ The reader will recall that the $|\Delta I| = \frac{1}{2}$ rule can be "explained" by invoking either neutral currents or *octet dominance* (Chap. 9).

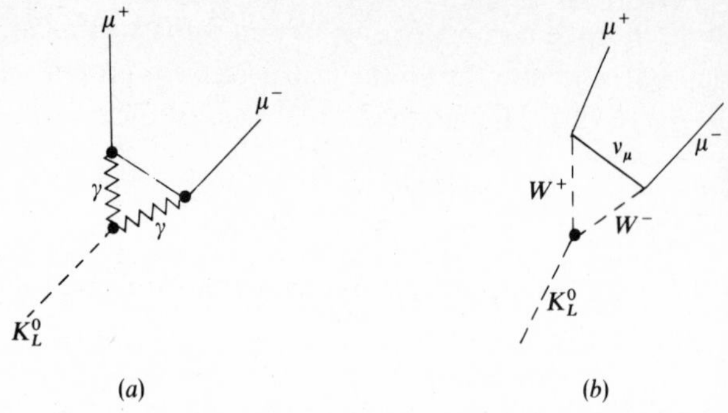

FIGURE 10.6

of a known rate for $K_L{}^0 \to 2\gamma$. Thus, as illustrated in Fig. 10.6*a*, the two virtual γ rays emitted by $K_L{}^0$ convert to a $\mu^+\mu^-$ pair in a way which can be calculated from standard quantum electrodynamics. The result of a recent calculation based on rather general arguments (Qui 68) is that the branching ratio

$$\frac{K_L{}^0 \to \mu^+\mu^-}{K_L{}^0 \to \text{all}}$$

must be *greater* than 5×10^{-9} because of this diagram, provided that no other intermediate states (for example, $2\pi\gamma$, 3π) reduce the amplitude by destructive interference.

The transition $K_L{}^0 \to \mu^+\mu^-$ might also proceed via the *second-order* weak interaction, as shown in Fig. 10.6*b*, where an intermediate boson is assumed to mediate the weak interaction. Actually, the amplitude corresponding to this diagram is logarithmically divergent, unless one imposes an upper cutoff Λ on the four-momentum of the internal loop in the diagram. One finds that the theoretical ratio R, defined by

$$R = \frac{\Gamma(K_L{}^0 \to \mu^+\mu^-)}{\Gamma(K_L{}^0 \to \gamma\gamma)} \qquad (10.108)$$

is given by $R = [G\Lambda^2/(4\pi)^2]^2$ (Iof 68, Moh 68). The experimental upper limit then yields $\Lambda < 5$ GeV.

REFERENCES

Aac 68 Aachen-Bari-Bergen-CERN-Nijmegan-Orsay Collaboration, Nuovo Cimento, **56A**:1106 (1968).

Abr 66 Abrams, R. J., A. Abashian, D. W. Carpenter, B. M. W. Nefkens, J. H. Smith, R. C. Talcher, C. W. Verhey, and H. Wattemberg: *Proc. 12th Internatl. Conf. High-Energy Physics, Berkeley* (1966).

Ade 64 Ademollo, M., and R. Gatto: *Phys. Rev. Letters,* **13**:264 (1964).

Aue 66 Auerbach, L. B., et al.: *Phys. Rev. Letters,* **17**:980 (1966); *Phys. Rev.,* **149**:1052 (1966).

Bas 68 Basile, P., J. W. Cronin, B. Thevenet, R. Turlay, S. Zylberajch, and A. Zylbersztejn: *Phys. Letters,* **26B**:542 (1968).

Bet 68 Bettels, J., et al.: *Nuovo Cimento,* **56A**:1106 (1968).

Cab 64a Cabibbo, N., and A. Maksymovicz: *Phys. Letters,* **9**:352 (1964).

Cab 64b Cabibbo, N., and A. Maksymovicz: *Phys. Letters,* **11**:360 (1964).

Cab 65 Cabibbo, N., and A. Maksymovicz: *Phys. Rev.,* **137B**:438 (1965).

Cal 69 Callan, C.: *Proc. Topical Conf. Weak Interactions, CERN, Geneva,* 263 (1969).

Cla 71 Clark, A., T. Elioff, R. Field, H. Frisch, R. Johnson, L. Kerth, and W. Wenzel: *Phys. Rev. Letters,* **26**:1667 (1971).

Cro 68 Cronin, J. W.: *Proc. 14th Internatl. Conf. High-Energy Physics, Vienna, CERN, Geneva,* 281 (1968).

Cut 68 Cutts, D., R. Stiening, C. Wiegand, and M. Deutsch: *Phys. Rev. Letters,* **20**:955 (1968).

Ely 68 Ely, R. P., et al.: *Univ. of Calif. Lawrence Rad. Lab. Rept.* UCRL-18626, 1968.

Fis 71a Fischbach, E., M. Nieto, H. Primakoff, C. Scott, and J. Smith: *Phys. Rev. Letters,* **27**:1403 (1971).

Fis 71b Fischbach, E., F. Iachello, A. Lande, M. Nieto, and C. Scott: *Phys. Rev. Letters,* **26**:1200, 1971.

Fri 71 Frisch, H. J.: doctoral dissertation, Univ. of Calif., Berkeley, *Univ. of Calif. Lawrence Rad. Lab. Rept.* UCRL-20264, 1971.

Gin 66 Ginsburg, E. S.: *Phys. Rev.,* **142**:1035 (1966).

Gin 67 Ginsburg, E. S.: *Phys. Rev.,* **162**:1570 (1967).

Gin 68 Ginsburg, E. S.: *Phys. Rev.,* **171**:1675 (1968).

Hel 68 Helland, J. A., M. A. Longo, and K. K. Young: *Phys. Rev. Letters,* **21**:257 (1968).

Iof 68 Ioffe, B. L., and E. P. Shabalin: *Soviet J. Nucl. Phys.,* **6**:603 (1968).

Kem 39 Kemmer, N.: *Proc. Roy. Soc. London A,* **173**:91 (1939).

Lit 69 Littenberg, L. S., J. H. Field, O. Piccioni, W. A. W. Mehlhop, S. S. Murty, and P. H. Bowles: paper submitted to Topical Conf. Weak Interactions, CERN, Geneva, 1969.

Moh 68 Mohapatra, R. N., J. Rao, and R. A. Marshak: *Phys. Rev.,* **171**:1502 (1968).

Qui 68 Quigg, C. and J. D. Jackson: *Univ. of Calif. Lawrence Rad. Lab. Rept.* UCRL-48487, 1968.

Rub 69 Rubbia, C.: *Proc. Topical Conf. Weak Interactions, CERN, Geneva,* 227 (1969).

Sti 68 Stiening, R. F.: *Univ. of Calif. Lawrence Rad. Lab. Rept.* UCRL-18231, 1968.

You 67 Young, K. K., M. J. Longo, and J.Helland: *Phys. Rev. Lett.,* **18**:806 (1967).

11

WEAK INTERACTIONS OF K MESONS, II; NONLEPTONIC KAON DECAYS; CP VIOLATION

11.1 $K_{\pi 2}$ DECAYS

Let us consider the decays

$$K^+ \to \pi^+\pi^0 \qquad (11.1)$$

$$K_S{}^0 \to \pi^+\pi^- \qquad (11.2)$$

$$K_S{}^0 \to \pi^0\pi^0 \qquad (11.3)$$

$$K_L{}^0 \to \pi^+\pi^- \qquad (11.4)$$

$$K_L{}^0 \to \pi^0\pi^0 \qquad (11.5)$$

As noted earlier (Chap. 7), the $K^+ \to \pi^+\pi^0$ decay cannot occur unless the $|\Delta I| = \frac{1}{2}$ rule is violated. However, this violation is a very "gentle" one in the sense that the $K^+ \to \pi^+\pi^0$ decay rate is small compared, for example, with the rates for $K_S{}^0 \to \pi^+\pi^-$, $\pi^0\pi^0$. From the experimental values of these rates one can establish that

$$\left|\frac{A_2}{A_0}\right| \approx 4.5\%$$

where

$$A_0 = \langle (2\pi)_{I=0}{}^S | H_w | K^0 \rangle$$

and

$$A_2 = \langle (2\pi)_{I=2}{}^S | H_w | K^0 \rangle$$

[See Eqs. (7.5) and (7.6).]

We also mentioned in Chap. 7 that the $|\Delta I| = \frac{1}{2}$ rule implies

$$R = \left. \frac{\Gamma(K_S{}^0 \to \pi^+\pi^-)}{\Gamma(K_S{}^0 \to \pi^0\pi^0)} \right|_{\text{theo}} = 2 \qquad (11.6)$$

Unfortunately, the experimental measurements of this ratio are not in complete agreement. Two recent results are

$$R = 2.285 \pm 0.055 \qquad \text{(Gob 69)} \qquad (11.7)$$

and

$$R = 2.10 \;\; \pm 0.06 \qquad \text{(Mor 69)} \qquad (11.8)$$

which yield an average value

$$R = 2.201 \pm 0.092 \qquad \text{(Barb 70)} \qquad (11.9)$$

Result (11.7) indicates a very clear-cut violation of the $|\Delta I| = \frac{1}{2}$ rule, but the situation is somewhat confused in the light of (11.8) and (11.9). The most we can state is that the $|\Delta I| = \frac{1}{2}$ rule *may* be violated in $K_S{}^0 \to 2\pi$ decays.

The decays $K_L{}^0 \to \pi^+\pi^-, \pi^0\pi^0$ violate CP invariance. In order to understand the CP violation in detail, we now undertake a rather detailed analysis of the $K^0\bar{K}^0$ system.

11.2 PHENOMENOLOGICAL ANALYSIS OF THE $K^0\bar{K}^0$ SYSTEM; THE EFFECTIVE HAMILTONIAN[1]

Suppose we have an arbitrary superposition of K^0 and $\bar{K}^0$ states with time-dependent coefficients A and B:

$$|\psi(t)\rangle = A(t)|K^0\rangle + B(t)|\bar{K}^0\rangle \qquad (11.10)$$

or simply

$$\psi(t) = \begin{pmatrix} A \\ B \end{pmatrix} \qquad (11.11)$$

The effective hamiltonian governing the time evolution of the system

$$H = H_{\text{strong}} + H_{\text{electromagnetic}} + H_{\text{weak}}$$
$$H = H_S + H_E + H_W \qquad (11.12)$$

satisfies the equation

$$H\psi = i\frac{d\psi}{dt} \qquad (11.13)$$

[1] The analysis discussed here was originally presented by Wu and Yang (Wu 64).

In the K^0, $\overline{K}^0$ basis H is represented by a 2×2 matrix

$$H = M - i\Gamma \tag{11.14}$$

where M and Γ are hermitian 2×2 matrices (the *mass* and *decay* matrices, respectively). H itself is not hermitian because the neutral kaons decay.

States $|K^0\rangle$ and $|\overline{K}^0\rangle$ are defined as eigenstates of $H_S + H_E$ with strangeness eigenvalues $+1$ and -1, respectively. Therefore, since the strong and electromagnetic interactions conserve strangeness, we have

$$\langle \overline{K}^0 | H_S + H_E | K^0 \rangle = 0 \tag{11.15}$$

If we pretend for a moment that $H_W = 0$, then the K^0, $\overline{K}^0$ mesons would not decay and Γ would be zero. In this case, in light of (11.15), the H matrix would reduce to

$$H \to M = \begin{pmatrix} m_{K^0} & 0 \\ 0 & m_{\overline{K}^0} \end{pmatrix} \tag{11.16}$$

where

$$m_{K^0} \equiv \langle K^0 | H_S + H_E | K^0 \rangle \tag{11.17}$$

and

$$m_{\overline{K}^0} \equiv \langle \overline{K}^0 | H_S + H_E | \overline{K}^0 \rangle \tag{11.18}$$

CPT invariance of H_S and H_E implies that $m_K = m_{\overline{K}^0}$.

Now, in fact, the weak interaction hamiltonian is not zero, and so $\Gamma \neq 0$; also M contains nonzero off-diagonal elements

$$M_{11} = m_{K^0} + \langle K^0 | H_W | K^0 \rangle + \sum_n{}' \frac{|\langle K^0 | H_W | n \rangle|^2}{m_{K^0} - \omega_n} + \cdots \tag{11.19}$$

$$M_{22} = m_{\overline{K}^0} + \langle \overline{K}^0 | H_W | \overline{K}^0 \rangle + \sum_n{}' \frac{|\langle \overline{K}^0 | H_W | n \rangle|^2}{m_{\overline{K}^0} - \omega_n} + \cdots \tag{11.20}$$

and

$$M_{21} = M_{12}^* = \langle \overline{K}^0 | H_W | K^0 \rangle + \sum_n{}' \frac{\langle \overline{K}^0 | H_W | n \rangle \langle n | H_W | K^0 \rangle}{m_{K^0} - \omega_n} + \cdots \tag{11.21}$$

In the discussion which follows we shall find that the second-order terms $\sum_n'$ in (11.19) to (11.21) are important and must be retained, but we shall assume that it is legitimate to neglect all terms of higher order. Furthermore, since first-order weak interactions obey the selection rule $|\Delta S| \leq 1$, the term $\langle \overline{K}^0 | H_W | K^0 \rangle$ in (11.21) is zero. Also, CPT invariance of H_W in addition to that of $H_S + H_E$ implies that $M_{11} = M_{22}$.

Let us now turn our attention to matrix Γ. Conditions on its elements are derived from the equation which describes conservation of probability:

$$\frac{d}{dt} \langle \psi | \psi \rangle + 2\pi \sum_F \rho(F) |\langle F | H_W | \psi(t) \rangle|^2 = 0 \tag{11.22}$$

Here $|F\rangle$ denotes a particular final state resulting from K^0, $\bar{K}^0$ weak decay, for example, $\pi^+ l^- \nu$ or $\pi^+\pi^-$, and so on; and $\rho(F)$ is the density of final states per unit energy. We have for an arbitrary linear combination

$$|\psi\rangle = A|K^0\rangle + B|\bar{K}^0\rangle$$

$$\begin{aligned}\frac{d}{dt}\langle\psi|\psi\rangle &= \langle\dot{\psi}|\psi\rangle + \langle\psi|\dot{\psi}\rangle \\ &= -\langle\psi|\Gamma - iM|\psi\rangle - \langle\psi|\Gamma + iM|\psi\rangle \\ &= -2\langle\psi|\Gamma|\psi\rangle \qquad (11.23)\end{aligned}$$

Therefore, since A and B are arbitrary, we obtain

$$\Gamma_{11} = \langle K^0|\Gamma|K^0\rangle = \pi\sum_F \rho(F)|\langle F|H_W|K^0\rangle|^2 \qquad (11.24)$$

$$\Gamma_{22} = \langle \bar{K}^0|\Gamma|\bar{K}^0\rangle = \pi\sum_F \rho(F)|\langle F|H_W|\bar{K}^0\rangle|^2 \qquad (11.25)$$

and

$$\begin{aligned}\Gamma_{12}^* = \Gamma_{21} &= \langle \bar{K}^0|\Gamma|K^0\rangle \\ &= \pi\sum_F \rho(F)\langle \bar{K}^0|H_W|F\rangle\langle F|H_W|K^0\rangle \qquad (11.26)\end{aligned}$$

The matrix elements Γ_{11}, Γ_{22} are thus expressed as sums over terms corresponding to distinct final states F. In principle it is possible to determine, by the measurement of partial decay rates, the contribution of each of these terms to Γ_{11}, Γ_{22} (and some restrictions can also be derived for Γ_{12}). However, no such separation is possible for the various contributions to M_{11}, M_{22} [Eqs. (11.19) and (11.20)].

If CPT invariance holds for H_W, then

$$\begin{aligned}\langle F|H_W|K^0\rangle &= \langle F|(\text{CPT})^{-1}H_W(\text{CPT})|K^0\rangle \\ &= -\langle \bar{K}^0|H_W|\bar{F}'\rangle = -\langle \bar{F}'|H_W|\bar{K}^0\rangle^*\end{aligned}$$

where the prime denotes reversed spins. However, in (11.24) and (11.25), the sums are taken over all possible final states F; and so CPT invariance implies $\Gamma_{11} = \Gamma_{22}$. If T invariance holds for H_W, then we can always adjust phases so that $\langle F|H_W|K^0\rangle$ and $\langle F|H_W|\bar{K}^0\rangle$ are relatively real. In this case it follows from (11.21) and (11.26) that

$$\left.\begin{aligned}\Gamma_{12} = \Gamma_{21} = \Gamma_{12}^* \\ \text{and} \qquad M_{12} = M_{21} = M_{12}^*\end{aligned}\right\} \quad \text{T invariance}$$

11.3 EIGENVALUES AND EIGENVECTORS OF $\Gamma + iM$

To examine the eigenvalues and eigenvectors of $\Gamma + iM$, we write

$$\Gamma + iM = DI + i\mathbf{E}\cdot\boldsymbol{\sigma} \qquad (11.27)$$

where I and $\boldsymbol{\sigma}$ are the 2×2 identity and Pauli matrices, respectively, and D and $\mathbf{E}$ are a complex number and a complex vector, respectively:

$$\mathbf{E} = E_1\hat{\imath} + E_2\hat{\jmath} + E_3\hat{k} \qquad (11.28)$$

It is convenient to write

$$E_1 = E\sin\theta\cos\phi$$
$$E_2 = E\sin\theta\sin\phi$$
$$E_3 = E\cos\theta$$

where θ and ϕ are in general complex. Thus we have

$$\Gamma + iM = \begin{pmatrix} D + iE_3 & i(E_1 - iE_2) \\ i(E_1 + iE_2) & D - iE_3 \end{pmatrix}$$
$$= \begin{pmatrix} D + iE\cos\theta & iE\sin\theta e^{-i\phi} \\ iE\sin\theta^{i\phi} & D - iE\cos\theta \end{pmatrix} \qquad (11.29)$$

which is readily diagonalized with the aid of some matrix C to yield

$$C(\Gamma + iM)C^{-1} = \begin{pmatrix} \lambda_S = D + iE & 0 \\ 0 & \lambda_L = D - iE \end{pmatrix} \qquad (11.30)$$

The eigenvalues of $\Gamma + iM$ are thus

$$D + iE \equiv \lambda_S \equiv \frac{\gamma_S}{2} + im_S \qquad (11.31)$$

$$D - iE \equiv \lambda_L \equiv \frac{\gamma_L}{2} + im_L \qquad (11.32)$$

where γ_S, γ_L, m_S, and m_L are real quantities defined by Eqs. (11.31) and (11.32) and corresponding to the eigenvectors $|K_S{}^0\rangle$, $|K_L{}^0\rangle$:

$$(\Gamma + iM)|K_S{}^0\rangle = \lambda_S|K_S{}^0\rangle = (D + iE)|K_S{}^0\rangle \qquad (11.33)$$

and

$$(\Gamma + iM)|K_L{}^0\rangle = \lambda_L|K_L{}^0\rangle = (D - iE)|K_L{}^0\rangle \qquad (11.34)$$

Thus, states $|K_S{}^0\rangle$ and $|K_L{}^0\rangle$ each have definite lifetimes $\tau_S = 1/\gamma_S$ and $\tau_L = 1/\gamma_L$ for weak decay:

$$|K_S{}^0(t)\rangle = |K_S{}^0(0)\rangle e^{-im_S t} e^{-(\gamma_S/2)t} \qquad (11.35)$$
$$|K_L{}^0(t)\rangle = |K_L{}^0(0)\rangle e^{-im_L t} e^{-(\gamma_L/2)t} \qquad (11.36)$$

Let us now express the eigenvectors $|K_S{}^0\rangle$, $|K_L{}^0\rangle$ of $\Gamma + iM$ as linear combinations of the basis states $|K^0\rangle$, $|\bar{K}^0\rangle$. To this end we write

$$|K_S{}^0\rangle = \frac{1}{\sqrt{|r|^2 + |s|^2}} [r|K^0\rangle + s|\bar{K}^0\rangle] \tag{11.37}$$

$$|K_L{}^0\rangle = \frac{1}{\sqrt{|p|^2 + |q|^2}} [p|K^0\rangle + q|\bar{K}^0\rangle] \tag{11.38}$$

Then we obtain from (11.29), (11.33), and (11.34)

$$iE(\cos\theta - 1)r + iE\sin\theta e^{-i\phi}s = 0$$

and

$$iE(\cos\theta + 1)p + iE\sin\theta e^{-i\phi}q = 0$$

or

$$\frac{s}{r} = \frac{1 - \cos\theta}{\sin\theta} e^{i\phi} = \tan\frac{\theta}{2} e^{i\phi} \tag{11.39}$$

and

$$\frac{q}{p} = -\cot\frac{\theta}{2} e^{i\phi} \tag{11.40}$$

It is customary to define the new complex quantities ε_1, ε_2 and $\varepsilon = (\varepsilon_1 + \varepsilon_2)/2$, $\delta = (\varepsilon_1 - \varepsilon_2)/2$ by

$$\frac{1 - \varepsilon_1}{1 + \varepsilon_1} \equiv -\frac{s}{r} = -\tan\frac{\theta}{2} e^{i\phi} \tag{11.41}$$

and

$$\frac{1 - \varepsilon_2}{1 + \varepsilon_2} = \frac{q}{p} = -\cot\frac{\theta}{2} e^{i\phi} \tag{11.42}$$

In terms of ε_1 and ε_2, (11.37) and (11.38) become

$$|K_S{}^0\rangle = \frac{1}{\sqrt{2(1 + |\varepsilon_1|^2)}} [(1 + \varepsilon_1)|K^0\rangle - (1 - \varepsilon_1)|\bar{K}^0\rangle] \tag{11.43}$$

and

$$|K_L{}^0\rangle = \frac{1}{\sqrt{2(1 + |\varepsilon_2|^2)}} [(1 + \varepsilon_2)|K^0\rangle + (1 - \varepsilon_2)|\bar{K}^0\rangle] \tag{11.44}$$

If CPT invariance holds, then $(\Gamma + iM)_{11} = (\Gamma + iM)_{22}$. Thus, from (11.29), $E_3 = 0$ and $\theta = -\pi/2$. Therefore,

$$\frac{1 - \varepsilon_1}{1 + \varepsilon_1} = \frac{1 - \varepsilon_2}{1 + \varepsilon_2} = e^{i\phi}$$

and

$$\left.\begin{aligned} \varepsilon_1 = \varepsilon_2 &= \varepsilon \\ \delta &= 0 \end{aligned}\right\} \quad \text{CPT invariance} \tag{11.45}$$

Consequently, CPT invariance implies

$$|K_S{}^0\rangle = \frac{1}{\sqrt{2(1+|\varepsilon|^2)}}[(1+\varepsilon)|K^0\rangle - (1-\varepsilon)|\bar{K}^0\rangle]$$
$$|K_L{}^0\rangle = \frac{1}{\sqrt{2(1+|\varepsilon|^2)}}[(1+\varepsilon)|K^0\rangle + (1-\varepsilon)|\bar{K}^0\rangle] \tag{11.46}$$

If T invariance holds, then $(\Gamma + iM)_{12} = (\Gamma + iM)_{21}$. Thus, from (11.29), $e^{i\phi} = e^{-i\phi} = 1$, $\phi = 0$, and

$$\frac{1-\varepsilon_1}{1+\varepsilon_1} = -\tan\frac{\theta}{2} = \frac{1+\varepsilon_2}{1-\varepsilon_2}$$

or

$$\left.\begin{aligned} \varepsilon_1 &= -\varepsilon_2 \\ \varepsilon &= 0 \\ \delta &= \varepsilon_1 = -\varepsilon_2 \end{aligned}\right\} \text{T invariance} \tag{11.47}$$

Consequently, T invariance implies

$$|K_S{}^0\rangle = \frac{1}{\sqrt{2(1+|\delta|^2)}}[(1+\delta)|K^0\rangle - (1-\delta)|\bar{K}^0\rangle]$$
$$|K_L{}^0\rangle = \frac{1}{\sqrt{2(1+|\delta|^2)}}[(1-\delta)|K^0\rangle + (1+\delta)|\bar{K}^0\rangle] \tag{11.48}$$

Obviously, if T invariance and CPT invariance both hold, then CP invariance is also valid and $\varepsilon = \delta = 0$. Of course, we know that CP violation is an extremely small effect, and so $|\varepsilon|$, $|\delta|$, $|\varepsilon_1|$, $|\varepsilon_2|$ are all very small compared to unity. It follows that $\phi \ll 1$ and that $\theta \geq -\pi/2$, even without the assumptions of CPT or T invariance.

The existence of nonzero ε_1 and ε_2 implies that $|K_S{}^0\rangle$ and $|K_L{}^0\rangle$ are not orthogonal states. From (11.43) and (11.44) we have

$$\langle K_S{}^0 | K_L{}^0\rangle = \frac{(1+\varepsilon_1^*)(1+\varepsilon_2) - (1-\varepsilon_1)^*(1-\varepsilon_2)}{2(1+|\varepsilon_1|^2)^{1/2}(1+|\varepsilon_2|^2)^{1/2}}$$

Neglecting $|\varepsilon_1|^2$, $|\varepsilon_2|^2$, and other second-order small quantities, this reduces to

$$\langle K_S{}^0 | K_L{}^0\rangle \approx \varepsilon_1^* + \varepsilon_2 = 2(\text{Re}\,\varepsilon - i\,\text{Im}\,\delta) \tag{11.49}$$

Therefore, if CPT invariance holds, $\langle K_S{}^0 | K_L{}^0\rangle$ is real:

$$\langle K_S{}^0 | K_L{}^0\rangle = 2\,\text{Re}\,\varepsilon \qquad \text{CPT invariance} \tag{11.50}$$

But if T invariance holds, $\langle K_S{}^0 | K_L{}^0 \rangle$ is imaginary:

$$\langle K_S{}^0 | K_L{}^0 \rangle = -2i \operatorname{Im} \delta \qquad \text{T invariance} \tag{11.51}$$

11.4 REGENERATION

In Sec. 10.2 we indicated briefly how a pure $K_L{}^0$ beam may be converted into a mixture of $K_L{}^0$ and $K_S{}^0$ when the beam passes through matter (phenomenon of regeneration). Let us now analyze regeneration in more detail. This will help us to understand the experiments in which the mass difference $\Delta m = m_L - m_S$ is determined.

Let ψ denote a two-component neutral kaon state, as usual:

$$\psi = \begin{pmatrix} A & (t) \\ B & (t) \end{pmatrix}$$

or

$$|\psi\rangle = A(t)|K^0\rangle + B(t)|\bar{K}^0\rangle$$

We wish to describe the change of ψ with respect to z (distance along the beam direction) as the beam propagates through matter. There will be two contributions to $d\psi/dz$, a part which would exist even in vacuum and a part arising from scattering with the matter nuclei:

$$\frac{d\psi}{dz} = \frac{d\psi}{dz}_{\text{vac}} + \frac{d\psi}{dz}_{\text{nuc}} \tag{11.52}$$

This can equally well be written

$$\frac{d\psi}{d\tau} = \frac{d\psi}{d\tau}_{\text{vac}} + \frac{d\psi}{d\tau}_{\text{nuc}} \tag{11.53}$$

where τ is the proper time and

$$\frac{dz}{d\tau} = \frac{v}{(1 - v^2)^{1/2}} \tag{11.54}$$

where v is the kaon velocity in the medium.

Let us define the indices of refraction n, $\bar{n}$ in the matter for K^0 and $\bar{K}^0$ components, respectively. These are related to the forward-scattering amplitudes $f(0)$ and $\overline{f(0)}$ for K^0, $\bar{K}^0$, respectively, by

$$n = 1 + \frac{2\pi N}{k^2} f(0) \tag{11.55}$$

and

$$\bar{n} = 1 + \frac{2\pi N}{k^2} \overline{f(0)} \tag{11.56}$$

where N is the number of scattering centers per unit volume and k is the wave number. By the very meaning of the term *index of refraction*, we have

$$\left.\frac{d\psi}{dz}\right)_{\text{nuc}} = ik\begin{pmatrix} n-1 & 0 \\ 0 & \bar{n}-1 \end{pmatrix}\psi$$

$$= \frac{2\pi iN}{k}\begin{pmatrix} f(0) & 0 \\ 0 & \overline{f(0)} \end{pmatrix}\psi \qquad (11.57)$$

or

$$\left.\frac{d\psi}{d\tau}\right)_{\text{nuc}} = \frac{2\pi Nvi}{k(1-v^2)^{1/2}}\begin{pmatrix} f(0) & 0 \\ 0 & \overline{f(0)} \end{pmatrix}\psi \qquad (11.58)$$

Now,

$$\left.\frac{d\psi}{d\tau}\right)_{\text{vac}} = -(\Gamma + iM)\psi$$

We define new matrices Γ', M' by

$$\frac{d\psi}{d\tau} = -(\Gamma' + iM')\psi$$

$$= -(\Gamma + iM)\psi + \frac{2\pi Nvi}{k(1-v^2)^{1/2}}\begin{pmatrix} f & 0 \\ 0 & \bar{f} \end{pmatrix}\psi$$

or

$$\Gamma' + iM' = \Gamma + iM - i\alpha_0\begin{pmatrix} f & 0 \\ 0 & \bar{f} \end{pmatrix} \qquad (11.59)$$

where

$$\alpha_0 = \frac{2\pi Nv}{k(1-v^2)^{1/2}}$$

From now on we assume CPT invariance but not CP invariance. In this case,

$$\Gamma + iM = D + iE_1\sigma_1 + iE_2\sigma_2 \qquad (11.60)$$

It is then straightforward to diagonalize the matrix $\Gamma' + iM'$ and find its eigenvalues and eigenvectors. Defining these by

$$(\Gamma' + iM')|K_S'\rangle \equiv \lambda_S'|K_S'\rangle \equiv \left(im_S' + \frac{\tau_S'}{2}\right)|K_S'\rangle$$

$$(\Gamma' + iM')|K_L'\rangle \equiv \lambda_L'|K_L'\rangle \equiv \left(im_L' + \frac{\gamma_L'}{2}\right)|K_L'\rangle$$

we find that

$$|K_S'\rangle = \frac{1}{\sqrt{2(1+|\varepsilon_1'|^2)}}\{(1+\varepsilon_1')|K^0\rangle - (1-\varepsilon_1')|\overline{K}^0\rangle\} \qquad (11.61)$$

and

$$|K_L'\rangle = \frac{1}{\sqrt{2(1+|\varepsilon_2'|^2)}}\{(1+\varepsilon_2')|K^0\rangle + (1-\varepsilon_2')|\overline{K}^0\rangle\} \qquad (11.62)$$

where
$$\varepsilon_1' = \varepsilon - \frac{i\pi N\Lambda}{k}\frac{f-\bar{f}}{\frac{1}{2}+i\mu} \tag{11.63}$$

$$\varepsilon_2' = \varepsilon + \frac{i\pi N\Lambda}{k}\frac{f-\bar{f}}{\frac{1}{2}+i\mu} \tag{11.64}$$

$$\mu = \frac{m_S' - m_L'}{\gamma_S' - \gamma_L'} \tag{11.65}$$

and
$$\Lambda = \frac{v}{(1-v^2)^{1/2}}\frac{1}{(\gamma_S' - \gamma_L')} \tag{11.66}$$

In Eqs. (11.61) to (11.66) all quadratic terms in ε_1', ε_2', and $\alpha_0(f \pm \bar{f})$ are neglected. Of course, $\gamma_S' \gg \gamma_L'$ since γ_S', γ_L' are not very different from their respective values in vacuum. Therefore, to a good approximation, Λ is simply the mean decay length of the short-lived kaon in matter.

Let us express the states $|K_S'\rangle$, $|K_L'\rangle$ of (11.61) and (11.62) in terms of the vacuum states $|K_S\rangle$, $|K_L\rangle$. Neglecting quadratic terms in ε, ε_1', and ε_2', we have

$$|K_S'\rangle \approx \frac{1}{\sqrt{2}}\{(1+\varepsilon-r)|K^0\rangle - (1-\varepsilon+r)|\bar{K}^0\rangle\}$$

$$= \frac{1}{\sqrt{2}}\{(1+\varepsilon)|K^0\rangle - (1-\varepsilon)|\bar{K}^0\rangle - r(|K^0\rangle + |\bar{K}^0\rangle)\}$$

$$|K_S'\rangle \approx |K_S\rangle - r|K_L\rangle \tag{11.67}$$

and similarly

$$|K_L'\rangle = |K_L\rangle + r|K_S\rangle \tag{11.68}$$

where
$$r(k) = \frac{-i\pi N\Lambda[f(0) - \bar{f}(0)]}{k(i\mu + \frac{1}{2})} \tag{11.69}$$

The complex number r (called the *regeneration parameter*) characterizes the regenerating power of a given medium for given k. It is important to notice that $r(k)$ is independent of the thickness of the regenerator. In all real cases, $|r| \ll 1$ and terms quadratic in r may be neglected when analyzing regeneration effects.

11.5 TRANSITION AMPLITUDES FOR $K_S^0 \to \pi\pi$, $K_L^0 \to \pi\pi$

We now calculate the transition amplitudes for the 2π decay modes of K_S^0 and K_L^0 in vacuum, assuming that CPT invariance is valid. To begin, we consider the two-pion final state in any of these decays. It must be completely symmetric, and therefore the possible final isospin is $I = 2$ or $I = 0$. For the moment, let us ignore $\pi\pi$ final-state interaction. In this case, the states $\langle\pi^+\pi^-|$ and $\langle\pi^0\pi^0|$ can be expressed very simply as linear combinations of the states $\langle(2\pi)_{I=0}{}^S|$ and $\langle(2)_{I=2}{}^S|$, where, as before, the latter refer to the $I = 0$, $I_3 = 0$ and $I = 2$, $I_3 = 0$ stationary isospin states, respectively. (*Stationary* means there is no final-state interaction.)

Since

$$|(2\pi)_{I=2}{}^S\rangle = \frac{1}{\sqrt{6}}\,|\pi_1{}^+\pi_2{}^- + \pi_1{}^-\pi_2{}^+ + 2\pi_1{}^0\pi_2{}^0\rangle^S$$

$$|(2\pi)_{I=0}{}^S\rangle = \frac{1}{\sqrt{3}}\,|\pi_1{}^+\pi_2{}^- + \pi_1{}^-\pi_2{}^+ - \pi_1{}^0\pi_2{}^0\rangle^S$$

and

$$|\pi^+\pi^-\rangle^S = \frac{1}{\sqrt{2}}\,|\pi_1{}^+\pi_2{}^- + \pi_1{}^-\pi_2{}^+\rangle^S$$

we have

$$\langle\pi^+\pi^-|^S = \frac{1}{\sqrt{3}}\{\langle(2\pi)_{I=2}{}^S| + \sqrt{2}\langle(2\pi)_{I=0}{}^S|\}$$

and

$$\langle\pi^0\pi^0|^S = \frac{1}{\sqrt{3}}\{\sqrt{2}\,\langle(2\pi)_{I=2}{}^S| - \langle(2\pi)_{I=0}{}^S|\}$$

Then, writing $e^{i\delta_0}$ and $e^{i\delta_2}$ as the final-state $\pi\pi$ phase shifts for $I = 0$ and $I = 2$, respectively, we have

$$\langle\pi^+\pi^-| = \frac{1}{\sqrt{3}}\{e^{i\delta_2}\langle(2\pi)_{I=2}{}^S| + \sqrt{2}\,e^{i\delta_0}\langle(2\pi)_{I=0}{}^S|\} \tag{11.70}$$

and

$$\langle\pi^0\pi^0| = \frac{1}{\sqrt{3}}\{\sqrt{2}\,e^{i\delta_2}\langle(2\pi)_{I=2}{}^S| - e^{i\delta_0}\langle(2\pi)_{I=0}{}^S|\} \tag{11.71}$$

Now we recall the amplitudes

$$A_0 = \langle(2\pi)_{I=0}{}^S|H_W|K^0\rangle \tag{11.72}$$

and

$$A_2 = \langle(2\pi)_{I=2}{}^S|H_W|K^0\rangle \tag{11.73}$$

Under a CPT transformation,

$$|K^0\rangle \rightarrow -\langle\bar{K}^0|$$

$$\langle(2\pi)_{I=0}{}^S| \rightarrow |(2\pi)_{I=0}{}^S\rangle$$

and

$$\langle(2\pi)_{I=2}{}^S| \rightarrow |(2\pi)_{I=2}{}^S\rangle$$

Therefore, if H_W is CPT-invariant, we have

$$\langle (2\pi)_{I=0}{}^S | H_W | \bar{K}^0 \rangle = -A_0^* \tag{11.74}$$

and

$$\langle (2\pi)_{I=2}{}^S | H_W | \bar{K}^0 \rangle = -A_2^* \tag{11.75}$$

Actually, we have the freedom to adjust the relative phase of $|K^0\rangle$ and $|\bar{K}^0\rangle$ in such a way that A_0 is real, although, in general, A_2 will be complex:

$$A_0 = A_0^* \tag{11.76}$$

Now taking into account Eqs. (11.46) and (11.70) to (11.76), we obtain

$$\langle \pi^+\pi^- | H_W | K_S{}^0 \rangle = \frac{1}{3^{1/2}\sqrt{2(1+|\varepsilon|^2)}} \times \left[(A_2 + A_2^*)e^{i\delta_2} + \frac{4}{\sqrt{2}} A_0 e^{i\delta_0} + (A_2 - A_2^*)e^{i\delta_2}\varepsilon \right] \tag{11.77}$$

$$\langle \pi^0\pi^0 | H_W | K_S{}^0 \rangle = \left(\frac{2}{3}\right)^{1/2} \frac{1}{\sqrt{2(1+|\varepsilon|^2)}} \times [(A_2 + A_2^*)e^{i\delta_2} - \sqrt{2}\, A_0 e^{i\delta_0} + (A_2 - A_2^*)e^{i\delta_2}\varepsilon] \tag{11.78}$$

$$\langle \pi^+\pi^- | H_W | K_L{}^0 \rangle = \frac{1}{3^{1/2}\sqrt{2(1+|\varepsilon|^2)}} \times \left\{ (A_2 - A_2^*)e^{i\delta_2} + \varepsilon \left[(A_2 + A_2^*)e^{i\delta_2} + \frac{4}{\sqrt{2}} A_0 e^{i\delta_0} \right] \right\} \tag{11.79}$$

$$\langle \pi^0\pi^0 | H_W | K_L{}^0 \rangle = \left(\frac{2}{3}\right)^{1/2} \frac{1}{\sqrt{2(1+|\varepsilon|^2)}} \times \{(A_2 - A_2^*)e^{i\delta_2} + \varepsilon[(A_2 + A_2^*)e^{i\delta_2} - 2^{1/2} A_0 e^{i\delta_0}]\} \tag{11.80}$$

From a previous discussion of $K^+ \to \pi^+\pi^0$ we know that the amplitude A_2 (which would give $|\Delta I| = \frac{3}{2}, \frac{5}{2}$) is about twenty times smaller in absolute value than the $|\Delta I| = \frac{1}{2}$ amplitude A_0. Thus we can safely neglect all terms of order $\varepsilon A_2/A_0$, and we can also neglect terms of order ε^2 in Eqs. (11.77) to (11.80). Thus defining

$$\eta^{+-} = \frac{\langle \pi^+\pi^- | H_W | K_L{}^0 \rangle}{\langle \pi^+\pi^- | H_W | K_S{}^0 \rangle} \tag{11.81}$$

$$\eta^{00} = \frac{\langle \pi^0\pi^0 | H_W | K_L{}^0 \rangle}{\langle \pi^0\pi^0 | H_W | K_S{}^0 \rangle} \tag{11.82}$$

and

$$\varepsilon' = \frac{1}{\sqrt{2}} \left(\frac{\operatorname{Im} A_2}{A_0} \right) e^{i(\pi/2 + \delta_2 - \delta_0)} \tag{11.83}$$

we finally obtain

$$\left.\begin{aligned} \eta^{+-} &\approx \varepsilon + \varepsilon' \qquad (11.84) \\ \eta^{00} &\approx \varepsilon - 2\varepsilon' \qquad (11.85) \end{aligned}\right\} \text{CPT invariance}$$

It is to be noted that if Im $A_2 = 0$, ε' vanishes and $\eta^{+-} = \eta^{00}$.

We may summarize the foregoing discussion by stating that, with the assumption of CPT invariance, CP violation in $K_L{}^0 \to 2\pi$ decay is characterized phenomenologically by the four complex numbers η^{+-}, η^{00}, ε, and ε', which are connected by the two relations (11.84) and (11.85). We now turn to a brief discussion of the crucial experiments in which the mass difference Δm, η^{+-}, η^{00}, and ε are determined.

11.6 DETERMINATION OF $|\eta^{+-}|$

From Eq. (11.63) we have

$$|\eta^{+-}|^2 = \frac{\text{rate } [K_L{}^0 \to \pi^+\pi^-]}{\text{rate } [K_S{}^0 \to \pi^+\pi^-]} \qquad (11.86)$$

Since the rate for $K_S{}^0 \to \pi^+\pi^-$ is well known, $|\eta^{+-}|^2$ is determined from observation of the branching ratio

$$\frac{\text{rate } (K_L{}^0 \to \pi^+\pi^-)}{\text{rate } (K_L{}^0 \to \text{all})} \qquad (11.87)$$

The decay $K_L{}^0 \to \pi^+\pi^-$ was first observed in the famous experiment of Christenson, Cronin, Fitch, and Turlay (Chr 64), and their results were subsequently verified at other laboratories over a wide range of $K_L{}^0$ laboratory momenta. Very briefly, their method was as follows.

K^0 mesons are produced by proton bombardment of a beryllium target. The K's are formed into a beam by collimators, etc., and charged particles are removed from the beam by sweeping magnets. The beam path from production point to observation point is long enough so that the $K_S{}^0$ component decays away, leaving an essentially pure $K_L{}^0$ beam (with the exception of a contaminating neutron component). The decays to be observed take place in a region filled with helium, and so the effects of regeneration are negligible (see Fig. 11.1). The charged pion detectors consist of two spectrometers, each composed of two spark chambers which are separated by a strong magnetic field. Most of the observed $K_L{}^0$ decays are to $\pi^+\pi^-\pi^0$, $\mu^\pm\pi^\mp\nu$, and $e^\pm\pi^\mp\nu$. However, we distinguish $K_L{}^0 \to \pi^+\pi^-$ decays from the others by requiring that the pion momenta add vectorally to give a resultant along the beam direction, and that the invariant mass m^* equals the $K_L{}^0$ mass:

$$m^* = [(\sqrt{p_{\pi+}{}^2 + m_{\pi+}{}^2} + \sqrt{p_{\pi-}{}^2 - m_{\pi-}{}^2})^2 - (\mathbf{p}_{\pi+} + \mathbf{p}_{\pi-})^2]^{1/2} = m_{K^0} \qquad (11.88)$$

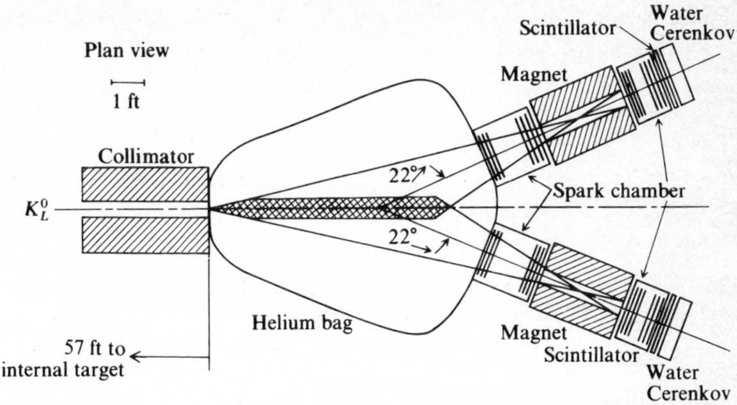

FIGURE 11.1
Experimental arrangement for observation of the decay $K_L{}^0 \to \pi^+\pi^-$ (Chr 64).

The "best" value of $|\eta^{+-}|$ based on all world results to date is (Cro 68)

$$|\eta^{+-}| = (1.90 \pm 0.05) \times 10^{-3} \qquad (11.89)$$

This value is in good agreement with the original result of Christenson, Cronin, Fitch, and Turlay (Chr 64).

11.7 THE MASS DIFFERENCE Δm

Rather than give a comprehensive review of the many experiments[1] which have been performed to measure Δm, we shall merely describe the principle of a recent, very precise and elegant experiment performed at Chicago (Aro 70), the method for which originated several years earlier at Princeton (Chr 65). The apparatus (Fig. 11.2) includes two regenerators R_1 and R_2 of the same material and with thicknesses L_1 and L_2, respectively. These are separated by an air

[1] See, for example, Alf 66, Bot 66, Bal 65, Pic 65, Chr 65, and Fit 61.

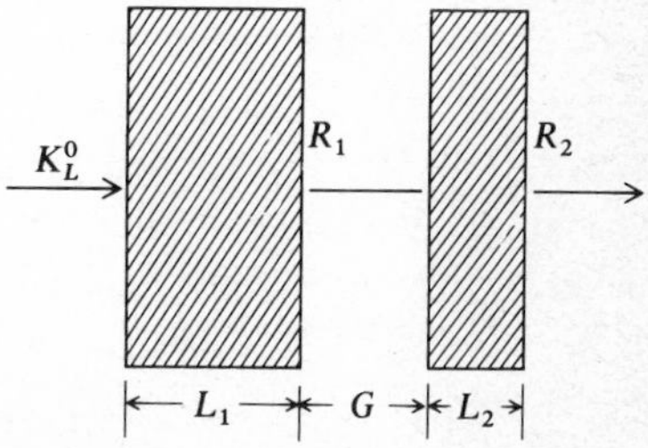

FIGURE 11.2
Experiment for determination of Δm by the gap method (Aro 70).

gap of variable length G. Let us analyze the wave function ψ at various points along the beam, retaining only terms of first order in the regeneration parameter r.

1 Before entering R_1, the system is in a pure $K_L{}^0$ state,

$$\psi = K_L{}^0 = K_L' - rK_S'$$

2 After a distance $L_1 = vl_1/(1 - v^2)^{1/2}$ in regenerator R_1,

$$\begin{aligned}\psi &= K_L' e^{-\lambda_L' l_1} - rK_S' e^{-\lambda_S' l_1} \\ &= (K_L + rK_S)e^{-\lambda_L' l_1} - r(K_S - rK_L)e^{-\lambda_S' l_1} \\ &\approx K_L e^{-\lambda_L' l_1} + rK_S(e^{-\lambda_L' l_1} - e^{-\lambda_S' l_1})\end{aligned}$$

3 After a gap distance $G = vg/(1 - v^2)^{1/2}$,

$$\psi = K_L e^{-\lambda_L' l_1} e^{-\lambda_L g} + rK_S(e^{-\lambda_L' l_1} - e^{-\lambda_S' l_1})e^{-\lambda_S g}$$

$$\approx K_L' e^{-\lambda_L' l_1} e^{-\lambda_L g} + rK_S'[e^{-\lambda_S g}(e^{-\lambda_L' l_1} - e^{-\lambda_S' l_1}) - e^{-\lambda_L g} e^{-\lambda_L' l_1}]$$

4 After a distance $L_2 = vl_2/(1 - v^2)^{1/2}$ in regenerator R_2,

$$\begin{aligned}\psi = K_L' e^{-\lambda_L' l_1} e^{-\lambda_L g} e^{-\lambda_L' l_2} &+ rK_S' e^{-\lambda_S' l_2} \\ &\times [e^{-\lambda_S g}(e^{-\lambda_L' l_1} - e^{-\lambda_S' l_1}) - e^{-\lambda_L g} e^{-\lambda_L' l_1}]\end{aligned}$$

or

$$\psi = A_0 e^{-\lambda_L g}\{K_L + rK_S[B_0 + C_0 e^{-(\lambda_S - \lambda_L)g}]\} \tag{11.90}$$

where A_0, B_0, and C_0 are all constants:

$$A_0 = e^{-(\lambda_L' l_1 + \lambda_L' l_2)} \tag{11.91}$$

$$B_0 = 1 - e^{(\lambda_L' - \lambda_S')l_2} \tag{11.92}$$

$$C_0 = e^{(\lambda_L' - \lambda_S')l_2}(1 - e^{(\lambda_L' - \lambda_S')l_1}) \tag{11.93}$$

It is now easy to see why a measurement of the $\pi^+\pi^-$ decay rate depends sensitively on g and leads to a determination of the mass difference Δm. Since $K_L \to \pi^+\pi^-$ violates CP invariance, the overwhelming contribution to the $\pi^+\pi^-$ decay rate comes from the K_S term on the right-hand side of (11.90), which is proportional to $B_0 + C_0 e^{-(\lambda_S - \lambda_L)g}$. The complex numbers B_0 and C_0 are represented in Fig. 11.3, and for zero gap size $g = 0$, the K_S decay amplitude is proportional to the vector sum of B_0 and C_0. As the gap size increases, the vector $C_0 e^{-(\lambda_S - \lambda_L)g}$ decreases in length at a rate determined by $(\gamma_S - \gamma_L)$ and rotates in a sense determined by the sign of the mass difference and at a rate determined by its magnitude. (Note that for the appropriate choice of lengths L_1 and L_2, B_0 and C_0 are such that for a convenient gap length g the K_S amplitude is reduced to zero. Thus one finds a sharp minimum in the $\pi^+\pi^-$ decay rate as a function of increasing gap size. The method is obviously relatively

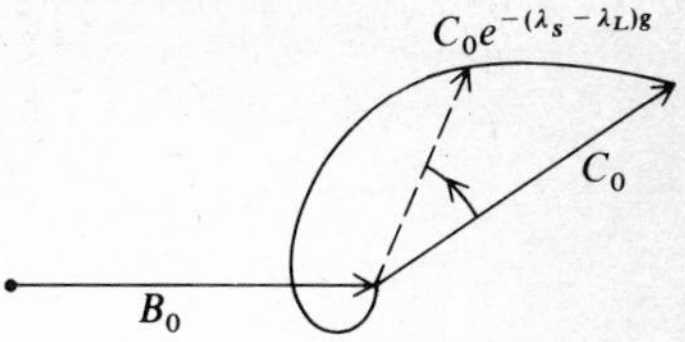

FIGURE 11.3

insensitive to the magnitude and phase of r, and if η^{+-} were zero, r could be disregarded.) The Chicago experiment (Aro 70) leads to the result

$$\Delta m = m_L - m_S = (0.542 \pm 0.006) \times 10^{10}\ \text{s}^{-1} \tag{11.94}$$

which is in agreement with earlier, less precise values, i.e.,

$$\Delta m = (0.542 \pm 0.013) \times 10^{10}\ \text{s}^{-1} \qquad \text{(Dar 69)} \tag{11.95}$$

11.8 THE PHASE OF η^{+-}

The phase ϕ^{+-}, defined by $\eta^{+-} = |\eta^{+-}|e^{i\phi^{+-}}$, is determined from observations of the interference between $K_S{}^0 \to \pi^+\pi^-$ and $K_L{}^0 \to \pi^+\pi^-$ decay amplitudes. We consider once more the time development of a neutral kaon state, which at $t = 0$ is pure $|K^0\rangle$. To first order in ε, at time $t = 0$,

$$|\psi(0)\rangle = |K^0\rangle = \frac{(1-\varepsilon)}{\sqrt{2}}\,[|K_S{}^0\rangle + |K_L{}^0\rangle] \tag{11.96}$$

At proper time t,

$$|\psi(t)\rangle = \frac{(1-\varepsilon)}{\sqrt{2}}\,[|K_S{}^0\rangle e^{-\lambda_S t} + |K_L{}^0\rangle e^{-\lambda_L t}] \tag{11.97}$$

Therefore, the decay rate to $\pi^+\pi^-$ at proper time t is proportional to

$$\begin{aligned}
I_{+-}(t) = \{&|\langle\pi^+\pi^-|H_W|K_S{}^0\rangle|^2 e^{-\gamma_S t} + |\langle\pi^+\pi^-|H_W|K_L{}^0\rangle|^2 e^{-\gamma_L t} \\
&+ e^{-(\gamma_S+\gamma_L)t/2}[\langle\pi^+\pi^-|H_W|K_S{}^0\rangle\langle\pi^+\pi^-|H_W|K_L{}^0\rangle^* e^{-i(m_S-m_L)t} \\
&\qquad + \langle\pi^+\pi^-|H_W|K_L{}^0\rangle\langle\pi^+\pi^-|H_W|K_S{}^0\rangle^* e^{i(m_S-m_L)t}]\} \\
= {}&|\langle\pi^+\pi^-|H_W|K_S{}^0\rangle|^2 \\
&\times [e^{-\gamma_S t} + |\eta^{+-}|^2 e^{-\gamma_L t} + 2e^{-(\gamma_S+\gamma_L)t/2}|\eta^{+-}|\cos\theta^{+-}]
\end{aligned} \tag{11.98}$$

where $\theta^{+-} = \phi^{+-} - \Delta m t$.

Equation (11.98) shows that the $\pi^+\pi^-$ decay rate is not exponential but contains an oscillatory interference term in addition to $e^{-\gamma_S t} + |\eta^{+-}|^2 e^{-\gamma_L t}$.

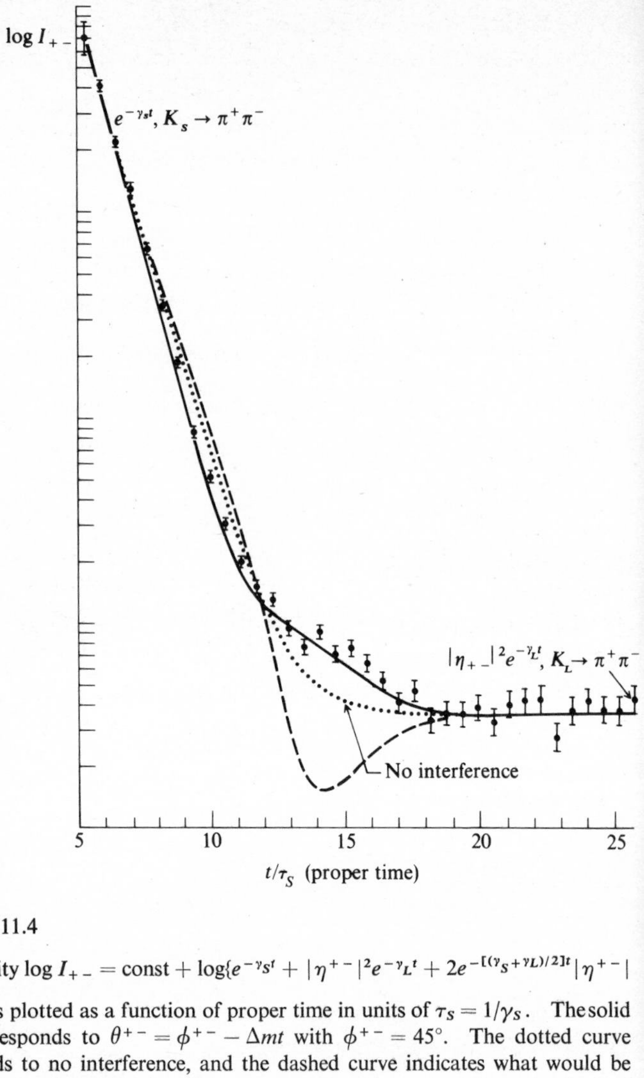

FIGURE 11.4

The quantity $\log I_{+-} = \text{const} + \log\{e^{-\gamma_S t} + |\eta^{+-}|^2 e^{-\gamma_L t} + 2e^{-[(\gamma_S+\gamma_L)/2]t}|\eta^{+-}|$ $\cos\theta^{+-}\}$ is plotted as a function of proper time in units of $\tau_S = 1/\gamma_S$. The solid curve corresponds to $\theta^{+-} = \phi^{+-} - \Delta m t$ with $\phi^{+-} = 45°$. The dotted curve corresponds to no interference, and the dashed curve indicates what would be expected if the sign of the interference term were reversed. [*From the thesis of D. Jensen, University of Chicago (unpublished). We are grateful to Professor V. Telegdi for his permission to use this illustration.*]

Observations of this interference term allow determination of ϕ^{+-} since Δm is already known from independent measurements (see Fig. 11.4). Relative to $e^{-\gamma_S t} + |\eta^{+-}|^2 e^{-\gamma_L t}$, the interference term is largest when

$$e^{+(\gamma_S+\gamma_L)t/2}|\eta^{+-}| \approx 1$$

i.e., when

$$t \approx 12\tau_S$$

for $|\eta^{+-}| \approx 2 \times 10^{-3}$. Therefore it is most advantageous to observe the interference effect a distance of 6 to 18 $K_S{}^0$ decay lengths downstream from the point of K^0 production. Unfortunately, the K^0 beam is then badly attenuated, and measurements of ϕ^{+-} are very difficult. Nevertheless, in recent experiments of great sophistication at Chicago (Jen 69) and CERN (Böh 69), ϕ^{+-} has indeed been determined in this manner:

$$\phi^{+-} = (44.7 \pm 4.0)^0 \qquad \text{(Jen 69)} \qquad (11.99)$$

$$\phi^{+-} = (49 \pm 12)^0 \qquad \text{(Böh 69)} \qquad (11.100)$$

Combining these results and taking into account the uncertainties in Δm, one finds

$$\phi^{+-} = (45.2 \pm 4.0)^0 \qquad (11.101)$$

In earlier experiments at CERN, ϕ^{+-} was determined by observation of the $\pi^+\pi^-$ decay time distribution following a regenerator (Alf 66, Bot 66, Dar 69). The results for ϕ^{+-} were rendered somewhat uncertain by the presence of a regenerator phase in θ^{+-}, which had to be determined from independent measurements of the time dependence of K_{l3} decays following the regenerator. The *vacuum regeneration* results (11.99) to (11.101) represent a substantial improvement over this older method.

11.9 THE REAL PART OF ε

Let us recall the decays K_{l3}, which are described by the following hadronic amplitudes, assuming CPT invariance:

$$f = \langle \pi^- | \mathcal{V}_1{}^\lambda | K^0 \rangle \qquad f^* = \langle \pi^+ | \mathcal{V}_1{}^{\lambda\dagger} | \bar{K}^0 \rangle$$

$$g = \langle \pi^+ | \mathcal{V}_1{}^{\lambda\dagger} | K^0 \rangle \qquad g^* = \langle \pi^- | \mathcal{V}_1{}^\lambda | \bar{K}^0 \rangle$$

We now calculate the ratio of decay rates for $K_L{}^0 \to \pi^- l^+ \nu$ and $K_L{}^0 \to \pi^+ l^- \nu$:

$$R = \frac{\Gamma(K_L{}^0 \to \pi^- l^+ \nu)}{\Gamma(K_L{}^0 \to \pi^+ l^- \nu)} = \frac{|(1+\varepsilon)f + (1-\varepsilon)g^*|^2}{|(1+\varepsilon)g + (1-\varepsilon)f^*|^2}$$

Writing $x = -g^*/f$ and rearranging terms, this becomes

$$R = \frac{\{1 - \varepsilon[(1+x)/(1-x)]\}\{1 + \varepsilon^*[(1+x)/(1-x)]^*\}}{\{1 + \varepsilon[(1+x)/(1-x)]^*\}\{1 - \varepsilon^*[(1+x)/(1-x)]^*\}}$$

$$\approx 1 + (\varepsilon + \varepsilon^*)\left(\frac{1+x^*}{1-x^*} + \frac{1-x}{1-x}\right)$$

where, in the last step, we ignore terms of second order in ε. Thus,

$$R \approx 1 + 4 \operatorname{Re} \varepsilon \left(\frac{1 - |x|^2}{|1 - x|^2} \right) \tag{11.102}$$

Quantity x is a measure of violation of the $\Delta S = \Delta Q$ rule: If $\Delta S = \Delta Q$ is valid, then $x = 0$. In this case

$$R = 1 + 4 \operatorname{Re}(\varepsilon) \qquad \Delta S = \Delta Q \tag{11.103}$$

Thus CP violation leads to a difference in the rates for $K_L{}^0 \to \pi^- e^+ \nu_e$ versus $K_L{}^0 \to \pi^+ e^- \bar{\nu}_e$ or $K_L{}^0 \to \pi^- \mu^+ \nu_\mu$ versus $K_L{}^0 \to \pi^+ \mu^- \bar{\nu}_\mu$. It is customary to define the *charge asymmetry*

$$\delta_l = \frac{\Gamma(K_L{}^0 \to \pi^- l^+ \nu) - \Gamma(K_L{}^0 \to \pi^+ l^- \bar{\nu})}{\Gamma(K_L{}^0 \to \pi^- l^+ \nu) + \Gamma(K_L{}^0 \to \pi^+ l^- \bar{\nu})} = 2 \operatorname{Re}(\varepsilon) \left(\frac{1 - |x|^2}{|1 - x|^2} \right) \tag{11.104}$$

The factor $[(1 - |x|^2)/(|1 - x|^2)]$ can actually be measured independently (Ben 68):

$$\left[\frac{1 - |x|^2}{|1 - x|^2} \right] = 0.95 \pm 0.05 \tag{11.105}$$

(See also Sec. 10.3.2.) Observations of δ_l have been carried out for K_{e3} and $K_{\mu 3}$ decays:

$$\delta_e = (2.24 \pm 0.36) \times 10^{-3} \qquad \text{(Ben 67)} \tag{11.106}$$

$$\delta_\mu = (4.05 \pm 1.35) \times 10^{-3} \qquad \text{(Dor 67)} \tag{11.107}$$

These results may be combined to yield

$$\delta = (2.7 \pm 0.3) \times 10^{-3} \tag{11.108}$$

or

$$\operatorname{Re} \varepsilon = (1.42 \pm 0.17) \times 10^{-3} \tag{11.109}$$

It is important to note that the charge asymmetry δ is the only experimental evidence for CP violation aside from the decays $K_L{}^0 \to 2\pi$.

11.10 THE PHASE OF ε

From Eqs. (11.29), (11.31), and (11.32) it is easy to show that

$$\varepsilon \approx \frac{(\Gamma_{12}^* - \Gamma_{12}) + i(M_{12}^* - M_{12})}{(\gamma_S - \gamma_L) + 2i(m_S - m_L)} \tag{11.110}$$

$$\approx \frac{(\Gamma_{12}^* - \Gamma_{12})}{\gamma_S[1 - 2i(m_L - m_S)/\gamma_S]} + \frac{i(M_{12}^* - M_{12})}{\gamma_S[1 - 2i(m_L - m_S)/\gamma_S]} \tag{11.111}$$

We shall establish that the first term in (11.111) is very small in absolute value compared with the second term, whose numerator is real. Therefore it will follow that the phase of ε is

$$\phi_\varepsilon \approx \arctan \frac{2(m_L - m_S)}{\gamma_S} = (43.2 \pm 0.4)^0 \qquad (11.112)$$

In order to demonstrate that the first term in (11.111) is negligible, we recall the unitarity relation (11.26) from which

$$\Gamma_{12}^* - \Gamma_{12} = \pi \sum_F \rho(F) | \langle \bar{K}^0 | H_W | F \rangle \langle F | H_W | K^0 \rangle - \langle K^0 | H_W | F \rangle \langle F | H_W | \bar{K}^0 \rangle | \qquad (11.113)$$

At this point we examine the separate contributions of the final states 2π, 3π, and $\pi l \nu$ to the sum (11.113). These are the only significant final states.

11.10.1 2π Final State

Since A_0 is real, the contribution to (11.113) from $|2\pi, I = 0\rangle$ is zero. For $|2\pi, I = 2\rangle$ we have

$$\begin{aligned} \frac{\pi \rho(F)(A_2{}^2 - A_2^{*2})}{\gamma_S(1 - 2i\, \Delta m/\gamma_S)} &\approx \frac{4}{\sqrt{2}} \pi \rho(F) \frac{|\mathrm{Im}\, A_2 \,\mathrm{Re}\, A_2|}{\gamma_S} \\ &\approx \sqrt{2} \frac{|\mathrm{Im}\, A_2 \,\mathrm{Re}\, A_2|}{|A_0{}^2 + |A_2|^2|} \\ &\leq 2|\varepsilon'| \left| \frac{A_2}{A_0} \right| \approx 0.1 |\varepsilon'| \end{aligned}$$

where the last step follows from the known rate of $K^+ \to \pi^+ \pi^0$ decay. Now, as shall appear in subsequent paragraphs, virtually all experimental results are consistent with $|\varepsilon'| \ll |\varepsilon|$. Therefore, the $|2\pi, I = 2\rangle$ contribution to the first term in (11.111) is negligible compared with the sum of the two terms.

11.10.2 3π Final States

Let $\Gamma_{12}{}^F$ denote a particular term in the sum over final states:

$$\Gamma_{12} = \sum_F \Gamma_{12}{}^F$$

Then, since Γ is positive definite,

$$|\Gamma_{12}{}^{F*} - \Gamma_{12}{}^{F}| \leq 2|\Gamma_{12}{}^{F}| \leq 2(\Gamma_{11}{}^{F}\Gamma_{22}{}^{F})^{1/2} \leq (\Gamma_{11}{}^{F} + \Gamma_{22}{}^{F}) \qquad (11.114)$$

Also,

$$\Gamma_{11}{}^{F} \approx \frac{\pi\rho_F}{2}\,|(1+\varepsilon)^{-1}|\langle F|H_W|K_L\rangle + \langle F|H_W|K_S\rangle||^2$$

and

$$\Gamma_{22}{}^{F} \approx \frac{\pi\rho_F}{2}\,|(1-\varepsilon)^{-1}|\langle F|H_W|K_L\rangle - \langle F|H_W|K_S\rangle||^2$$

Thus

$$\Gamma_{11}{}^{F} + \Gamma_{22}{}^{F} = \tfrac{1}{2}(\gamma_S{}^{F} + \gamma_L{}^{F}) + \mathcal{O}(\varepsilon)$$

where $\gamma_S{}^F$ and $\gamma_L{}^F$ are the partial decay rates to final state F from K_S and K_L respectively. Neglecting the term of order ε, we have, from (11.114),

$$|\Gamma_{12}{}^{F*} - \Gamma_{12}{}^{F}| \leq \tfrac{1}{2}(\gamma_S{}^{F} + \gamma_L{}^{F}) \qquad (11.115)$$

Now, in $K^0 \to 3\pi$ decays, we shall demonstrate subsequently that the only transition not hindered by centrifugal barrier effects, approximate CP invariance, and the $|\Delta I| = \frac{1}{2}$ rule is $K_L{}^0 \to 3\pi(I=1)$. The $K_S{}^0 \to 3\pi$ transition is presumably strongly hindered and has not been observed. Experimentally one can say that

$$\gamma_S(K_S{}^0 \to 3\pi) \leq \tfrac{1}{2}\gamma_L(K_L{}^0 \to 3\pi)$$

Therefore in (11.115) we are justified in putting

$$|\Gamma_{12}{}^{3\pi*} - \Gamma_{12}{}^{3\pi}| \leq \gamma(K_L \to 3\pi,\, I=1)$$

Consequently in (11.111) the 3π contribution to the first term is less than $\gamma(K_L \to 3\pi)/\gamma_S$, which is considerably smaller than the lower bound on $|\varepsilon|$ established from result (11.109) for the real part of $|\varepsilon|$.

11.10.3 $\pi l \nu$ Final States

If the $\Delta S = \Delta Q$ rule is strictly valid, there is no contribution whatever to Γ_{12} or Γ_{12}^* from $K^0 \to \pi l \nu$ decays. Accepting the present experimental limits on $\Delta S = -\Delta Q$ amplitudes in $K^0 \to \pi l \nu$ decay, we find the contribution to the first term of (11.111) to be quite negligible. Therefore we conclude that expression

(11.112) for the phase of ε is indeed justified. Combining this value with result (11.109) for the real part of ε we find

$$|\varepsilon| = \frac{\text{Re } \varepsilon}{\cos 43.2^0} = (1.95 \pm 0.25) \times 10^{-3} \qquad (11.116)$$

11.11 DETERMINATION OF η^{00}

Observation of the decay $K_L{}^0 \to \pi^0\pi^0$ is extremely difficult—far more so than $K_L{}^0 \to \pi^+\pi^-$—since now the pions are neutral and their presence must be inferred from detection of γ rays of presumed origin: $\pi^0 \to 2\gamma$. The possible sources of background in such experiments are the γ's from

$$K_L{}^0 \to \pi^+\pi^-\pi^0,\ \pi^0\pi^0\pi^0$$

and γ's from charged particles striking the walls of experimental apparatus. It is a very delicate matter to separate the legitimate γ's from the spurious ones. Thus it is perhaps not surprising that the experimental values of $|\eta^{00}|$ have shown large fluctuations over the years. (See Table 11.1 for a summary of $|\eta^{00}|$ measurements.)

It is important to note that at least one often-discussed theoretical model of CP violation—the *superweak hypothesis* (to be described later)—predicts $\eta^{+-} = \eta^{00}$. In any event, $\eta^{+-} = \eta^{00}$ is required if Im $A_2 = 0$ [Eqs. (11.84, 11.85)]. It is therefore a very interesting experimental question, still not completely settled, whether $|\eta^{+-}| = |\eta^{00}|$.

Very preliminary work has been done to determine ϕ^{00} in $\eta^{00} = |\eta^{00}|e^{i\phi^{00}}$ by observation of interference between $K_L{}^0$, $K_S{}^0 \to 2\pi^0$ decay amplitudes following regeneration in copper. The result (Cho 70) is that

$$\phi^{00} = (51 \pm 30)^0 \qquad (11.117)$$

Table 11.1
EXPERIMENTAL VALUES OF $|\eta^{00}|$

$\lvert\eta^{00}\rvert \times 10^{-3}$	Reference
3.6 ± 0.5	Gai 68
2.24 ± 0.47	Bug 68
2.26 ± 0.29	Ban 68
3.36 ± 0.5	Cen 68
3.2 ± 0.8	Cho 69
2.02 ± 0.23	Bar 70

11.12 SUMMARY OF CP-VIOLATION EXPERIMENTAL DATA

In Table 11.2 we summarize the experimental results. It can be seen that, within experimental errors, $\eta^{+-} = \varepsilon$. This implies that $\varepsilon' \approx 0$ and $\eta^{00} \approx \varepsilon$ also. However, as we have noted, present results on η^{00} are still too uncertain to make this conclusion ironclad.

11.13 IS CP VIOLATION ACCOMPANIED BY T VIOLATION OR CPT VIOLATION IN $K_L{}^0$ DECAY?

Let us return to the unitarity relation [Eq. (11.22)]. Since it is valid for an arbitrary linear combination of $|K^0\rangle$, $|\bar{K}^0\rangle$ states, we have

$$(\lambda_S + \lambda_L^*)\langle K_L{}^0 | K_S{}^0\rangle = 2\pi \sum_F \rho(F)\langle K_L{}^0 | H_W | F\rangle\langle F | H_W | K_S{}^0\rangle$$

or

$$\langle K_L{}^0 | K_S{}^0\rangle = \frac{2}{\gamma_S} \cos\phi_\varepsilon e^{i\phi_\varepsilon} 2\pi \sum_F \rho(F)\langle F | H_W | K_L{}^0\rangle^* \langle F | H_W | K_S{}^0\rangle \tag{11.118}$$

We recall that for CPT invariance $\langle K_L{}^0 | K_S{}^0\rangle$ is real, whereas for T invariance $\langle K_L{}^0 | K_S{}^0\rangle$ is imaginary [Eqs. (11.50) and (11.51)]. Therefore, CPT invariance implies that

$$\frac{1}{\gamma_S} \operatorname{Im}\left[2\pi e^{i\phi_\varepsilon} \sum_F \rho(F)\langle F | H_W | K_L{}^0\rangle^* \langle F | H_W | K_S{}^0\rangle\right] = 0 \tag{11.119}$$

Let us consider separately the terms in the sum $\sum_F$ over final states which correspond to $\pi^+\pi^-$, $\pi^0\pi^0$, on the one hand, and all other possible final states (K_{l3}, $\pi^+\pi^-\pi^0$, $\pi^0\pi^0\pi^0$, and so on), on the other hand:

$$\sum_F = \sum_{2\pi} + \sum_{F'}$$

Table 11.2 SUMMARY OF EXPERIMENTAL OBSERVATIONS OF CP VIOLATION IN $K_L{}^0 \to 2\pi$ DECAY

Parameter	Value
$\lvert\eta^{+-}\rvert$	$(1.90 \pm 0.05) \times 10^{-3}$
ϕ^{+-}	$(45.2 \pm 4.0)^0$
Re ε	$(1.42 \pm 0.17) \times 10^{-3}$
$\phi_\varepsilon = \arctan \dfrac{2\Delta m}{\gamma_S}$	$(43.2 \pm 0.4)^0$
Therefore $\lvert\varepsilon\rvert$	$(1.95 \pm 0.25) \times 10^{-3}$
$\lvert\eta^{00}\rvert$	$1.8 \times 10^{-3} < \lvert\eta^{00}\rvert < 3 \times 10^{-3}$?
ϕ^{00}	$(51 \pm 30)^0$

Then (11.119) becomes

$$\frac{1}{\gamma_S}\,\mathrm{Im}\Big[2\pi e^{i\phi_\varepsilon}\sum_{\pi^+\pi^-}\rho(\pi^+\pi^-)\eta^{+-*}|\langle\pi^+\pi^-|H_W|K_S{}^0\rangle|^2 + 2\pi e^{i\phi_\varepsilon}\sum_{\pi^0\pi^0}\rho(\pi^0\pi^0)\eta^{00*}|\langle\pi^0\pi^0|H_W|K_S{}^0\rangle|^2\Big]$$
$$= -\frac{1}{\gamma_S}\,\mathrm{Im}\Big[2\pi e^{i\phi_\varepsilon}\sum_{F'}\rho(F')\langle F'|H_W|K_L{}^0\rangle^*\langle F|H_W|K_S{}^0\rangle\Big] \qquad (11.120)$$

After a simple manipulation of the left-hand side, (11.120) becomes

$$\left[\sin(\phi^{+-}-\phi_\varepsilon) + R\left|\frac{\eta^{00}}{\eta^{+-}}\right|\sin(\phi^{00}-\phi_\varepsilon)\right]$$
$$= \frac{R+1}{\gamma_S|\eta^{+-}|}\,\mathrm{Im}\Big[e^{i\phi_\varepsilon}2\pi\sum_{F'}\rho(F')\langle F'|H_W|K_L{}^0\rangle^*\langle F|H_W|K_S{}^0\rangle\Big] \qquad (11.121)$$

where
$$R = \frac{\text{rate}\,(K_S{}^0\to\pi^0\pi^0)}{\text{rate}\,(K_S{}^0\to\pi^+\pi^-)} \simeq 0.48$$

Now, no CP violation has been observed in any of the channels $K^0\to\pi^+\pi^-\pi^0$, $K^0\to\pi l\nu$,‡ and an upper limit on the amplitudes for CP violation in these decays yields

$$\frac{R+1}{\gamma_S|\eta^{+-}|}\left|2\pi\sum_{F'}\rho(F')\langle F'|H_W|K_L{}^0\rangle^*\langle F'|H_W|K_S{}^0\rangle\right| < 0.2 \qquad (11.122)$$

One must admit that little is known about possible CP violation in $K^0\to\pi^0\pi^0\pi^0$, but any reasonable assumptions still lead to the conclusion that the left-hand side of (11.122) is consistent with zero.

As for the left-hand side of (11.121), it can be determined from the observations listed in Table 11.2, with the conclusion that it is also very small (consistent with zero). Therefore, Eq. (11.121) is borne out by observation and we have no contradiction of CPT invariance. On the other hand, T invariance implies that

$$\frac{1}{\gamma_S}\,\mathrm{Re}\Big[e^{i\phi_\varepsilon}2\pi\sum_{\pi^+\pi^-}\rho(\pi^+\pi^-)\eta^{+-*}|\langle\pi^+\pi^-|H_W|K_S{}^0\rangle|^2 + e^{i\phi_\varepsilon}2\pi\sum_{\pi^0\pi^0}\rho(\pi^0\pi^0)\eta^{00*}|\langle\pi^0\pi^0|H_W|K_S{}^0\rangle|^2\Big]$$
$$= -\frac{1}{\gamma_S}\,\mathrm{Re}\Big[e^{i\phi_\varepsilon}2\pi\sum_{F'}\rho(F')\langle F'|H_W|K_L{}^0\rangle^*\langle F'|H_W|K_S{}^0\rangle\Big] \qquad (11.123)$$

‡ For further discussion of $K\to 3\pi$ decays, see Secs. 11.14 and 11.15.

or
$$\left[\cos(\phi^{+-}-\phi_\varepsilon)+\cos(\phi^{00}-\phi_\varepsilon)R\left|\frac{\eta^{00}}{\eta^{+-}}\right|\right]$$
$$=-\frac{(R+1)}{|\eta^{+-}|\gamma_S}\,\mathrm{Re}\left[e^{i\phi_\varepsilon}2\pi\sum_{F'}\rho(F')\langle F'|H_W|K_L{}^0\rangle^*\langle F|H_W|K_S{}^0\rangle\right] \qquad (11.124)$$

Although the absolute value of the right-hand side of (11.124) is very small, as already noted, the left-hand side is large: 1.5 ± 0.3. We thus contradict the assumption of T invariance. Although one requires better knowledge of CP-violating effects in $K^0 \to \pi^0\pi^0\pi^0$ to clinch the argument, it seems quite safe to state that the CP violation in $K_L{}^0$ decay is consistent with T violation and CPT invariance.

11.14 SPECULATIONS ON THE ORIGIN OF CP VIOLATION IN $K_L{}^0$ DECAYS

Unfortunately, the experimental results of Table 11.2 do not tell us very much about the possible origin of CP violation. Many speculations have been offered since 1964, and it is impossible for us to do anything more than summarize the few most frequently discussed possibilities. CP violation in neutral K decay could be caused by one or more of the following effects:

1. A small T or C violation (approximately 2×10^{-3}) in the strong interaction
2. A large T or C violation in the electromagnetic interaction of hadrons
3. A small T violation (approximately 2×10^{-3}) in the weak interaction
4. Existence of a CP-violating superweak interaction (coupling strength about 10^{-8} times that of the weak interaction or even smaller) obeying the selection rule $|\Delta S| = 2$

Let us now consider each of these possibilities briefly.

11.14.1 Small C and T Violation in the Strong Interaction?

Here it is imagined that $K_L{}^0 \to 2\pi$ decay occurs via a two-step process

$$K_L{}^0 \to M \to 2\pi \qquad (11.125)$$

where M is some intermediate hadronic state and the first step is supposed to be due to the ordinary $|\Delta S| = 1$ CP-conserving weak interaction, resulting in the quantum numbers for M: $B = 0$, $S = 0$, $CP = -1$. The second step proceeds via a $\Delta S = 0$ CP-violating strong interaction. The total amplitude for this second-order process must be of order $2 \times 10^{-3}G$ in order to account for observed $K_L{}^0 \to 2\pi$ decay. The amplitude is of order

$$\frac{Gf}{m_K - m_M} \approx 2 \times 10^{-3}G \qquad (11.126)$$

where f is the coupling strength of the new hypothetical interaction. Since the energy denominator is about unity, we have $f \approx 2 \times 10^{-3}$, which may be compared to the ordinary strong coupling of order unity.

Now, CPT invariance of the various interactions implies that the masses, decay lifetimes, and gyromagnetic ratios of a particle and its corresponding antiparticle are equal. Of the various equalities of this kind established experimentally, the most precise is that of m_{K^0} and $m_{\bar{K}^0}$, which are the same to about two parts in 10^{16} (see Table 11.3). This indicates that if the strong-interaction hamiltonian contains a CPT-noninvariant part, it must be less than about 10^{-15} of the CPT-invariant part. We may therefore safely assume that the strong interaction is CPT-invariant for all practical purposes. A similar argument establishes that the electromagnetic interaction of hadrons is CPT-invariant to at least one part in 10^{11} or 10^{12}.

It may also be established that the strong and hadronic electromagnetic interactions conserve parity to a very high degree. As discussed in detail in

Table 11.3 PARTICLE-ANTIPARTICLE MASSES, LIFETIMES, AND GYROMAGNETIC RATIOS

Particle	Quantity	Value
e^+, e^-	$\frac{1}{2}(g^+ - g^-)$	$(1.5 \pm 2)\alpha^2/\pi^2$
μ^+, μ^-	$\frac{1}{2}(g^+ - g^-)$	$(0 \pm 1.5)\alpha^2/\pi^2$
μ^+, μ^-	$(\tau_+/\tau_-) - 1$	0.000 ± 0.001
π^+, π^-	$(\tau_+/\tau_-) - 1$	-0.0009 ± 0.0008
K^+, K^-	$(\tau_+/\tau_-) - 1$	-0.0009 ± 0.0008
e^-	m_-	0.511006(2) MeV
e^+	m_+	0.510976(7) MeV
μ^+, μ^-	$(m_+/m_-) - 1$	10^{-4}
π^+	m_+	139.60 ± 0.05 MeV
π^-	m_-	139.578 ± 0.017 MeV
K^+	m_+	493.78 ± 0.17 MeV
K^-	m_-	493.7 ± 0.3 MeV
$K^0, \bar{K}^0$	Masses equal to 2 parts in 10^{16}	
p	m_p	938.256 ± 0.005 MeV
$\bar{p}$	$m_{\bar{p}}$	$1.008 \pm 0.005\ m_p$
d	$(m_d/m_{\bar{d}}) - 1$	$\pm 3\%$
Λ^0	m_Λ	1115.44 ± 0.12 MeV
	$Q_\Lambda(\Lambda \to p\pi^-)$	37.60 ± 0.12 MeV
	$Q_{\bar{\Lambda}}$	$35\ {}^{+2.6}_{-0.9}$ MeV

Chap. 12, very small parity violations have been observed in nuclear forces (as manifested by circular polarization of γ rays emitted from unpolarized nuclei and asymmetries in γ emission from polarized nuclei). However, such violations are attributed to weak interactions, and the experimental results demonstrate that parity is conserved in the strong and electromagnetic interactions to at least one part in 10^5.

It follows from the CPT and P invariance of the strong and electromagnetic interactions that a CP violation in either or both would have to come about through C and T violation. If this occurs in the strong interaction, we require a C- and T-violating coupling of order 10^{-3} of the ordinary strong coupling to explain the decays $K_L{}^0 \to \pi\pi$. T invariance in the strong interaction[1] is tested directly by comparison of the differential cross sections for certain nuclear reactions and their inverses, e.g.,

$$d + \mathrm{Mg}^{24} \rightleftharpoons p + \mathrm{Mg}^{25} \qquad \text{(Bod 66)} \qquad (11.127)$$

$$\alpha + \mathrm{Mg}^{24} \rightleftharpoons p + \mathrm{Al}^{27} \qquad \text{(VonW 66)} \qquad (11.128)$$

and

$$d + \mathrm{O}^{16} \rightleftharpoons \alpha + \mathrm{N}^{14} \qquad \text{(Tho 68)} \qquad (11.129)$$

These experiments show that time-reversal-odd amplitudes must be less than about 3×10^{-3} of corresponding time-reversal-even amplitudes.

C invariance of the strong interaction is tested by experiments involving proton-antiproton annihilation (Pai 59, Dob 66). Here the reaction

$$p\bar{p} \to \bar{K}^0 K^+ \pi^+$$

is compared with the reaction $p\bar{p} \to K^0 K^- \pi^+$, or

$$p\bar{p} \to K^0 K^+ \pi^- \pi^0$$

with $p\bar{p} \to K^0 K^- \pi^+ \pi^0$, for example. Such reactions are all of the type

$$p\bar{p} \to 1 + 2 + X \qquad (11.130)$$

and

$$p\bar{p} \to \bar{1} + \bar{2} + \bar{X} \qquad (11.131)$$

where X refers to any group of particles and only particles 1 and 2 in the final state are singled out for observation. The test of C invariance may be understood in a very simple way, as follows.

Let us define the transition probability $W(p_1, \theta_1, p_2, \theta_2, \phi_{12})$ for observing particle 1(2) with momentum $p_1(p_2)$ at an angle $\theta_1(\theta_2)$ relative to the direction of $\bar{p}$, where ϕ_{12} is the azimuth of particle 2 relative to the plane $\bar{p}1$. The various angles are defined in Fig. 11.5, and the reader will note that particle 2 does not necessarily lie in the plane $\bar{p}1$ since the final state contains three or more particles.

[1] E. M. Henley (Hen 69) presents an excellent review of parity and time-reversal invariance tests, and provides much more detail than is found here.

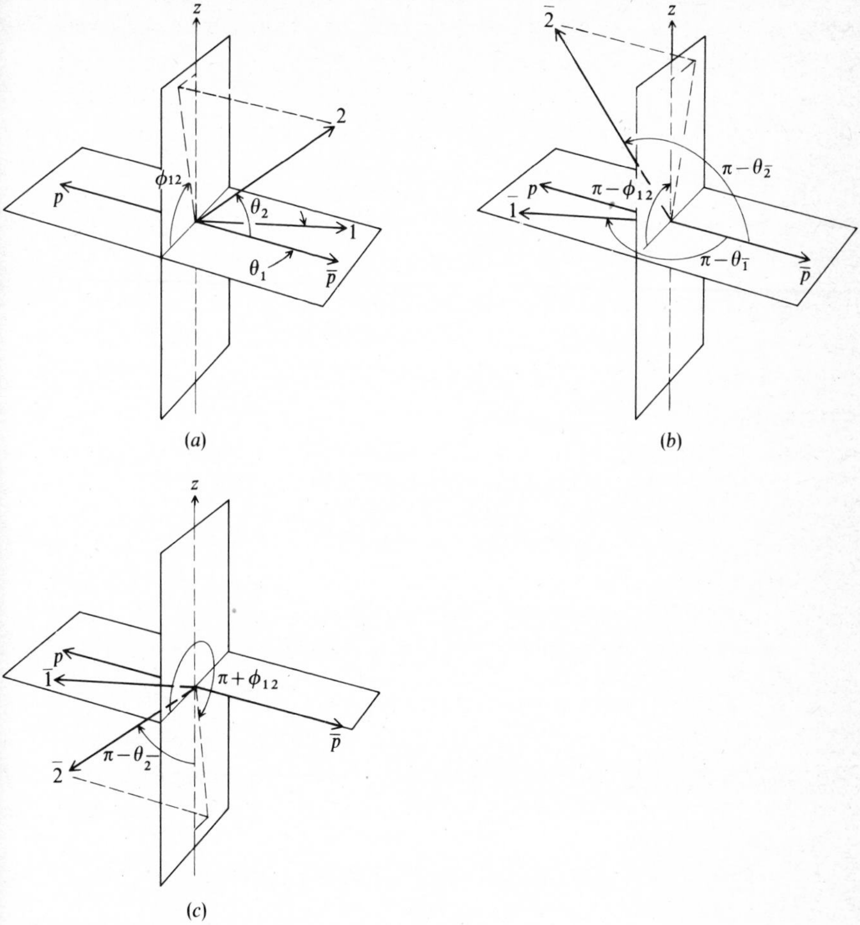

FIGURE 11.5
(*a*) Vectors which define the directions of particles p, $\bar{p}$, 1, and 2. (*b*) The result of a CR transformation [charge conjugation and a rotation of 180° about the z axis] on the state of (*a*). If the interaction is C invariant, the amplitudes for processes (*a*) and (*b*) must be equal. (*c*) The result of a CP transformation [charge conjugation and inversion of all vectors through the origin] on the state of (*a*). If the interaction is CP invariant, the amplitudes for processes (*a*) and (*c*) must be equal.

If we define a rotation R of 180° about the axis z, we see easily from Fig. 11.5a and b that invariance of strong interaction with respect to the operation CR implies

$$W(p_1, \theta_1, p_2, \theta_2, \phi_{12}) = W(p_{\bar{1}}, \pi - \theta_{\bar{1}}, p_{\bar{2}}, \pi - \theta_{\bar{2}}, \pi - \phi_{\bar{1}\bar{2}}) \qquad (11.132)$$

Of course, this is equivalent to C invariance since we assume rotational invariance. Incidentally, we see from Fig. 11.5 that CP invariance may also be tested in the same reactions. In this case we have

$$W(p_1, \theta_1, p_2, \theta_2, \phi_{12}) = W(p_{\bar{1}}, \pi - \theta_{\bar{1}}, p_{\bar{2}}, \pi - \theta_{\bar{2}}, \pi + \phi_{\bar{1}\bar{2}}) \qquad (11.133)$$

Experiments of this kind have established that C invariance holds in the strong interaction to at least one part in 10^2.

Thus the experimental results are not yet sufficiently precise to rule out with certainty a small C and T violation in the strong interactions. However, no direct evidence has been found to favor this hypothesis.

11.14.2 A Small C and T Violation in the Hadronic Electromagnetic Interaction?

A CT-violating electromagnetic interaction could give rise to CP violation in $K_L{}^0$ decay via a two-step process of the form (11.125), where the transition $M \to 2\pi$ is now imagined to be $|\Delta S| = 0$ electromagnetic and the coupling, of order 2×10^{-3}, is (perhaps accidentally) close to $\alpha/2\pi$ where α is the fine structure constant.

Interest in this hypothesis arose originally because the evidence for C invariance of the hadronic electromagnetic interaction is surprisingly poor. Elastic electron-nucleon scattering processes could not (in first order) reveal a C-noninvariant portion of the EM hamiltonian, even if it existed, because of current conservation. Although other hadronic electromagnetic processes such as η^0 or π^0 decay can in principle be used to test for C violation, various effects conspire to suppress the manifestations of a possible C-violating component. For example, in the decay $\eta^0 \to \pi^+\pi^-\pi^0$, an asymmetry in the kinetic energy distributions of π^+, π^- would be indicative of C violation. No clear evidence for such an asymmetry has been found in experiments reaching the 1 percent level of precision (Cno 66, Gor 68). On the other hand, no one has yet predicted firmly what kind of asymmetry should be expected if C is violated. The theoretical situation is murky owing to uncertain centrifugal barrier effects, etc.

There are a variety of experimental methods for testing T invariance in electromagnetic processes. These include direct (detailed balance) comparison of differential cross sections for photonuclear reactions with their inverses, e.g.,

$$\gamma + d \rightleftharpoons n + p \qquad (11.134)$$

with photon energies in the range of 300 MeV. The experimental results indicate no T violation at the 5 percent level of precision (Sch 71). Other high-energy inelastic EM reactions, e.g.,

$$e + p \to N^* + e \qquad (11.135)$$

make use of scattering from polarized protons to search for a correlation of the form

$$\boldsymbol{\sigma}_p \cdot \mathbf{p}_{ei} \times \mathbf{p}_{ef} \qquad (11.136)$$

No T violation is found, but the precision (about 5 percent) is rather crude (Che 68).

Time-reversal invariance requires that the relative phase between two competing amplitudes be zero or π in an electromagnetic transition between two nuclear levels (for example, an $E2$, $M1$ mixed transition). Recent Mössbauer observations of such "mixed" γ rays from Ru^{99} (Kis 67) and Ir^{193} (Ata 68) indicate that a T-odd amplitude, if any, must be less than about 3×10^{-3} of the corresponding T-even amplitude.

It seems plausible that if a CT violation does exist in the hadronic electromagnetic interaction, it may manifest itself in a permanent electric dipole moment (EDM) of the neutron and/or proton. Such a dipole moment cannot exist unless both P *and* T are violated (Lan 57). We thus imagine the EDM to arise from combined operation of the hypothesized C- and T-violating EM interaction and the C- and P-violating weak interaction. Here we can easily make a crude order-of-magnitude estimate of the EDM, as follows. If the CT violation of the EM interaction is large, the moment should be

$\mu \approx$ 1 electronic charge $\times$ nucleon compton wavelength $\times$ G $\times$ (mass of nucleon)^2

$$\approx e \frac{\hbar}{m_p C} \times G m_p^2$$

$$\approx 10^{-19} e \text{ cm}$$

More sophisticated estimates (see Mei 64, Sch 64, Gla 65, Bou 65, Fei 65a, b, Sal 65, Bab 67) vary from about 10^{-19} e cm to about 10^{-23} e cm, depending on the particular assumptions involved. However, if a CT violation of the EM interaction is invoked to explain $K_L{}^0 \to 2\pi$ decay, it seems likely that the neutron

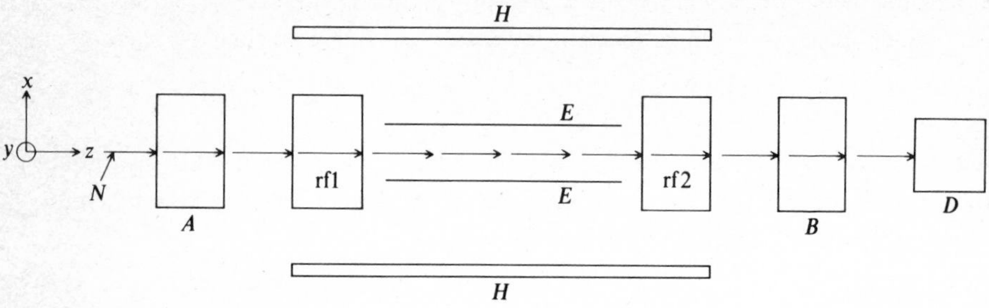

FIGURE 11.6
Schematic diagram of apparatus for measurement of neutron electric dipole moment (Bai 69).

A: Spin polarizer (magnetic mirror)
B: Spin analyzer
N: Thermal neutron beam
D: Neutron detector
rf1: Radio-frequency coil for neutron spin reorientation
rf2: Radio-frequency coil for neutron spin reorientation
E: Electrostatic plates
H: Homogeneous magnetic field pole pieces

Polarized neutrons pass through transverse static magnetic field H (along x) at each end of which is superimposed rf magnetic fields H_1 (along z). Spin flip transitions are induced when frequency of H_1 equals average neutron larmor frequency between coils. When strong electrostatic field E is applied in direction x, an additional precession would be introduced, proportional to the intrinsic electric dipole moment. Thus a finite electric dipole moment would result in a resonance shift.

EDM would not be smaller than 10^{-21} e cm. Recent experiments (Bai 69) utilizing slow polarized neutron beams and the atomic beam magnetic resonance method (Fig. 11.6) show that

$$\mu_n < 5 \times 10^{-23}\ e\ \text{cm} \qquad (11.137)$$

which would seem to argue strongly against the hypothesis of CT violation in EM interaction of hadrons.

11.14.3 Small T Violation in the Weak Interaction?

Here we imagine that the decay $K_L{}^0 \rightarrow 2\pi$ proceeds directly by a first-order weak interaction which violates T (and therefore CP). The coupling is then of order $2 \times 10^{-3}G$, and we are led to expect T violations of order 10^{-3} in other weak interactions. As discussed in Chap. 4,[1] we cannot test for T invariance directly

[1] See also Appendix 3.

in the weak interaction, but instead we are compelled to make an indirect test by searching for the existence of triple correlations, for example, $\hat{\boldsymbol{\sigma}}_n \cdot \mathbf{p}_e \times \mathbf{p}_\nu$ in neutron or Ne^{19} decay (Sec. 5.7), $\hat{\boldsymbol{\sigma}}_\mu \cdot \mathbf{p}_\mu \times \mathbf{p}_\pi$ in $K_{\mu 3}$ decay (Sec. 10.3), or $\mathbf{p}_p \cdot \hat{\boldsymbol{\sigma}}_p \times \hat{\boldsymbol{\sigma}}_\Lambda$ in $\Lambda^0 \to p\pi^-$ decay (Sec. 9.2).

In each of these tests we must be careful to correct for final-state interactions (which are particularly large in $\Lambda^0 \to p\pi^-$ decay) which can masquerade as T violation. The experiments fail to reveal any evidence for T violation at the 1 percent level of precision (neutron decay, $K_{\mu 3}$ decay) or the 3×10^{-3} level (Ne^{19} decay). Clearly, such precision is not sufficient to rule out the hypothesis. But even if the Ne^{19} experiment were improved by an order of magnitude, a null result in this $\Delta S = 0$ decay would not necessarily imply that T invariance holds in the $\Delta S = 1$ weak amplitude for $K_L{}^0 \to 2\pi$.

11.14.4 The Superweak Interaction?

Finally we discuss a very simple model proposed by Wolfenstein (Wol 64). Here we imagine the $K_L{}^0 \to 2\pi$ transition to be the result of a two-step process

$$K_L{}^0 \to K_S{}^0 \to 2\pi \qquad (11.138)$$

The first step involves a $|\Delta S| = 2$ transition via some new interaction with coupling constant f. The second step is just the ordinary allowed weak interaction with coupling constant G. The overall amplitude is

$$A \approx \frac{fG}{m_L - m_S} \approx 2 \times 10^{-3} G \qquad (11.139)$$

Thus

$$f \approx 2 \times 10^{-3} \Delta m \approx 2 \times 10^{-3} m_\pi{}^5 G^2$$
$$\approx 2 \times 10^{-8} m_\pi{}^5 G \qquad (11.140)$$

If we insert the mass of the pion, we obtain $f \approx 10^{-12} G$. Even if we allow the pion mass to be replaced by the nucleon mass (of order unity), we obtain

$$f \approx 10^{-8} G \qquad (11.141)$$

Obviously the numerical value of f is not to be taken too seriously. Rather, the essential point is that a minute coupling f (the *superweak* interaction) can manifest itself in an observable amplitude because of the huge amplification afforded by the tiny energy denominator Δm. Unfortunately the $K_L{}^0 K_S{}^0$ system is unique—we do not encounter similar pairs of particle states so close in energy elsewhere in nature. Therefore it seems extremely unlikely that the superweak interaction, if it indeed exists, could be observed except in the one case of $K_L{}^0 \to 2\pi$ decay.

The superweak hypothesis stands alone among the various speculations we have mentioned in that it makes definite experimental predictions about the phenomenological parameters of $K^0 \to 2\pi$ decay. This follows directly from the assumed form (11.138). Thus, according to the superweak hypothesis,

$$\eta^{+-} = \eta^{00} = \varepsilon \qquad (11.142)$$

and

$$\varepsilon' = 0 \qquad (11.143)$$

Moreover, the phase of ε should be given by

$$\phi_\varepsilon = \arctan \frac{2\,\Delta m}{\gamma_S} = 43^\circ \qquad (11.144)$$

The experimental results of $K_L{}^0$ decay, although not yet complete, are consistent with these predictions and give the superweak hypothesis support which increases with time.

11.15 $K_{\pi 3}$ DECAYS; GENERAL KINEMATIC CONSIDERATIONS

We now turn our attention to the $K_{\pi 3}$ decays:

$$K^+ \to \pi^+\pi^+\pi^- \qquad Q = 75.1 \text{ MeV} \qquad (11.145)$$

$$K^- \to \pi^-\pi^-\pi^+ \qquad Q = 75.1 \text{ MeV} \qquad (11.146)$$

$$K^\pm \to \pi^0\pi^0\pi^\pm \qquad Q = 84.3 \text{ MeV} \qquad (11.147)$$

$$K^0 \to \pi^0\pi^0\pi^0 \qquad Q = 92.8 \text{ MeV} \qquad (11.148)$$

$$K^0 \to \pi^+\pi^-\pi^0 \qquad Q = 79.0 \text{ MeV} \qquad (11.149)$$

In the K rest frame, let us denote the energies and three-momenta of the three pions by E_1, E_2, E_3 and $\mathbf{p}_1$, $\mathbf{p}_2$, $\mathbf{p}_3$, respectively. By conservation of energy,

$$m_K = E_1 + E_2 + E_3$$

$$= T_1 + T_2 + T_3 + m_1 + m_2 + m_3$$

or

$$Q = m_K - m_1 - m_2 - m_3 = T_1 + T_2 + T_3 \qquad (11.150)$$

where the T's are the pion kinetic energies [T_3 always stands for the kinetic energy of the last-named pion in reactions (11.145) to (11.149)].

It is very convenient to represent condition (11.150) on a *Dalitz* diagram (Fig. 11.7). Thus, consider an equilateral triangle of height Q and an arbitrary point P which represents a 3π final state with kinetic energies T_1, T_2, T_3. The sum of the distances T_1, T_2, T_3 from point P to the three sides of the triangle is equal to Q. (This can be seen from the fact that the area of triangle ABC is

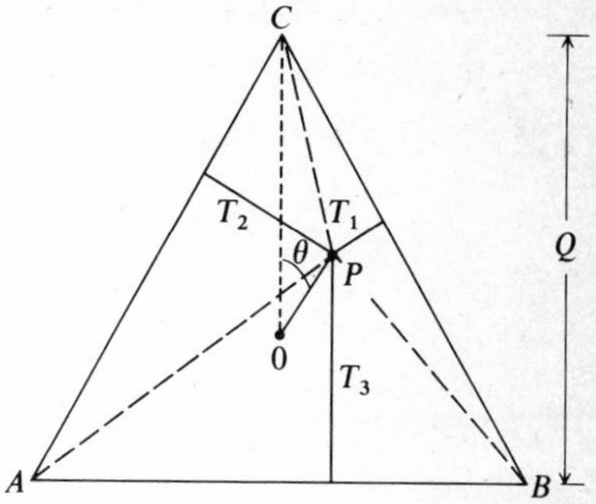

FIGURE 11.7

equal to the sum of the areas of triangles APB, BPC, and CPA.) Relative to the center of the triangle O, we assign point P the polar coordinates $Q/3r$, θ. Then it follows that

$$T_3 = \frac{Q}{3}(1 + r\cos\theta) \tag{11.151}$$

$$T_1 = \frac{Q}{3}\left[1 + r\cos\left(\theta + \frac{2\pi}{3}\right)\right] \tag{11.152}$$

and

$$T_2 = \frac{Q}{3}\left[1 + r\cos\left(\theta - \frac{2\pi}{3}\right)\right] \tag{11.153}$$

We see that $r = 0$ corresponds to $T_1 = T_2 = T_3 = Q/3$.

Although a given decay event $K \to 3\pi$ is characterized by definite T_1, T_2, T_3 and is represented by a single point in the plane, specified by r, θ, not all points within the triangle are accessible. From conservation of momentum,

$$\mathbf{p}_1 + \mathbf{p}_2 + \mathbf{p}_3 = 0 \tag{11.154}$$

we have
$$2\mathbf{p}_1 \cdot \mathbf{p}_2 = (E_3{}^2 - m_3{}^2) - (E_1{}^2 - m^2) - (E_2{}^2 - m^2)$$

where $m = m_1 = m_2$. Thus,

$$2(E_1{}^2 - m^2)^{1/2}(E_2{}^2 - m^2)^{1/2} \geq (E_3{}^2 - m_3{}^2) - (E_1{}^2 - m^2) - (E_2{}^2 - m^2) \tag{11.155}$$

Inequality (11.155) and Eq. (11.150) define a region of the triangle bounded by the closed curve

$$4(E_1{}^2 - m^2)(E_2{}^2 - m^2) = [m_K{}^2 - 2m_K(E_1 + E_2) + 2E_1E_2 + 2m^2 - m_3{}^2]^2 \tag{11.156}$$

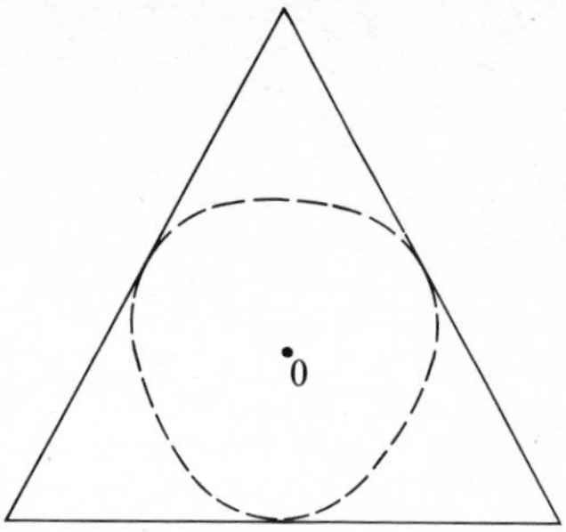

FIGURE 11.8

which may be expressed in polar coordinates as follows:

$$r(\theta) \leq r_0(\theta) \tag{11.157}$$

where

$$r_0{}^2 = (1 + \varepsilon_0)^{-1}(1 - \varepsilon_0 r_0{}^3 \cos 3\theta) \tag{11.158}$$

and

$$\varepsilon_0 = \frac{2Qm_K}{(2m_K - Q)^2} \tag{11.159}$$

The accessible region is shown in Fig. 11.8.

Next, we calculate the transition probability for any one of the decays (11.145) to (11.149). The amplitude A_i for the ith decay must be a function of the kinematic invariants which can be constructed from the four-momenta of the various particles. However, because of conservation of energy and momentum, the amplitude for any three-body decay can always be expressed as a function of the kinetic energies of two of the particles alone:

$$A_i = A_i(T_1, T_2) \tag{11.160}$$

or

$$A_i = A_i(r, \theta) \tag{11.161}$$

The transition probability is

$$dW_i = \frac{1}{(2\pi)^5} \frac{\delta(m_K - E_1 - E_2 - E_3)\delta^3(\mathbf{p}_1 + \mathbf{p}_2 + \mathbf{p}_3)}{16m_K E_1 E_2 E_3} \times d^3\mathbf{p}_1 d^3\mathbf{p}_2 d^3\mathbf{p}_3 |A_i(T_1, T_2)|^2 \tag{11.162}$$

Integrating over $\mathbf{p}_3$, we have

$$dW_i = \frac{1}{(2\pi)^5} \frac{1}{16m_K E_1 E_2 E_3} |A_i|^2 |\mathbf{p}_1| E_1\, dE_1 |\mathbf{p}_2| E_2\, dE_2 \times \delta(m_K - E_1 - E_2 - E_3)\, d\Omega_1\, d\Omega_2$$

where

$$E_3 = \sqrt{\mathbf{p}_1{}^2 + \mathbf{p}_2{}^2 + 2\mathbf{p}_1 \cdot \mathbf{p}_2 + m_3{}^2}$$

Then, integrating over $d\Omega_1\, d\Omega_2 = d\Omega_1\, d\Omega_{12}$, we have

$$\rho_i \equiv \frac{d^2 W_i}{dT_1\, dT_2} = \frac{d^2 W_i}{dE_1\, dE_2} = \frac{8\pi^2}{(2\pi)^5} \frac{|A_i|^2}{16 m_K E_3} p_1 p_2 \times \int \sin\theta_{12}\, d\theta_{12}\, \delta(m_K - E_1 - E_2 - E_3)$$

The integral is

$$I = \int \sin\theta_{12}\, d\theta_{12}\, \delta\left(m_K - E_1 - E_2 - \sqrt{\mathbf{p}_1^2 + \mathbf{p}_2^2 + 2|\mathbf{p}_1||\mathbf{p}_2|\cos\theta_{12} + m_3^2}\right) = \frac{E_3}{|\mathbf{p}_1||\mathbf{p}_2|}$$

Thus,

$$\rho_i = \frac{d^2 W}{dT_i\, dT_2} = \frac{1}{64\pi^3 m_K} |A_i(T_1, T_2)|^2 \qquad (11.163)$$

The total decay rate is thus

$$\Gamma_i = \int_{\substack{\text{accessible}\\ \text{region}}} \rho\, dT_1\, dT_2 = \frac{1}{64\pi^3 m_K} \int |A_i(T_1, T_2)|^2 dT_1\, dT_2 \qquad (11.164)$$

It is easy to convert this into an integral over polar coordinates r, θ:

$$\Gamma_i = \frac{1}{64\pi^3 m_K} \int |A_i|^2 \left| \frac{\partial(T_1, T_2)}{\partial(r, \theta)} \right| dr\, d\theta$$

where the Jacobian is

$$\left| \frac{\partial(T_1, T_2)}{\partial(r, \theta)} \right| = \frac{\sqrt{3}}{18} Q^2 r$$

Thus,

$$\Gamma_i = \frac{\sqrt{3}}{18} \frac{Q^2}{64\pi^3 m_K} \int |A_i(r, \theta)|^2 r\, dr\, d\theta \qquad (11.165)$$

This means that if we perform an experiment to observe a particular $K_{\pi 3}$ decay and plot each observed event as a point in the Dalitz diagram, then the density of events in the neighborhood of a given point with coordinates r, θ must be proportional to $|A_i(r, \theta)|^2$. Although it is beyond our powers to calculate the $A_i(r, \theta)$, we can establish a number of useful restrictions on these amplitudes from (approximate) CP invariance, conservation of angular momentum, Bose–Einstein statistics, and the $|\Delta I| = \frac{1}{2}$ rule.

11.16 SYMMETRY RESTRICTIONS ON $K_{\pi 3}$ AMPLITUDES

11.16.1 $A_i(r, \theta) = A_i(r, -\theta)$

For the decays (11.145) to (11.148) the first two pions are identical. The final three-pion state must therefore be symmetric with respect to exchange of these two pions, and we have $A_i(r, \theta) = A_i(r, -\theta)$. For $K^0 \to \pi^+\pi^-\pi^0$ decay, we shall establish subsequently that CP invariance (which holds to good approximation) implies $K_L{}^0 \to \pi^+\pi^-\pi^0$ is allowed, $K_S{}^0 \to \pi^+\pi^-\pi^0$ is forbidden, and the amplitude $A_{+-0}(r, \theta)$ for $K_L{}^0 \to \pi^+\pi^-\pi^0$ satisfies $A(r, \theta) = A(r, -\theta)$.

Assuming then that each amplitude is an even function of θ, we expand A_i as a function of T_1, T_2, T_3 about the origin. Taking into account that $A_i(T_1, T_2, T_3) = A_i(T_2, T_1, T_3)$ and keeping only first-order terms since $Q/3 \ll m_\pi$, we have

$$A_i(T_1, T_2, T_3) \approx A_i(0) + \frac{\partial A_i}{\partial T_1}\bigg|_0 \left(T_1 - \frac{Q}{3}\right) + \frac{\partial A_i}{\partial T_2}\bigg|_0 \left(T_2 - \frac{Q}{3}\right) + \frac{\partial A_i}{\partial T_3}\bigg|_0 \left(T_3 - \frac{Q}{3}\right)$$

$$= A_i(0) + a_i\left(r \cos\left(\theta + \frac{2\pi}{3}\right) + r \cos\left(\theta - \frac{2\pi}{3}\right)\right) + b_i r \cos\theta$$

where a_i and b_i are two constants. Thus introducing a new constant, the *slope parameter* σ_i, and the quantity $y = r\cos\theta$, we have

$$A_i \approx A_i(0)\left[1 + \frac{2m_K Q}{3m_\pi{}^2}\sigma_i y\right] \tag{11.166}$$

11.16.2 Possible Isospin States of Three Pions

Let us consider two pions π_a and π_b. These may be coupled to form isospin states $I_{ab} = 2$, 1, or 0. When the isospin of the third pion is included, we get one $I = 3$ state, two $I = 2$ states, three $I = 1$ states, and, if the total charge of the three pions is zero, one $I = 0$ state. Of these the $I = 3$ state is completely symmetric:

$$|I = 3, I_3 = 0\rangle \equiv |3, 0\rangle$$

$$= \frac{1}{\sqrt{10}}\,|(+0-) + (-+0) + (0-+) + (0+-)$$

$$+ (+-0) + (-0+) + 2(000)\rangle \tag{11.167}$$

and

$$|3, 1\rangle = \frac{1}{\sqrt{15}} |(++-) + (+-+) + (-++) + 2(+00) + 2(0+0) + 2(00+)\rangle \quad (11.168)$$

where $|(+0-)\rangle = |\pi_1{}^+\pi_2{}^0\pi_3{}^-\rangle$, and so on.

The $I = 0$ state is completely antisymmetric, and the $I = 2$ and $I = 1$ states, as obtained by the procedure described, are of mixed symmetry. However, an appropriate linear combination of $I = 1$ states corresponding to $I_{ab} = 0$ and 2 yields a completely symmetric $I = 1$ state:

$$\Big|1, 0\rangle = \frac{1}{\sqrt{15}}\Big|(+0-) + (-+0) + (0-+) + (0+-) + (+-0) + (-0+) - 3(000)\rangle \quad (11.169)$$

$$\Big|1, 1\rangle = \frac{1}{\sqrt{15}}\Big|2(++-) + 2(+-+) + 2(-++) - (+00) - (0+0) - (00+)\rangle \quad (11.170)$$

For convenience in what follows, we also define

$$\Big|+-0\rangle = \frac{1}{\sqrt{6}}\Big|(+0-) + (-+0) + (0-+) + (0+-) + (+-0) + (-0+)\rangle \quad (11.171)$$

$$|000\rangle = |(000)\rangle \quad (11.172)$$

$$\Big|++-\rangle = \frac{1}{\sqrt{3}}\Big|(++-) + (+-+) + (-++)\rangle \quad (11.173)$$

and

$$\Big|+00\rangle = \frac{1}{\sqrt{3}}\Big|(+00) + (0+0) + (00+)\rangle \quad (11.174)$$

It is then easy to verify that

$$\Big|+-0\rangle = \sqrt{\frac{2}{5}}\Big|1, 0\rangle + \sqrt{\frac{3}{5}}\Big|3, 0\rangle \quad (11.175)$$

$$\Big|000\rangle = -\sqrt{\frac{3}{5}}\Big|1, 0\rangle + \sqrt{\frac{2}{5}}\Big|3, 0\rangle \quad (11.176)$$

$$\Big|++-\rangle = \frac{2}{\sqrt{5}}\Big|1, 1\rangle + \frac{1}{\sqrt{5}}\Big|3, 1\rangle \quad (11.177)$$

$$\Big|+00\rangle = -\frac{1}{\sqrt{5}}\Big|1, 1\rangle + \frac{2}{\sqrt{5}}\Big|3, 1\rangle \quad (11.178)$$

Since pions and kaons have zero spin, the zero angular momentum of the three-pion final state is compounded of the orbital angular momentum l of two pions (dipion) about their CM and the orbital angular momentum L of the third pion about the three-pion CM:

$$\mathbf{L} + \mathbf{l} = 0$$

It follows that $l = L$. Since pions are bosons, the total three-pion wave function must be symmetric in spatial coordinates and charge coordinates with respect to interchange of any two pions. Since $I = 1, 3$ isospin states are symmetric, these must correspond to a symmetric spatial wave function for which $l = L = 0, 2, 4, \ldots$. On the other hand, the $I = 0$ state is totally antisymmetric and corresponds to $l = L = 1, 3, \ldots$. In the latter case the three-pion decay amplitude is sharply reduced by the centrifugal barrier effect; i.e., the three-pion spatial wave function is zero at the origin of coordinates. For the $I = 2$ case, with mixed symmetry, the centrifugal barrier reduction also occurs but to a lesser extent.

Although CP violation occurs in $K_L{}^0 \to 2\pi$ decay, it is a very small effect and presumably can be ignored in discussing $K_{\pi 3}$ decays in the first approximation. Thus with CP invariance assumed, $K_S{}^0$ and $K_L{}^0$ can decay to three-pion final states only with CP $= +1$ and -1, respectively. Now, the neutral three-pion $J = 0$ final states have $P = -1$ and $C = (-1)^l$; thus CP $= (-1)^{l+1}$. It follows that, for $K_S{}^0 \to 3\pi$, only $I = 0$ and 2 are permitted on CP invariance grounds alone, whereas for $K_L{}^0 \to 3\pi$ only $I = 1$ and $I = 3$ are permitted. In principle, CP violation can manifest itself in $K_L{}^0 \to \pi^+\pi^-\pi^0$ decay through an asymmetry in the π^+, π^- momentum distributions. This would be expressed quantitatively by a nonzero coefficient $\sigma_\pm$ in the formula for the amplitude

$$A \approx A(0)\left[1 + 2\sigma(T_0 - T_{\max})\frac{m_K}{m_{\pi^2}} + 2\sigma_\pm(T_+ - T_-)\frac{m_K}{m_{\pi^2}}\right]$$

which replaces (11.155). However, no evidence for such an asymmetry is found experimentally (Bla 69):

$$\sigma_\pm = +0.001 \pm 0.004$$

CP invariance leads to no restrictions on $K_{3\pi}{}^\pm$ decays.

Finally we encounter the $|\Delta I| = \frac{1}{2}$ rule, which allows transitions from the $I = \frac{1}{2}$ initial state to $I = 0$ or 1 but not to $I = 2$ or $I = 3$ final states. When all these restrictions are combined (see Table 11.4), we conclude that the only allowed transitions are $K^\pm \to (I = 1)$ and $K_L{}^0 \to (I = 1)$. Let us state somewhat

Table 11.4 SELECTION RULES ON $K \to 3\pi$ DECAYS

Particle	$K^{\pm}$			$K_S{}^0$			$K_L{}^0$		
Selection rule	$\lvert \Delta I \rvert = \frac{1}{2}$	CP invariance	Centrifugal barrier	$\lvert \Delta I \rvert = \frac{1}{2}$	CP invariance	Centrifugal barrier	$\lvert \Delta I \rvert = \frac{1}{2}$	CP invariance	Centrifugal barrier
$I = 0$				√	√	×	√	×	×
1	√	√	√	√	×	√	√	√	√
2	×	√	×	×	√	×	×	×	×
3	×	√	√	×	×	√	×	√	√

† √ = allowed; × = forbidden.

more precisely the implications of the $|\Delta I| = \frac{1}{2}$ rule by considering the amplitudes for the transitions

$$K^+ \to \pi^+\pi^+\pi^-$$
$$K^+ \to \pi^0\pi^0\pi^+$$
$$K^0 \to \pi^+\pi^-\pi^0$$

and

$$K^0 \to \pi^0\pi^0\pi^0$$

assuming that only $I = 1$ and $I = 3$ symmetric states contribute. According to Eqs. (11.175) to (11.178), these amplitudes are written

$$\langle ++-|H_W|K^+\rangle = \frac{1}{\sqrt{5}}\langle 1, 1|H_W|K^+\rangle + \frac{1}{\sqrt{5}}\langle 3, 1|H_W|K^+\rangle \tag{11.179}$$

$$\langle +00|H_W|K^+\rangle = -\frac{1}{\sqrt{5}}\langle 1, 1|H_W|K^+\rangle + \frac{2}{\sqrt{5}}\langle 3, 1|H_W|K^+\rangle \tag{11.180}$$

$$\langle +-0|H_W|K^0\rangle = \sqrt{\frac{2}{5}}\langle 1, 0|H_W|K^0\rangle + \sqrt{\frac{3}{5}}\langle 3, 0|H_W|K^0\rangle \tag{11.181}$$

$$\langle 000|H_W|K^0\rangle = -\sqrt{\frac{3}{5}}\langle 1, 0|H_W|K^0\rangle + \sqrt{\frac{2}{5}}\langle 3, 0|H_W|K^0\rangle \tag{11.182}$$

Suppose that the $|\Delta I| = \frac{5}{2}, \frac{7}{2}$ amplitudes are zero. Then, defining $\gamma_i = \Gamma_i/\Phi_i$ where Γ_i is the decay rate and Φ_i is the invariant phase-space factor for the ith decay, we predict from (11.179) and (11.180)

$$\frac{\gamma(++-)}{\gamma(+00)} = 4 \tag{11.183}$$

and from (11.181) and (11.182)

$$\frac{\gamma(K_L \to +-0)}{\gamma(K_L \to 000)} = \frac{2}{3} \tag{11.184}$$

To impose a further restriction, suppose that the $|\Delta I| = \frac{1}{2}$ rule itself is valid. This implies that H_W transforms like an isospinor. Thus, according to the Wigner–Eckhart theorem,

$$\langle ++-|H_W|K^+\rangle = \frac{2}{\sqrt{5}}\langle \tfrac{1}{2}, \tfrac{1}{2}; \tfrac{1}{2}, \tfrac{1}{2}|1, 1\rangle \mathcal{M}_0 = \frac{2}{\sqrt{5}}\mathcal{M}_0$$

$$\langle +00|H_W|K^+\rangle = -\frac{1}{\sqrt{5}}\langle \tfrac{1}{2}, \tfrac{1}{2}; \tfrac{1}{2}, \tfrac{1}{2}|1, 1\rangle \mathcal{M}_0 = -\frac{1}{\sqrt{5}}\mathcal{M}_0$$

$$\langle +-0|H_W|K^0\rangle = \sqrt{\frac{2}{5}}\langle \tfrac{1}{2}, -\tfrac{1}{2}; \tfrac{1}{2}, \tfrac{1}{2}|1, 0\rangle \mathcal{M}_0 = -\frac{1}{\sqrt{5}}\mathcal{M}_0$$

$$\langle 000|H_W|K^0\rangle = -\sqrt{\frac{3}{5}}\langle \tfrac{1}{2}, -\tfrac{1}{2}; \tfrac{1}{2}, \tfrac{1}{2}|1, 0\rangle \mathcal{M}_0 = \sqrt{\frac{3}{10}}\mathcal{M}_0$$

where $\mathcal{M}_0$ is a reduced matrix element. Then, taking into account that

$$K^0 \approx \frac{1}{\sqrt{2}} |K_L{}^0 + K_S{}^0\rangle$$

we predict

$$\frac{\gamma(K_L \to +-0)}{\gamma(K^+ \to 00+)} = 2 \qquad (11.185)$$

and

$$\frac{\gamma(K_L \to 000)}{\gamma(K^+ \to ++-) - \gamma(K^+ \to 00+)} = 1 \qquad (11.186)$$

It is also possible to show that the slope parameters σ are related by the $|\Delta I| = \frac{1}{2}$ rule (Wei 60):

$$\frac{\sigma(K^+ \to 00+)}{\sigma(K^+ \to ++-)} = -2 \qquad (11.187)$$

$$\frac{\sigma(K^+ \to 00+)}{\sigma(K_L{}^0 \to +-0)} = +1 \qquad (11.188)$$

Let us now compare these results with experiment. As shown in Table 11.5, the observations are consistent with absence of $|\Delta I| = \frac{5}{2}, \frac{7}{2}$ amplitudes but

Table 11.5 TESTS OF THE $|\Delta I| = \frac{1}{2}$ RULE IN $K_{\pi 3}$ DECAYS

Test	Prediction	Observation	Reference
	$\lvert\Delta I\rvert = \frac{1}{2}$		
$\dfrac{\gamma(K_L \to +-0)}{2\gamma(K^+ \to +00)}$	1	0.85 ± 0.04	Wil 68
$\dfrac{\gamma(K_L \to 000)}{\gamma(K^+ \to ++-) - \gamma(K^+ \to +00)}$	1	0.95 ± 0.05	Wil 68
$\dfrac{\sigma(K^+ \to +00)}{\sigma(K^+ \to ++-)}$	-2	-2.63 ± 0.18	Gra 69
$\dfrac{\sigma(K_L \to +-0)}{\sigma(K^+ \to +00)}$	1	0.99 ± 0.09	Buc 70
Tests for existence of $\lvert\Delta I\rvert = \frac{5}{2}, \frac{7}{2}$ components			
	$\lvert\Delta I\rvert \le \frac{3}{2}$		
$\dfrac{\gamma(K^+ \to ++-)}{4\gamma(K^+ \to +00)}$	1	1.02 ± 0.03	Wil 68
$\dfrac{\gamma(K_L \to 000)}{\frac{3}{2}\gamma(K_L \to +-0)}$	1	1.15 ± 0.11	Wil 68

appear to violate the $|\Delta I| = \frac{1}{2}$ rule. However, the predictions (11.185) and (11.186) are ambiguous to a certain extent because of electromagnetic effects (Bou 69). Let us consider as an example the decay $K_L{}^0 \to \pi^+\pi^-\pi^0$. The slope parameter in this case is quite large, and we find that the density of points in the Dalitz plot is four times as large for $T_{\min}(\pi^0)$ as it is for $T_{\max}(\pi^0)$. The prediction (11.185) concerns the total rate of $K_L{}^0 \to \pi^+\pi^-\pi^0$ decay and thus refers to the square of the matrix element taken at $r = 0$. However, the question arises as to what we ought to take for $r = 0$. Is it that point corresponding to equal physical pion kinetic energies T_1, T_2, T_3, or does it refer to the Dalitz plot we would have if the pion masses were all equal (before the electromagnetic interaction is switched on)? Since the mass difference is of order 4 percent, the prediction of the $|\Delta I| = \frac{1}{2}$ rule is ambiguous by this amount also. In spite of this argument, one tends to believe that there is some violation of the $|\Delta I| = \frac{1}{2}$ rule in $K_{\pi 3}$ decays, as indicated in Table 11.5, and that, as in $K^+ \to \pi^+\pi^0$ decays, the ratio of $I = \frac{3}{2}$ to $\frac{1}{2}$ amplitudes is about

$$\left|\frac{A_{3/2}}{A_{1/2}}\right| \approx 5\% \qquad (11.189)$$

Finally we consider very briefly an application of the *zero-energy*, or *soft-pion*, theorem of Sec. 10.4 to the $K \to 2\pi$, 3π decays. There it was demonstrated that the matrix element of an operator Q between two states α and $\beta + \pi_i$ may be reduced in the limit of vanishing pion four-momentum to the matrix element of $[F_i{}^5, Q]$ between α and β:

$$\lim_{q_i \to 0} \frac{f\pi}{\sqrt{2}} \langle \beta, \pi^i | Q | \alpha \rangle = i \cos \Theta \langle \beta | [F_i{}^5, Q] | \alpha \rangle \qquad (11.190)$$

We take for Q that portion of the weak lagrangian which refers to nonleptonic K decays:

$$Q = \mathfrak{L} = \frac{G}{\sqrt{2}} (\mathfrak{J}_0 \mathfrak{J}_1{}^\dagger + \mathfrak{J}_1 \mathfrak{J}_0{}^\dagger)$$

The term $1/\sqrt{2}\,\mathfrak{J}_0 \mathfrak{J}_1{}^\dagger$ contains a $|\Delta I| = \frac{1}{2}$ part and a $|\Delta I| = \frac{3}{2}$ part, both of which correspond to $\Delta I_3 = -\frac{1}{2}$:

$$\frac{G}{\sqrt{2}} \mathfrak{J}_0 \mathfrak{J}_1{}^\dagger = \mathfrak{L}(\tfrac{1}{2}, -\tfrac{1}{2}) + \mathfrak{L}(\tfrac{3}{2}, -\tfrac{1}{2}) \qquad (11.191)$$

Similarly, we have

$$\frac{G}{\sqrt{2}} \mathfrak{J}_1 \mathfrak{J}_0{}^\dagger = \mathfrak{L}(\tfrac{1}{2}, \tfrac{1}{2}) + \mathfrak{L}(\tfrac{3}{2}, \tfrac{1}{2}) \qquad (11.192)$$

Now let us consider the commutator of F_i^5 with each of these portions of the lagrangian. Since each weak current contains a vector and an axial vector part, it is easy to see, taking into account Eqs. (7.87) and (7.88), that

$$[F_i^5, \mathcal{L}(I, I_3)] = [F_i, \mathcal{L}(I, I_3)]$$

for each piece of the lagrangian.

Now we may apply the reduction formula (11.190) repeatedly to obtain

$$\lim_{q_i, q_j, q_k \to 0} \left(\frac{f_\pi}{\sqrt{2}}\right)^3 \langle \pi_i \pi_j \pi_k | \mathcal{L}(\tfrac{3}{2}, \pm\tfrac{1}{2}) | K \rangle$$
$$= -i \cos^3 \Theta \langle 0 | [F_i, [F_j, [F_k, \mathcal{L}(\tfrac{3}{2}, \pm\tfrac{1}{2})]]] | K \rangle \qquad (11.193)$$

The right-hand side of (11.193) contains a succession of commutators of the operators I_+, I_-, and/or I_3 with $\mathcal{L}(\frac{3}{2}, \pm\frac{1}{2})$, which can only yield a linear combination of the operators $\mathcal{L}(\frac{3}{2}, m_\alpha)$, with $m_\alpha = -\frac{3}{2}, \ldots \frac{3}{2}$:

$$[F_i, [F_j, [F_k, \mathcal{L}(\tfrac{3}{2}, \pm\tfrac{1}{2})]]] = \sum_{m_\alpha} a_\pm(m_\alpha) \mathcal{L}(\tfrac{3}{2}, m_\alpha)$$

where the $a_\pm$ are numerical coefficients. Thus the right-hand side of (11.193) is zero since the K mesons are members of $I = \frac{1}{2}$ multiplets. Therefore, the $|\Delta I| = \frac{1}{2}$ rule for $K \to 2\pi$, 3π decays is seen to be an automatic consequence of the current algebra assumptions *in the unphysical limit* of zero pion four-momenta.

If we then neglect $\mathcal{L}(\frac{3}{2}, \pm\frac{1}{2})$ contributions in (11.191) and (11.192) and note that

$$[F_+, \mathcal{L}(\tfrac{1}{2}, -\tfrac{1}{2})] = \mathcal{L}(\tfrac{1}{2}, \tfrac{1}{2})$$
$$[F_-, \mathcal{L}(\tfrac{1}{2}, -\tfrac{1}{2})] = 0$$

and

$$[F_3, \mathcal{L}(\tfrac{1}{2}, -\tfrac{1}{2})] = -\tfrac{1}{2}\mathcal{L}(\tfrac{1}{2}, -\tfrac{1}{2})$$

[analogous relations hold for $\mathcal{L}(\frac{1}{2}, \frac{1}{2})$], it is straightforward to apply the zero-energy theorem (11.190) to $\mathcal{L}(\frac{1}{2}, \pm\frac{1}{2})$ to find relations connecting the amplitudes for $K \to 3\pi$ and $K \to 2\pi$ decays [Hara and Nambu (Har 66)]. For example, one obtains

$$\frac{m_\pi^2}{3c} A(K_S^0 \to \pi^+\pi^-)\left\{1 + \mathcal{O}\left(\frac{m_\pi^2}{m_K^2}\right)\right\}$$
$$= A(K_L^0 \to \pi^0\pi^+\pi^-)$$
$$= \tfrac{1}{3}A(K_L^0 \to 3\pi^0)$$
$$= -\tfrac{1}{2}A(K^+ \to \pi^-\pi^+\pi^+) \qquad (11.194)$$

where A means the average amplitude (center of Dalitz plot). Relation (11.194) is in good agreement with experimental results, as are similar relations derived for the slope parameters σ.

REFERENCES

Alf 66 Alff-Steinberger, C., W. Heuer, K. Kleinknecht, C. Rubbia, A. Scribano, J. Steinberger, M. J. Tannenbaum, and K. Tittel: *Phys. Letters,* **20**:207 (1966); *ibid.*, **21**:595 (1966).

Aro 70 Aronson, S. H., R. D. Ehrlich, H. Hofer, D. A. Jensen, R. A. Swanson, V. L. Telegdi, H. Goldberg, and J. Solomon: *Phys. Rev. Letters,* **25**:1057 (1970).

Ata 60 Atac, M., B. Chrisman, P. Debrunner, and H. Frauenfelder: *Phys. Rev. Letters,* **20**: 691 (1968).

Bab 67 Babu, P., and M. Suzuki: *Phys. Rev.,* **162**:1359 (1967).

Bai 69 Baird, J. K., P. D. Miller, W. D. Dress, and N. F. Ramsey: *Phys. Rev.,* **179**:1285 (1969).

Bal 65 Baldo Ceolin, M., E. Calimani, S. Campolillo, C. Filippi-Filosofo, H. Huzita, F. Mattioli, and G. Miari: *Nuovo Cimento,* **38**:684 (1965).

Ban 68 Banner, M., J. W. Cronin, J. K. Liu, and J. E. Pilcher: *Phys. Rev. Letters,* **21**:1107 (1968).

Bar 70 Barmin, V. V., et al.: *Phys. Letters,* **33B**:377 (1970).

Barb 70 Barbaro-Galtieri, A., et al.: *Rev. Mod. Phys.,* **42**:87 (1970).

Ben 67 Bennett, S., D. Nygren, H. Saal, J. Steinberger, and J. Sunderland: *Phys. Rev. Letters,* **19**:993 (1967).

Ben 68 Bennett, S., D. Nygren, and J. Steinberger: *Phys. Letters,* **27B**:244 (1968); *ibid.*, **27B**:248 (1968).

Ben 69 Bennett, S., D. Nygren, H. Saal, and J. Steinberger: *Phys. Letters,* **29B**:317 (1969).

Bla 69 Blanpied, W. A., L. B. Levit, E. Engels, M. Gotein, T. Kuk, D. G. Ryan, and D. G. Stairs: *Phys. Rev. Letters,* **21**:1650 (1969).

Bod 66 Bodansky, D., W. J. Braithwaite, D. C. Shreve, D. W. Storm, and W. G. Weitkamp: *Phys. Rev. Letters,* **17**:589 (1966). See also: Weitkamp et al., *Phys. Rev.,* **165**:1233 (1968).

Böh 69 Böhm, A., P. Darriulat, C. Grosso, V. Kaftanov, K. Kleinknecht, H. Lynch, C. Rubbia, H. Ticho, and K. Tittel: *Nucl. Phys.,* **B9**:606 (1969).

Bot 66 Bott-Bodenhausen, M., X. de Bouard, D. G. Cassell, D. Dekkers, R. Felst, R. Mermod, I. Savin, P. Scharff, M. Vivargent, T. Willits and K. Winter: *Phys. Letters,* **20**:212 (1966); *ibid.,* **23**:277 (1966).

Bou 65 Boulware, G. D.: *Nuovo Cimento,* **40A**:1041 (1965).

Bou 69 Bouchiat, C., and M. Veltman: *Proc. Topical Conf. Weak Inter-, actions, CERN, Geneva,* 225 (1969).

Buc 70	Buchanan, C. D., et al.: *Phys. Letters,* **33B**:623 (1970).
Bug 68	Bugadov, I. A., et al.: *Phys. Letters,* **28B**:215 (1968).
Cen 68, 69	Cence, R. J., B. D. Jones, V. Z. Peterson, V. J. Stenger, J. Wilson, D. I. Cheng, R. D. Eandi, R. W. Kenney, I. Linsott, W. P. Oliver, S. Parker, and C. Rey: paper 476 contributed to the 14th Internatl. Conf. High-Energy Physics, Vienna (1968) [CERN, Geneva (1968)]; *Phys. Rev. Letters,* **22**:1210 (1969).
Che 68	Chen, J. R., et al.: *Phys. Rev. Letters,* **21**:1279 (1968).
Cho 69	Chollet, J. C., J. M. Gaillard, M. R. Jane, T. J. Ratcliffe, and J. P. Repellin: CERN, **69–7**:309 (1969).
Cho 70	Chollet, J. C., J. M. Gaillard, M. R. Jane, T. J. Ratcliffe, J. P. Repellin, K. R. Schubert, and B. Wolff: *Phys. Letters,* **31B**:658 (1970).
Chr 64	Christenson, J. H., J. W. Cronin, V. L. Fitch, and R. Turlay: *Phys. Rev. Letters,* **13**:138 (1964).
Chr 65	Christenson, J. H., J. W. Cronin, V. L. Fitch, and R. Turlay: *Phys. Rev.,* **140**:B74 (1965).
Cno 66	Cnops, A. M., et al.: *Phys. Letters,* **22**:546 (1966).
Cro 68	Cronin, J. W.: *Proc. 14th Internatl. Conf. High-Energy Physics, Vienna,* 281 (1968) [CERN, Geneva (1968)].
Dar 69	Darriulat, P., K. Kleinknecht, C. Rubbia, T. Sandweiss, H. Foeth, A. Staude, K. Tittel, M. I. Ferrero, and C. Grosso: *Phys. Letters,* **30B**:209 (1969).
Dob 66	Dobrzynski, L., et al.: *Phys. Letters,* **22**:105 (1966).
Dor 67	Dorfau, D., J. Enstrom, D. Raymond, M. Schwartz, W. Wojcicki, D. H. Miller, and M. Paciotti: *Phys. Rev. Letters,* **19**:987 (1967).
Fei 65a	Feinberg, G., and H. S. Mani: *Phys. Rev.,* **137**:637 (1965).
Fei 65b	Feinberg, G.: *Phys. Rev.,* **140**:B1402 (1965).
Fit 61	Fitch, V. L., P. A. Pirone, and R. B. Perkins: *Nuovo Cimento,* **22**:1160 (1961).
Gai 68	Gaillard, J. M., et al.: paper 139 contributed to the 14th Internatl. Conf. High-Energy Physics, Vienna (1968) [CERN, Geneva (1968)].
Gla 65	Glashow, S. L.: *Phys. Rev. Letters,* **14**:35 (1965).
Gob 69	Gobbi, B., D. Green, W. Hakel, R. Moffett, and J. Rosen: *Phys. Rev. Letters,* **22**:682 (1969).
Gor 68	Gormley, M., et al.: *Phys. Letters,* **21**:399, 402 (1968).
Gra 69	Grauman, J., S. Taylor, E. Koller, D. Pandoulas, S. Hoffmaster, O. Raths, L. Romano, P. Stamer, A. Kanofsky, and V. Mainkar: *Phys. Rev. Letters,* **23**:737 (1969).
Har 66	Hara, Y., and Y. Nambu: *Phys. Rev. Letters,* **16**:875 (1966).
Hen 69	Henley, E. M.: *Ann. Rev. Nucl. Sci.,* **19**:367 (1969).
Jen 69	Jensen, D. A., S. H. Aronson, R. D. Ehrich, D. Fryberger, C. Nissim-Sabat, V. L. Telegdi, H. Goldberg, and J. Solomon: *Phys. Rev. Letters,* **23**:615 (1969).

Kis 67 Kistner, O. C.: *Phys. Rev. Letters,* **19**:872 (1967).

Lan 57 Landau, L.: *Nucl. Phys.,* **3**:127 (1957).

Mei 64 Meister, N. T., and T. K. Radha: *Phys. Rev.,* **135**:B769 (1964).

Mor 69 Morfin and Sinclair: *Phys. Rev. Letters,* **23**:660 (1969).

Pai 59 Pais, A.: *Phys. Rev. Letters,* **3**:242 (1959).

Pic 65 Piccioni, O.: *Proc. Internatl. Conf. Weak Interactions,* p. 230, Argonne Natl. Lab. (1965).

Sal 65 Salzman, G., and F. Salzman: *Phys. Letters,* **15**:91 (1965).

Sch 64 Schwinger, J.: *Phys. Rev.* **136**:B1821 (1964).

Sch 71 Schrock, B. L., et al.: *Phys. Rev. Letters,* **26**:1659 (1971).

Tho 68 Thornton, S. T., C. M. Jones, J. K. Bair, M. D. Mancusi, and H. B. Willard: *Phys. Rev. Letters,* **21**:447 (1968).

VonW 66 Von Witsch, W., A. Richter, and P. Von Brentano: *Phys. Letters,* **22**:631 (1966); *Phys. Rev. Letters,* **19**:524 (1967); *Phys. Rev.,* **169**:923 (1968).

Wei 60 Weinberg, S.: *Phys. Rev. Letters,* **4**:87, 585(E) (1960).

Wil 68 Willis, W. J.: *Proc. Heidelberg Internatl. Conf. Elementary Particle Phys.,* p. 273, North-Holland (1968).

Wol 64 Wolfenstein, L.: *Phys. Rev. Letters,* **13**:562 (1964).

Wu 64 Wu, T. T., and C. N. Yang: *Phys. Rev. Letters,* **13**:380 (1964).

12

PARITY VIOLATION IN NUCLEAR FORCES

12.1 GENERAL REMARKS

The symmetric current-current lagrangian $\mathfrak{L}_W{}^S$ contains terms of the form

$$\frac{1}{2}\frac{G}{\sqrt{2}}(\mathfrak{J}_0\mathfrak{J}_0{}^\dagger + \mathfrak{J}_0{}^\dagger\mathfrak{J}_0) \qquad (12.1)$$

and

$$\frac{1}{2}\frac{G}{\sqrt{2}}(\mathfrak{J}_1\mathfrak{J}_1{}^\dagger + \mathfrak{J}_1{}^\dagger\mathfrak{J}_1) \qquad (12.2)$$

each of which yields nonzero matrix elements for the two-nucleon interaction

$$n + p \to p + n \qquad (12.3)$$

Thus we are led to expect that there exists a weak nuclear force, which is described in the nonrelativistic limit by a two-nucleon potential ϕ_W. Since the weak interaction violates parity, we may express ϕ_W as the sum of a scalar part ϕ_{WS} and a pseudoscalar part ϕ_W'. The former arises from terms of the form vector × vector and axial vector × axial vector, while the latter is associated with vector × axial vector terms.

The weak potential coexists with the ordinary potential ϕ_0 (a scalar) of strong (and electromagnetic) interactions, but it is extraordinarily feeble compared with the latter. Therefore, for all practical purposes, ϕ_{WS} is unobservable and may be neglected. Thus we can write for the total potential

$$\phi_{\text{total}} \approx \phi_0 + \phi'_W \tag{12.4}$$

Even the simplest of arguments shows that

$$\frac{\phi'_W}{\phi_0} \approx G m_\pi{}^2 \approx 2 \times 10^{-7} \tag{12.5}$$

but quantitative estimates of ϕ'_W are difficult and uncertain (see Sec. 12.4).

Since, with the inclusion of ϕ'_W, the total potential is not inversion-symmetric, the eigenstates of the nuclear hamiltonian H are no longer eigenstates of parity. Suppose we let H_0 be the hamiltonian without the weak nuclear force, and ψ_n, χ_m be its eigenstates with parities $+P$ and $-P$, respectively. Then, when the weak nuclear force is included, ψ_n becomes

$$\begin{aligned} \psi'_n &= \psi_n + \sum_m \frac{\langle \chi_m | \phi'_W | \psi_n \rangle}{E_{0n} - E_m} \chi_m \\ &= \psi_n + \sum_m F_{nm} \chi_m \end{aligned} \tag{12.6}$$

by first-order perturbation theory. In the simplest case, only one term in m contributes and

$$\psi' = \psi + F\chi \tag{12.7}$$

with

$$F = \frac{\langle \chi | \phi'_W | \psi \rangle}{E(\psi) - E(\chi)} \approx 10^{-7} - 10^{-6} \tag{12.8}$$

Of the two terms in the weak lagrangian which contribute to ϕ'_W, $\mathfrak{J}_0 \mathfrak{J}_0{}^\dagger + \mathfrak{J}_0{}^\dagger \mathfrak{J}_0$ obeys the isospin selection rule $|\Delta I| = 2,\ 0$ while $\mathfrak{J}_1 \mathfrak{J}_1{}^\dagger + \mathfrak{J}_1{}^\dagger \mathfrak{J}_1$ satisfies the condition $|\Delta I| = 1$. This follows immediately from the fact that $\mathfrak{J}_0$ transforms as an isovector while $\mathfrak{J}_1$ transforms as an isospinor.[1]

12.2 PARITY-FORBIDDEN TRANSITIONS

We can perform two kinds of experiments to investigate the parity-violating potential ϕ'_W. In the first case, we seek to observe a transition from an initial nuclear state ψ_i to a final state ψ_f, which would be strictly forbidden if ϕ'_W were zero.

[1]The rule $|\Delta I| = 2, 0$ for $\mathfrak{J}_0 \mathfrak{J}_0{}^\dagger + \mathfrak{J}_0{}^\dagger \mathfrak{J}_0$ holds strictly only if $\mathfrak{J}_0$ contains no second-class component. Otherwise, $\mathfrak{J}_0 \mathfrak{J}_0{}^\dagger + \mathfrak{J}_0{}^\dagger \mathfrak{J}_0$ could yield $|\Delta I| = 1$ as well (Bli 66)

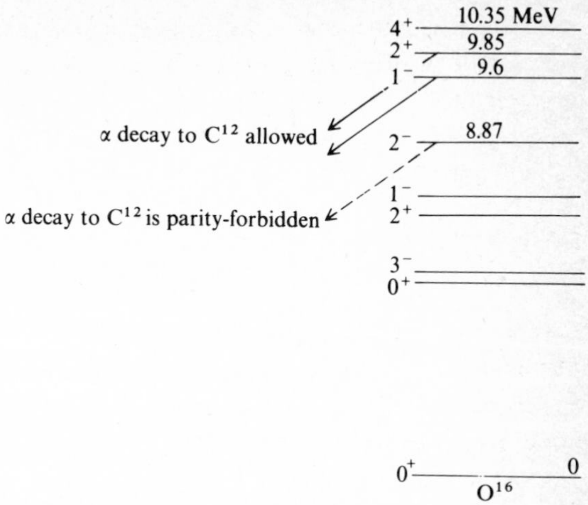

FIGURE 12.1
Energy levels of O^{16}. The α transitions from 2^+ and 1^- levels are allowed, but from the 2^- level they are forbidden.

For example, consider the α decay from excited states in O^{16} to C^{12} (see Fig. 12.1). Since ${}_6C^{12}$ and ${}_2He^4$ both have even nominal parity and zero spin, the total angular momentum of the $C^{12}\alpha$ final state is l, and the total parity is $(-1)^l$, where l is the orbital angular momentum of the C^{12}—He^4 system. The α transitions conserve angular momentum and parity; thus α decay from the 2^+ and 1^- levels of O^{16} is allowed, but α decay from the 8.88 MeV 2^- level is forbidden and can only proceed at a very slow rate through parity mixing of the 2^- level, via ϕ'_W, with nearby 2^+ levels. (Parity mixing in the ground states of C^{12} and He^4 is negligible.)

Let H_T be the transition operator. Then, in general, the transition rate is proportional to

$$\begin{aligned} I &= |\langle\psi'_f|H_T|\psi'_i\rangle|^2 \\ &= |\langle\psi_f + \sum F_f \chi_f|H_T|\psi_i + \sum F_i \chi_i\rangle|^2 \end{aligned}$$

Since, by assumption, $\langle\psi_f|H_T|\psi_i\rangle = 0$, we have, to lowest order in the F's,

$$I = |\sum F_f \langle\chi_f|H_T|\psi_i\rangle|^2 + |\sum F_i \langle\psi_f|H_T|\chi_i\rangle|^2 \qquad (12.9)$$

Since each F is of order 10^{-6} to 10^{-7} and appears quadratically in (12.9), we expect the parity-forbidden transition [for example, α decay of $O^{16}(2^-)$] to be inhibited by a factor 10^{12} to 10^{14}, as compared with the allowed transitions.

Much effort has been expended in searching for the forbidden α decay from $O^{16}(2^-)$. Experiments show that

$$\Gamma_\alpha(2^-) = 2.2 \pm 0.8 \times 10^{-10}\text{eV} \qquad \text{(Hüt 70)} \qquad (12.10)$$

in good agreement with theoretical estimates:

$$\Gamma_\alpha(2^-)_{\text{theo}} \approx 1 \text{ to } 2 \times 10^{-10}\text{eV} \qquad \text{(Gar 69, Hen 69a)} \qquad (12.11)$$

Thus, one finds $|F^2| \approx 10^{-13}$.

12.3 CIRCULAR POLARIZATION AND ASYMMETRY OF GAMMA RAYS

The second method for observing parity violation in nuclear forces consists in detecting the small component of circular polarization in the γ emission accompanying electromagnetic transitions of unpolarized nuclei or the asymmetry in γ emission from polarized nuclei. As before, the transition intensity is proportional to

$$I = |\langle\psi_f + \sum F_f \chi_f | H_T | \psi_i + \Sigma F_i \chi_i\rangle|^2 \qquad (12.12)$$

but this time $\langle\psi_f|H_T|\psi_i\rangle$ does not vanish and, in fact, is the matrix element for the "normal" multipole transitions. Expanding (12.12) and keeping only the terms of lowest order in F_i or F_f, we obtain

$$I = |\langle\psi_f|H_T|\psi_i\rangle|^2 \left\{1 + \sum F_f \frac{\langle\chi_f|H_T|\psi_i\rangle}{\langle\psi_f|H_T|\psi_i\rangle} + cc + \sum F_i \frac{\langle\psi_f|H_T|\chi_i\rangle}{\langle\psi_f|H_T|\psi_i\rangle} + cc\right\}$$

This time the parity violation manifests itself in interference terms linear in the F's, a typical one of which is

$$\sum F_i \frac{\langle\psi_f|H_T|\chi_i\rangle}{\langle\psi_f|H_T|\psi_i\rangle} \qquad (12.13)$$

For the purpose of the present discussion, we consider only this term and neglect the others. It is convenient, in dealing with electromagnetic transitions, to write H_T as the sum of electric and magnetic multipole operators:

$$H_T = \sum_L \mathcal{O}_E(L) + \sum_L \mathcal{O}_M(L) \qquad (12.14)$$

Hence, (12.13) becomes

$$\sum F_i \frac{\langle \psi_f | \mathcal{O}_{M(E)}(L) | \chi_i \rangle}{\langle \psi_f | \mathcal{O}_{E(M)}(L) | \psi_i \rangle} \qquad (12.15)$$

since $\mathcal{O}_E(L)$ has parity $(-1)^L$ and $\mathcal{O}_M(L)$ has parity $(-1)^{L+1}$. Thus, for example, if the normal transition proceeds by emission of an $E(1)$ photon (electric dipole), then parity violation mixes in some magnetic dipole amplitude. We may then show that the circular polarization is

$$\mathcal{P} = 2\overline{FR} \qquad (12.16)$$

where

$$\overline{FR} = \sum F_i R_i$$

and

$$R_i = \frac{\langle \psi_f | \mathcal{O}_M(1) | \chi_i \rangle}{\langle \psi_f | \mathcal{O}_E(1) | \psi_i \rangle} \qquad (12.17)$$

Analogous expressions hold for higher multipoles L. Similarly, if photons are emitted in direction $\hat{k}$ from nuclei with polarization $\mathbf{P}$, the angular distribution of the radiation is given by the formula

$$I \approx 1 + \alpha_0 \mathbf{P} \cdot \hat{k} \qquad (12.18)$$

where the asymmetry coefficient is

$$\alpha_0 = 2\overline{FR}A \qquad (12.19)$$

and A is a numerical coefficient of order unity which depends on the initial and final nuclear spins.

It is important to note that there are circumstances in which the *enhancement* factor

$$R = \sum \frac{\langle \psi_f | H_T | \chi_i \rangle}{\langle \psi_f | H_T | \psi_i \rangle}$$

may be as large as 10^2 or more. This occurs when the normal transition $\psi_i \rightarrow \psi_f$ is hindered by the operation of some approximate selection rule, while the abnormal transition is fully allowed. In such a case, the fractional circular polarization $\mathcal{P}$ or asymmetry α_0 may be as large as 10^{-5} or 10^{-4} even though F is only about 2×10^{-7}.

12.3.1 Observations of Circular Polarization

In order to detect circular polarizations of the order 10^{-5} to 10^{-6}, a huge number of photon counts is required to accumulate sufficient statistical accuracy. The difficulties are severely compounded by the very low efficiencies inherent in the only practical method for detecting γ-ray circular polarizations, namely, spin-dependent Compton scattering of photons from partially polarized electrons in iron magnetized to saturation.

In very elegant experiments, V. M. Lobashov and coworkers have observed circular polarizations in the following transitions:

Nucleus	Transition		Normal (abnormal) multipole	γ	Reference
K^{41}	$\frac{7}{2}^- \to \frac{3}{2}^+$	(1.293 MeV)	$M2$ $(E2)$	$(1.9 \pm 0.3) \times 10^{-5}$	Lob 69
Lu^{175}	$\frac{9}{2}^- \to \frac{7}{2}^+$	(0.396 MeV)	$E1 + M2\,(M1)$	$(4 \pm 1) \times 10^{-5}$	Lob 66
Ta^{181}	$\frac{5}{2}^- \to \frac{7}{2}^+$	(0.482 MeV)	$M1 + E2\,(E1)$	$-(6 \pm 1) \times 10^{-6}$	Lob 67

and they have placed an upper limit of 3×10^{-6} on the circular polarization of γ's accompanying neutron capture by protons (Lob 69).

In each case the γ rays of interest are passed through a magnetized iron polarimeter and the magnetization is reversed automatically, e.g., once each second. The transmitted γ rays are not counted individually, but rather they are detected continuously by a scintillation crystal and photomultiplier, which yields a current with a signal component at 0.5 cps. This is passed through a narrow-band amplifier, which is coupled to a precision pendulum with resonance frequency 0.5 cps. The phase and amplitude of pendulum oscillations are used to determine the sign and magnitude of the circular polarization.

Let us now discuss a number of theoretical features of interest in connection with the experiments of Lobashov. The simplest and most fundamental transition which can be used to study parity violation is

$$n + p \to d + \gamma \qquad (12.20)$$

Danilov (Dan 65) has calculated the γ circular polarization from capture of unpolarized neutrons, and the γ asymmetry from capture of polarized neutrons, with results as follows:

$$10^{-7} \leq \mathcal{T} \leq 10^{-6} \qquad (12.21)$$

$$\alpha_0 \approx 10^{-8} \qquad (12.22)$$

It is interesting to note that $\mathcal{T}$ arises from the $\Delta I = 0$ portion of ϕ'_W, whereas α_0 is due to the $|\Delta I| = 1$ portion. This can be seen as follows.

The capture takes place with slow (thermal) neutrons. The initial state is thus an s wave, and the normal transition is from ${}^1S_0$ to the (predominantly) ${}^3S_1$ deuteron ground state, with the emission of an $M1$ photon. When ϕ'_W is taken into account, the following amplitudes also contribute:

$$(a)\quad {}^1S_0\ (I = 1) \to {}^1P_1\ (I = 0) \qquad E1$$

$$(b)\quad {}^3S_1\ (I = 0) \to {}^3P_1\ (I = 1) \qquad E1$$

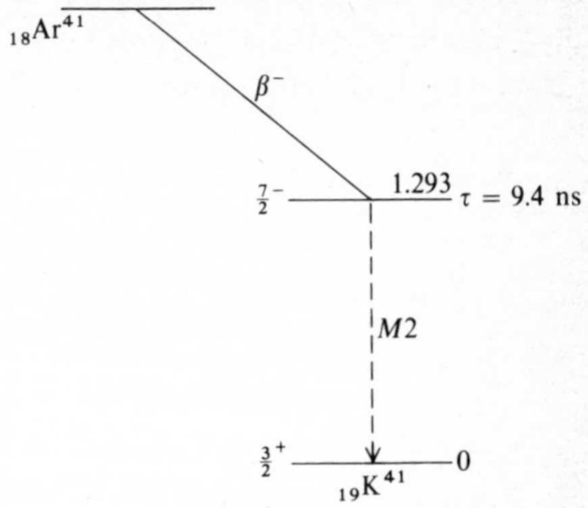

FIGURE 12.2
Energy levels of K^{41}.

Clearly, the abnormal amplitudes (*a*) and (*b*) occur because of $\Delta I = 0$ and $\Delta I = 1$ portions of ϕ'_W, respectively. Now, for unpolarized incoming neutrons, the $^1S_0$ and $^3S_1$ initial states are incoherent and the circular polarization can arise only through interference of abnormal amplitude (*a*) with the allowed amplitude. For polarized neutrons, no contribution to the asymmetry comes from amplitude (*a*) since the initial state in this case is $^1S_0$. However, for polarized neutrons, the initial $^1S_0$, $^3S_1$ mixture is partially coherent. This can be seen easily if we express the neutron and proton spinors in terms of the usual eigenstates with respect to the *z* axis. If the neutron spin is up (α_n) while half the proton spins are up (α_p) and half are down (β_p), the initial spin states are

$$\alpha_n \alpha_p$$

and

$$\alpha_n \beta_p = \tfrac{1}{2}(\alpha_n \beta_p + \beta_n \alpha_p) + \tfrac{1}{2}(\alpha_n \beta_p - \beta_n \alpha_p) \qquad (12.23)$$

On the right-hand side of (12.23) the first term corresponds to $^3S_1$ and the second to $^1S_0$. Therefore the asymmetry arises from interference between abnormal amplitude (*b*) and the normal amplitude.

Next let us consider the 1.293-MeV γ transition $K^{41*}(\frac{7}{2}^-) \rightarrow K^{41}(\frac{3}{2}^+)$. (See Fig. 12.2.) The normal (abnormal) multipolarity is $M2(E2)$. In order to make an estimate of the circular polarization to be expected, we assume that in $\mathcal{T} = 2\overline{FR}$ we can separate the factors F and R and write $F = 2 \times 10^{-7}$. The problem then is to estimate R. For this purpose we employ the single-particle model, which gives unambiguous predictions for γ-transition rates in terms of the mass number A, the γ-ray energy E, the initial and final nuclear spins, and the multipolarity of the transition. From a straightforward application of perturbation theory and multipole expansion, we find the transition rates as follows:

Multipole	W, s^{-1} E, MeV
$E1$	$1.5 \times 10^{14} A^{2/3} E^3 S$
$M1$	$2.8 \times 10^{13} E^3 S$
$E2$	$1.6 \times 10^{8} A^{4/3} E^5 S$
$M2$	$1.2 \times 10^{8} A^{2/3} E^5 S$
$E3$	$1.1 \times 10^{2} A^{2} E^7 S$
$M3$	$1.8 \times 10^{2} A^{4/3} E^7 S$
$E4$	$5.0 \times 10^{-5} A^{8/3} E^9 S$
$M4$	$1.5 \times 10^{-4} A^{2} E^9 S$

where S is a statistical weight factor of the order of unity, depending on the initial and final spins. These rates were originally derived by Weisskopf (Wei 51).

We must emphasize that single-particle model calculations are extremely crude and could not possibly apply without major modifications to anything except single-proton transitions. Nevertheless, as a start they are useful, and in particular they may be applied with fair confidence to K^{41}, which possesses 19 protons (closed shell minus one) and 22 neutrons. Thus, we find from the above table that the predicted lifetime of the $\frac{7}{2}^-$ level of K^{41} for the $M2$ transition is 1×10^{-9} s. The actual mean life is 9.4 ns, which means that the actual $M2$ amplitude is smaller by a factor of $9.4^{1/2} = 3.1$ than the corresponding single-particle (SP) amplitude. Let us now assume that the $E2$ amplitude is correctly described by the SP model. We find

$$\left|\frac{E(2)}{M(2)}\right|_{\text{SP}} = \left|\frac{\langle\psi_f|H_T|\chi_i\rangle}{\langle\psi_f|H_T|\psi_i\rangle}\right| \approx 10$$

Therefore the enhancement factor, with these assumptions, is approximately $10 \times 3.1 \approx 30$, and the circular polarization is $\mathcal{P} = 2RF \approx 12 \times 10^{-5}$. This estimate, although very rough, compares reasonably well with the actual result obtained by Lobashov.

Among the various circular polarization measurements which have been suggested but not yet carried out, a proposal by E. Henley is especially interesting. It concerns an attempt to isolate the $|\Delta I| = 1$ part of ϕ'_W by observing the circular polarization in γ transitions of self-conjugate nuclei (Hen 68), e. g., the 1.080-MeV $E1\,0^- \to 1^+$ transition to the ground state of F^{18} or the 4.388-MeV $E1\,2^- \to 1^+$ transition from the 5.105-MeV level to the 0.717 level in B^{10}. In each case the initial and final levels are $I = 0$, yet the admixing of other $I = 0$ levels is shown to be negligible. By contrast, in F^{18} the $J^P = 0^+$, $I = 1$ state at 1.043 MeV is probably strongly admixed to the 0^-, $I = 0$, 1.080-MeV state.

It is estimated that $\mathcal{P}_\gamma \approx 10^{-5}$ or even larger, in this case. Unfortunately, there seems to be no simple way to produce large numbers of F^{18} nuclei in the 1.080-MeV state. Practical circular polarization measurements generally require a powerful beta-active source which decays to the excited state of interest, for example, $Ar^{41} \xrightarrow{\beta} K^{41}(\frac{7}{2}^-)$.

12.3.2 γ-Ray Asymmetries

As an example, we consider the capture of slow polarized neutrons by Cd^{113}. The reaction proceeds as follows:

$$
\begin{array}{lllllll}
 & n(\text{polarized}) + Cd^{113} & \rightarrow Cd^{114*} & \rightarrow Cd^{114} + \gamma & & & \\
\text{Spins} & \frac{1}{2} \quad + \frac{1}{2} & \rightarrow 1 & \rightarrow 0 + 1 & (9.05\ \text{MeV}) & & (12.24)\\
 & & & \rightarrow 2 + 1 & (8.48\ \text{MeV}) & &
\end{array}
$$

The probability per capture for the 9.05- and 8.48-MeV γ rays is 1.4×10^{-3} and 2.4×10^{-3}, respectively. The slow neutron capture proceeds from an s-wave initial state. The 9.05-MeV γ is nominally $M1$, and an $E1$ admixture arises from parity violation. Abov et al. (Abo 64, 65) observed an asymmetry

$$\alpha_0 = -(3.7 \pm 0.9) \times 10^{-4} \qquad (12.25)$$

and similar results were obtained by Warming et al. (War 67). The enhancement factor R is only about 10 in the present case, and so we may ask: Why the large asymmetry? The explanation is that the 1^+ state of Cd^{114*} lies above the neutron separation energy, and so there are *many* nearby 1^- states. Therefore, F itself is greatly enhanced, and, as estimates by Blin-Stoyle (Bli 60) show,

$$F_{\text{eff}} \approx 50 \frac{\phi'_W}{\phi_0} \approx 50 \times 2 \times 10^{-7} = 10^{-5}$$

Thus

$$\alpha_0 = 2F_{\text{eff}} R \approx 10^3 \frac{\phi'_W}{\phi_0} \approx 2 \times 10^{-4}$$

An extreme example of enhancement is reported by Krane et al. (Kra 71), who observed the angular distribution of 501-keV γ rays emitted from Hf^{180m} polarized at very low temperatures and obtained an asymmetry of $-(1.49 \pm 0.25)$ percent. This nucleus is highly distorted from spherical symmetry, and its energy levels must be characterized, not only be the nuclear spin J, but also by its projection K on the nuclear symmetry axis. The γ transition in question corresponds to a highly forbidden jump $\Delta K = 8$; thus the normal amplitude is very inhibited.

12.4 THEORETICAL ESTIMATES OF THE PARITY-VIOLATING POTENTIAL

It is obvious that one must have a dependable means of estimating ϕ'_W and F if the results of parity-violation experiments are to be at all useful in understanding the fundamental structure of the weak interaction. Unfortunately, we encounter here the same basic difficulties which plague theories of nuclear structure, namely, our inability to calculate forces between physical nucleons with any reliability. Thus, all detailed estimates of parity violation in nuclear forces are subject to great uncertainties and should probably be viewed with caution and skepticism.

Bearing all this in mind, we can still arrive at certain general conclusions about the nature of ϕ'. First of all, we write down that portion of the weak lagrangian relevant to parity violation in terms of Cabibbo's theory:

$$\mathfrak{L} = \frac{G}{2\sqrt{2}} \cos^2 \Theta[(\mathcal{F}_1 - i\mathcal{F}_2)(\mathcal{F}_1 + i\mathcal{F}_2) + (\mathcal{F}_1 + i\mathcal{F}_2)(\mathcal{F}_1 - i\mathcal{F}_2)]$$

$$+ \frac{G}{2\sqrt{2}} \sin^2 \Theta[(\mathcal{F}_4 - i\mathcal{F}_5)(F_4 + i\mathcal{F}_5) + (\mathcal{F}_4 + i\mathcal{F}_5)(\mathcal{F}_4 - i\mathcal{F}_5)] \qquad (12.26)$$

where $\mathcal{F}_i = j_i + g_i$, and the j_i and g_i are members of the vector and axial vector octets of current operators. The first term of the right-hand side of (12.26) corresponds to $\Delta I = 0, 2$, while the second term corresponds to $\Delta I = 1$. Naively one would expect that the first term dominates greatly over the second because $\sin^2 \Theta/\cos^2 \Theta \approx 0.07$. However, the following important factors modify this conclusion.

1 We can calculate the contribution to ϕ'_W which arises from the current-current interaction model for bare nucleons N_1 and N_2 to lowest order in p/m. This is easily shown to be (Mic 64)

$$\phi'_c(\mathbf{r}) = -\frac{G}{2^{3/2}} \frac{1}{m_p} |C_A| \{(1 + \mu)(i\boldsymbol{\sigma}_1 \times \boldsymbol{\sigma}_2) \cdot [\mathbf{p}, \delta(\mathbf{r}_1 - \mathbf{r}_2)]$$

$$+ (\boldsymbol{\sigma}_1 - \boldsymbol{\sigma}_2) \cdot [\mathbf{p}, \delta(\mathbf{r}_1 - \mathbf{r}_2)](\boldsymbol{\tau}_1 \boldsymbol{\tau}_2 - {\boldsymbol{\tau}_1}^3 {\boldsymbol{\tau}_2}^3)\} \qquad (12.27)$$

where μ is the difference in anomalous magnetic moments of proton and neutron (and arises from the weak-magnetism term in the vector matrix element), $\mathbf{p} \equiv \frac{1}{2}(\mathbf{p}_1 - \mathbf{p}_2)$, $[\cdots]$ indicates a commutator, and $\{\cdots\}$ indicates an anticommutator. The effect of ϕ'_c is negligible, however, because it represents a contact interaction, while, in fact, the nucleon-nucleon parity-conserving potential ϕ_0 has a hard core.

(a) (b)

FIGURE 12.3

2 Therefore, the main contribution to ϕ'_W must arise from the exchange of one or more mesons. Such exchanges may be represented schematically as in Fig. 12.3*a* and *b*. In this figure G represents a weak vertex and f a strong vertex. One might naturally suppose that the main contribution to the $\cos^2 \Theta$ term in (12.26) would arise from single-pion exchange. However, it is possible to prove that if we neglect CP violation, single-pion exchange does not contribute at all to the $\Delta I = 0$ term (Hen 69b, Bar 61).[1] This may be seen as follows.

The pion is pseudoscalar, and therefore the most general parity-violating effective hamiltonian we can write for the weak vertex in Fig. 12.3*a* is

$$H_{\text{eff}} = \frac{G}{\sqrt{2}} \int \{a_1 \bar{\psi} \tau_+ \psi \phi_- + b_1 \bar{\psi} \tau_- \psi \phi_+ + a_2 \bar{\psi} \gamma^\mu \tau_+ \psi \partial_\mu \phi_-$$
$$+ b_2 \bar{\psi} \gamma^\mu \tau_- \psi \partial_\mu \phi_+ + a_3 \bar{\psi} \sigma^{\mu\nu} \tau_+ \psi \partial_\mu \partial_\nu \phi_- + b_3 \bar{\psi} \sigma^{\mu\nu} \tau_- \psi \partial_\mu \partial_\nu \phi_+\} \, d^3 r$$

where the a_i and b_i are constants, ψ and $\bar{\psi}$ are nucleon destruction and creation operators, respectively, and $\phi^\mp$ are $\pi^\pm$ field operators, respectively. We require the coefficients a_i and b_i to be such that H_{eff} is P-odd but CP-even. Now suppose that H_{eff} transforms under isospin rotations like $I = 0$ or 2. Then, under a G-parity transformation,

$$G = e^{i\pi I_2} \hat{C}$$

we have $G H_{\text{eff}} G^{-1} = - H_{\text{eff}}$. It follows that $b_1 = - a_1$. Furthermore, the coefficients a_2, b_2, a_3, and b_3 must vanish, as can be shown easily. Therefore,

$$H_{\text{eff}} \propto \frac{G}{\sqrt{2}} \int \bar{\psi}(\tau_+ \rho_- - \tau_- \phi_+) \psi \, d^3\tau = \frac{G}{\sqrt{2}} \int \bar{\psi} (\tau \times \phi)^{(3)} \psi \, d^3 \tau$$

Therefore H_{eff} satisfies $|\Delta I| = 1$, which is a contradiction.

[1] However, see a recent discussion by E. Henley (Hen 71), who points out that this is strictly true only in the approximation that the nucleons are free. When nuclear binding effects are taken into account, one finds a small but nonzero $\Delta I = 0$ one-pion exchange contribution.

It follows that if we limit ourselves to single-meson exchanges, the simplest contribution to the $\cos^2 \Theta$ term comes from $\rho^\pm$ meson exchange. This can be described by a potential ϕ'_ρ in which the delta function in (12.24) is replaced by a Yukawa factor for the $\rho^\pm$:

$$m_\rho{}^2 \frac{e^{-m_\rho|\mathbf{r}_1-\mathbf{r}_2|}}{4\pi|\mathbf{r}_1-\mathbf{r}_2|} \tag{12.28}$$

One can then compare the static parts of ϕ'_ρ and ϕ_0 (the parity-conserving strong interaction potential) at various distances, e.g., the pion Compton wavelength and the hard-core radius. One concludes that $\overline{\phi'_\rho/\phi_0} \approx 10^{-6}$ to 10^{-7}, where the bar means average over nucleon volume. More detailed conclusions are hardly warranted. [It should be noted that 2π exchange diagrams lead to contributions to the potential comparable to ϕ'_ρ (Lac 68).]

3 One-pion exchange *can* contribute to the $\sin^2 \Theta$ term (McKel 67). A detailed calculation relates the parity-violating weak vertex factor for pion emission to the nonleptonic decays of hyperons via SU_3 and current algebra. It is found that

$$\phi'_\pi(r) = \frac{gf}{2^{3/2}m_p}(\boldsymbol{\sigma}_1 + \boldsymbol{\sigma}_2)\cdot\nabla\left(\frac{e^{-m_\pi|\mathbf{r}_1-\mathbf{r}_2|}}{4\pi|\mathbf{r}_1-\mathbf{r}_2|}(\boldsymbol{\tau}_1 \times \boldsymbol{\tau}_2)^{(3)}\right) \tag{12.29}$$

where the factor g is determined from the hyperon nonleptonic decays to be

$$g \approx 4.2 \pm 0.8 \times 10^{-8}$$

and $f^2/4\pi \approx 14.4$ represents the pion-nucleon coupling. The one-pion exchange thus turns out to give a contribution to ϕ' which is *comparable to* that of the $\cos^2 \Theta$ term!

Of course, the mere computation of an effective two-nucleon parity-violating potential is not sufficient for the purpose of determining F factors. We must average the two-nucleon potential over a spherical nucleon core to obtain an effective one-body potential of the form constant $\times\, \boldsymbol{\sigma}\cdot\mathbf{p}$. This potential must then be employed together with crude nuclear wave functions to calculate F and then $\mathcal{F}$ or α_0. A detailed discussion of these problems is given by Michel (Mic 64).

In summary, it is fair to state that good evidence now exists for parity violations in nuclear forces, and that the experimental results are consistent with the conventional current-current scheme. However, theoretical estimates of these effects are rendered uncertain by at least a factor 10 because of the complications of meson-nucleon interactions and nuclear structure.

REFERENCES

Abo 64 Abov, Kruptchitsky, and Oratovsky: *Phys. Letters,* **12**:25 (1964).

Abo 65 Abov, Kruptchitsky, and Oratovsky: *Sov. J. Nucl. Phys.,* **1**:341 (1965).

Bar 61 Barton, G.: *Nuovo Cimento,* **19**:512 (1961).

Bli 60 Blin-Stoyle, R. J.: *Phys. Rev.,* **118**:1605 (1960).

Bli 66 Blin-Stoyle, R. J., and P. Herczeg: *Phys. Letters,* **23**:376 (1966).

Dan 65 Danilov, G. S.: *Phys. Letters,* **18**:40 (1965).

Gar 69 Gari, M., and H. Kümmel: *Phys. Rev. Letters,* **23**:26 (1969).

Hen 68 Henley, E. M.: *Phys. Letters,* **28B**:1 (1968).

Hen 69a Henley, E. M., T. E. Keliher, and D. V. L. Yu: *Phys. Rev. Letters,* **23**:941 (1969).

Hen 69b Henley, E. M.: *Ann. Rev. Nucl. Sci.,* **19**:367 (1969).

Hen 71 Henley, E. M.: *Phys. Rev. Letters,* **27**:542 (1971).

Hüt 70 Hüttig, H., K. Hünchen, and H. Wäffler: *Phys. Rev. Letters,* **25**:941 (1970).

Kra 71 Krane, K. S., C. E. Olsen, J. R. Sites, and W. A. Steyert: *Phys. Rev. Letters,* **26**:1579 (1971).

Lac 68 Lacaze, R.: *Nucl. Phys.,* **B4**:657 (1968).

Lob 66 Lobashov, V. M., et al.: *Soviet Phys. JETP Letters,* **3**:173 (1966).

Lob 67 Lobashov, V. M., et al.: *Soviet Phys. JETP Letters,* **5**:59 (1967).

Lob 69 Lobashov, V. M., et al.: *Phys. Letters,* **30B**:39 (1969).

Lob 70 Lobashov, V. M., et al.: *Soviet Phys. JETP Letters,* **11**:76 (1970).

McK 67 McKellar, B. H. J.: *Phys. Letters,* **26B**:107 (1967).

Mic 64 Michel, F. C.: *Phys. Rev.,* **133**:B329 (1964).

War 67 Warming, E., F. Stecher-Rasmussen, W. Ratynski, and J. Kopecky: *Phys. Letters,* **25B**:200 (1967).

Wei 51 Weisskopf, V.: *Phys. Rev.,* **83**:1073 (1951).

13

HIGH-ENERGY NEUTRINO-NUCLEON COLLISIONS

13.1 INTRODUCTION

In preceding chapters virtually all of our attention has been devoted to the weak decays, which were indispensable for elucidating the nature of the weak interaction at low energies and small momentum transfers. As we have seen, this much of our subject can be accounted for more or less adequately by the universal first-order weak lagrangian.

However, the decays tell us essentially nothing about weak processes at high energies and large momentum transfers and yield no clue as to the possible existence of an intermediate vector boson. Clearly, progress along these lines can only come from weak-scattering experiments. For this purpose we might, in principle, investigate collisions of hadrons on hadrons, or muons or electrons on hadrons, or even charged lepton–charged lepton scattering. However, in these cases it would be necessary to extract the extraordinarily feeble effects of the weak interaction from scattering cross sections dominated by strong and/or electromagnetic couplings. It is only when we use neutrinos as projectiles that we free ourselves of such complications. Therefore, it is easy to see why

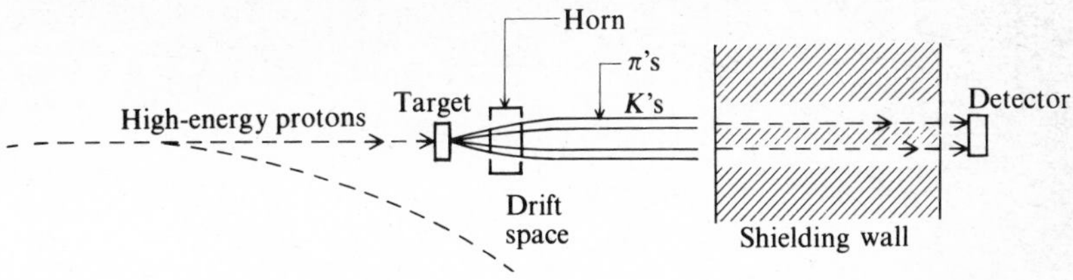

FIGURE 13.1
Schematic diagram for production of a high-energy neutrino beam.

scattering of high-energy neutrino beams has assumed such great importance in contemporary investigations of the weak interaction.[1]

A typical experimental arrangement for the production of a high-energy neutrino beam is shown schematically in Fig. 13.1. Protons from a large accelerator[2] strike a target such as boron carbide, aluminum, or beryllium, from which pions and kaons are emitted in great profusion. Those with the appropriate sign of charge and contained in a certain forward cone are focused into a beam by one or more powerful electromagnets (called neutrino *horns*, or lenses). These hadrons then decay in flight in a long drift space (CERN: 60 m; Batavia: 400 m). The decay products, also with high forward momentum in the laboratory frame, are overwhelmingly muons and muon neutrinos from the decays

$$\pi^{\pm} \to \mu^{\pm}\nu_{\mu}(\bar{\nu}_{\mu})$$
$$K^{\pm} \to \mu^{\pm}\nu_{\mu}(\nu_{\mu})$$

although a $\nu_e(\bar{\nu}_e)$ component of several percent originates from K_{e3} decay. Following the drift space an immense shielding wall of steel, concrete, and/or earth is provided to absorb the muons. Of course, the neutrino beam passes through this wall essentially unimpeded and emerges relatively pure on the far side. Next, interactions of the neutrino beam with nucleons are examined. The CERN experiments of 1967 and earlier years utilized a bubble chamber filled with freon or propane, while the bubble chamber intended for use at Batavia, starting in 1973, will operate with H_2 or D_2, and contain an expandable volume of about 100,000 l.

[1] The first ideas for accelerator neutrino experiments were proposed by Schwartz (Sch 60) and Pontecorvo (Pon 59). The first extended calculations of high-energy neutrino cross sections were published by Lee and Yang (Lee 60). Comprehensive reviews are given by Lederman (Led 67) and Perkins (Per 69).

[2] Maximum proton energies from the largest accelerators: Brookhaven, CERN: 30 GeV; Serphukov (USSR): 70 GeV; Batavia (NAL.): 200 to 500 GeV (projected).

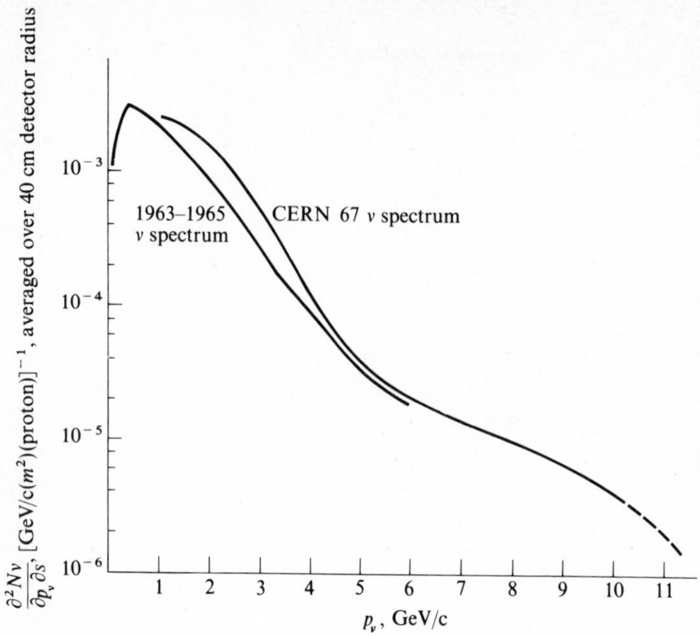

FIGURE 13.2
Neutrino ν_μ energy spectrum obtained at CERN during the 1967 experiments. Incident proton energy: 21 GeV. (*From H. Wachsmuth, Proc. Neutrino Meeting, CERN 69–28, Geneva, 1969.*)

The neutrino energy spectrum cannot be measured directly, but it can be inferred from the muon energy spectrum. The latter is measurable at various points along the shielding and can also be calculated from the incident proton energy, the pion and kaon production spectrum in the target (Hag 68), and the π and K orbits from target through focusing system to decay point (see, for example, Wac 69). Figure 13.2 gives the ν_μ spectrum obtained during the 1967 CERN experiments with a 21-GeV proton beam (Wac 69). Figure 13.3 gives the neutrino spectrum predicted for Batavia (Bur 70).

The earliest neutrino beam experiments were performed at the AGS at Brookhaven in 1961–1962 with neutrino event rates on free nucleons of a few times 10^{-4}/h. Since then, various technical improvements and increasing proton energies have resulted in enormous rate enhancements. For example, during the 1967 CERN experiments the rate was about 10^{-1}/h, while the expected rate at Batavia (by 1973–1974) for 200-GeV protons will be about 10^3/h.

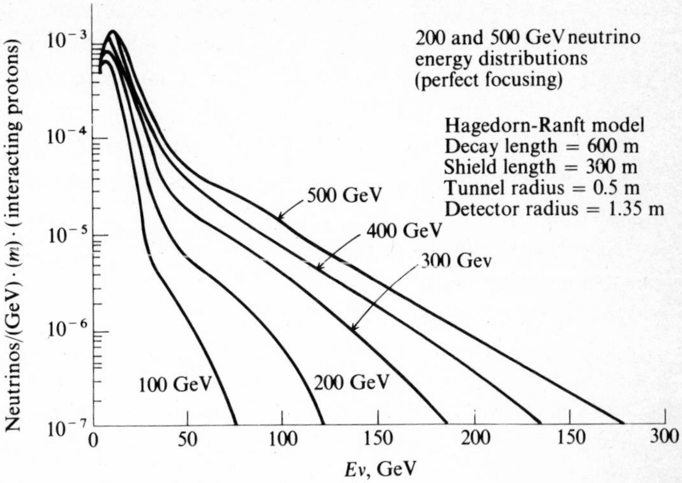

FIGURE 13.3
Predicted neutrino energy spectrum at the National Accelerator Laboratory (NAL) for various proton energies (Bur 70).

What can be learned from high-energy neutrino scattering experiments? Perhaps the most dramatic discovery so far was made in the first few years of work—this was the observation that ν_e and ν_μ are distinct particles (Dan 62, Bie 64; see also Chap. 1). Another very fundamental problem concerns the *diagonal* interactions

$$\nu_e + e \rightarrow e + \nu_e \tag{13.1}$$

$$\nu_\mu + \mu \rightarrow \mu + \nu_\mu \tag{13.2}$$

discussed in Chap. 3. Unfortunately it seems immensely difficult to investigate these reactions experimentally with high-energy neutrino beams. In addition to the fact that the cross sections are very small, rates for reaction (13.1) suffer from the feeble ν_e intensity of a typical accelerator neutrino beam, while the only practical way to obtain information about reaction (13.2) is to observe the related process of muon pair production in a coulomb field: $\nu_\mu + Z \rightarrow \mu^+ + \mu^- + \nu_\mu + Z$. We find (Czy 64) that

$$\begin{aligned}\sigma\text{ (pair)} &= 2.7 \times 10^{-41}\ \text{cm}^2/\text{Fe nucleus at } E_\nu = 10\ \text{GeV}\\ &= 1.6 \times 10^{-40}\ \text{cm}^2/\text{Fe nucleus at } E_\nu = 30\ \text{GeV}\\ &= 9 \times 10^{-40}\ \text{cm}^2/\text{Fe nucleus at } E_\nu = 100\ \text{GeV}\end{aligned}$$

Thus, for example, if we were to utilize 3×10^{18} protons on target at the 200- to 500-GeV Batavia facility (500,000 proton beam pulses) and employ a 100-ton neutrino detector, we would obtain only about fifty events for $\nu_e e$ scattering and one hundred events for the muon-pair process (Per 69). In addition, the experimental problem of distinguishing true events from background (mainly inelastic neutrino-nucleon scattering) would be truly monumental. Thus, prospects for investigation of the diagonal interactions are not encouraging.[1]

Still another problem of great importance concerns the existence of the intermediate vector boson. In Chap. 3 we discussed how one might produce this particle by a reaction of the form

$$\nu_\mu + Z \to W^+ + \mu^- + Z \tag{13.3}$$

Estimates show that if W has the properties commonly attributed to it and a mass $m_W \leq 8$ GeV, it should be observable with the Batavia neutrino beam (Per 69).

Our main concern in this chapter is with neutrino-nucleon scattering, however. Here we distinguish the *elastic* processes

$$\nu_\mu + n \to \mu^- + p \tag{13.4}$$

$$\bar{\nu}_\mu + p \to \mu^+ + n \tag{13.5}$$

(as well as the $|\Delta S| = 1$ transitions

$$\nu_\mu + n \to \mu^- + \Sigma^+$$

$$\bar{\nu}_\mu + p \to \mu^+ + \Lambda$$

etc.) from inelastic collisions, where two or more hadrons are in the final state:

$$\nu + N \to \mu + 2 \text{ or more hadrons} \tag{13.6}$$

In the latter case we have, at one extreme, single-pion production, e.g.,

$$\bar{\nu}_\mu + p \to p + \pi^+ + \mu^- \tag{13.7}$$

and at the other extreme, the so-called *deep* inelastic processes, where there is a very large energy transfer to the final hadron state, which may have a very large invariant mass. It has been observed that, in the deep inelastic case, there is a striking analogy between electron-proton collisions, which occur via the electromagnetic interaction by exchange of a single photon in the simplest approximation, and neutrino-nucleon weak collisions. In both cases the experimental results indicate in a preliminary way that the nucleon scatters as if it consisted of pointlike objects called *partons* (Fey 69, Bjo 69b).

[1] However, $\nu_\mu e^- \to \nu_e \mu^-$ seems feasible.

13.2 KINEMATICS OF THE SCATTERING PROCESS

Let us now consider some of the kinematic features common to all neutrino-nucleon scattering experiments, elastic or inelastic. We denote the four-momenta of initial and final leptons by $k = (\omega, k)$ and $k' = (\omega', k')$, respectively. The four-momentum of the initial nucleon is labeled $p = (E, p)$, and this reduces to $p_0 = (m, 0)$ in the laboratory frame. As for the final hadronic state, it may consist of one nucleon, as in elastic scattering, or two or more hadrons in the inelastic case. We denote its total four-momentum by $p' = (E', p')$ and its invariant mass by M. The energy transferred from the initial lepton to the hadronic system is

$$\nu = \omega - \omega'$$

and the four-momentum transfer is

$$q = k' - k = p - p'$$

Thus we have

$$q^2 = (k' - k)^2 = k'^2 + k^2 - 2k \cdot k'$$

Neglecting the final lepton mass, this becomes

$$\begin{aligned} q^2 &= 2\mathbf{k} \cdot \mathbf{k}' - 2\omega\omega' \\ &= -\, 2\omega\omega'(1 - \cos\theta) \\ q^2 &= -\, 4\omega\omega' \sin^2 \frac{\theta}{2} \end{aligned} \tag{13.8}$$

where θ is the lepton scattering angle. We also have

$$q^2 = (p - p')^2 = m^2 + M^2 - 2p \cdot p'$$

which becomes in the laboratory frame

$$q_{\text{lab}}{}^2 = M^2 - m^2 - 2m\nu \tag{13.9}$$

In practice, one does not observe the final hadron(s) but merely the final lepton energy and scattering angle. Thus, one determines q^2 and ν, from which M is obtained. Of course, for elastic scattering, q^2 and ν are not independent since in this case $M = m$ and Eq. (13.9) reduces to

$$q_{\text{lab}}{}^2 = -\, 2m\nu$$

These elementary kinematic considerations also apply to electron-proton scattering at high energies, where one can neglect the electron mass.

13.3 ELASTIC NEUTRINO-NUCLEON COLLISIONS

We now consider the reactions

$$\nu_\mu + n \to \mu^- + p \qquad (13.10)$$

$$\bar{\nu}_\mu + p \to \mu^+ + n \qquad (13.11)$$

which have been studied in some detail experimentally, especially at CERN (Bud 69a).

The cross section is given by the usual formula

$$d\sigma_{\nu,\bar{\nu}} = \frac{(2\pi)^4\delta^4(p + k - p' - k')}{4pk} \frac{d^3\mathbf{k}' \, d^3\mathbf{p}'}{(2\pi)^6 2\omega' 2E'} |\mathcal{M}_{\nu,\bar{\nu}}|^2 \qquad (13.12)$$

where $\mathcal{M}_\nu = \dfrac{G}{\sqrt{2}} \cos \Theta$

$$\times \, \bar{u}_p(p') \, [f_1\gamma^\lambda + if_2 \, \sigma^{\lambda\nu} q_\nu + f_3 q^\lambda + g_1\gamma^\lambda\gamma^5 - ig_2 \, \sigma^{\lambda\nu}\gamma^5 q_\nu + g_3 q^\lambda\gamma^5] u_n(p) L_\lambda \qquad (13.13)$$

for reaction (13.10),

$$\mathcal{M}_{\bar{\nu}} = \frac{G}{\sqrt{2}} \cos \Theta \bar{u}_n(p')[f_1^*\gamma^\lambda + if_2^*\sigma^{\lambda\nu} q_\nu - f_3^* q^\lambda + g_1^*\gamma^\lambda\gamma^5$$

$$+ ig_2^*\sigma^{\lambda\nu} q_\nu \gamma^5 + g_3^* q^\lambda\gamma^5] u_p(p) L_\lambda \qquad (13.14)$$

for reaction (13.11), and Θ is the Cabibbo angle. We also introduce the kinematic variables

$$s = (p + k)^2 = (p' + k')^2$$

and

$$u = (p - k')^2 = (p' - k)^2$$

It is then straightforward, if tedious, to show that

$$\frac{d\sigma(\nu, \bar{\nu})}{d(-q^2)} = \frac{G^2 \cos^2 \Theta}{8\pi\omega^2} [A(q^2) \mp B(q^2)(s - u) + C(q^2)(s - u)^2] \qquad (13.15)$$

where A, B, and C are functions of q^2, ${m_\mu}^2$, and the form factors f and g.

Rather than write out A, B, and C in full generality, we merely give their values in the ultrarelativistic limit in which m_μ may be neglected. Assuming no second-class currents ($f_3 = g_2 = 0$) and time-reversal invariance (all form factors real), we obtain (with $m_p = 1$)

$$\lim_{m_\mu \to 0} A(q^2) = \tfrac{1}{4}q^4 |g_1|^2 + \tfrac{1}{4}q^2(q^2 + 4)[{f_1}^2 - q^2{f_2}^2] + 2q^4 f_1 f_2$$

$$\lim_{m_\mu \to 0} B(q^2) = -q^2[g_1(f_1 + 2f_2)]$$

$$\lim_{m_\mu \to 0} C(q^2) = \tfrac{1}{4}({f_1}^2 + {g_1}^2) - \tfrac{1}{4}q^2{f_2}^2$$

For forward scattering, $q^2 = 0$ and thus $A = B = 0$, while $C = \frac{1}{4}(f_1{}^2 + g_1{}^2)$. Therefore

$$\frac{d\sigma_{\nu,\bar{\nu}}}{d(-q^2)}(\theta = 0, m_\mu \to 0) = \frac{G^2 \cos^2 \Theta}{8\pi\omega^2} \tfrac{1}{4}(f_1{}^2 + g_1{}^2)(s - u)^2$$

However, $(s - u) = 4\omega$ in this case, and so

$$\frac{d\sigma_{\nu,\bar{\nu}}}{d(-q^2)}(\theta = 0, m_\mu \to 0) = \frac{G^2 \cos^2 \Theta}{2\pi}(f_1{}^2 + g_1{}^2) \qquad (13.16)$$

Similarly, at very large ω, $s - u = 4\omega$, and so only the C term is important; thus

$$\lim_{w \to \infty} \frac{d\sigma_{\nu,\bar{\nu}}}{d(-q^2)} = \frac{G^2 \cos^2 \Theta}{3}\{f_1{}^2 + g_1{}^2 - q^2 f_2{}^2 - q^2 g_2{}^2\} \qquad (13.17)$$

In practice, neutrino-nucleon elastic scattering experiments are not yet sufficiently sensitive to uncover possible contributions from form factors f_3, g_2, or g_3 even if these are nonzero. (According to the CVC hypothesis and the selection rule *no second-class currents*, of course, we expect $f_3 = g_2 = 0$, but $g_3 \neq 0$.) If CVC continues to be valid at high energies, then $f_1(q^2)$ and $f_2(q^2)$ should be the same as their isovector electromagnetic counterparts, as found in scattering of electrons from nucleons, viz.,

$$\frac{f_1(q^2)}{f_1(0)} = \frac{f_2(q^2)}{f_2(0)} = \frac{1}{(1 + |q^2/M_V{}^2|)^2} \qquad (13.18)$$

with $M_V \approx 0.84$ GeV. Assuming this we attempt in practice to determine the q^2 dependence of g_1, generally assuming that the form

$$\frac{g_1(q^2)}{g_1(0)} = \frac{1}{(1 + |q^2/M_A{}^2|)^2} \qquad (13.19)$$

is valid. Thus the whole enterprise reduces to a determination of M_A.

The experimental difficulties are naturally formidable. For $\nu_\mu + n \to \mu^- + p$, we must, of course, use neutrons bound in nuclei. Thus we must take account of the Pauli principle and the resulting Fermi motion of the target neutron, scattering and/or absorption of the proton in traversing the nucleus, and background contributed by events of the type $\nu + N \to N' + \pi + \mu^-$, in which pions are absorbed by the nucleus. The 1967 CERN experiments show that (Bud 69a)

$$M_A = (0.7 \pm 0.2) \text{ GeV}/c^2 \qquad (13.20)$$

which is consistent with (13.19), although, of course, the uncertainty is large.

Even though the results until now are barely sufficient to determine M_A, we may anticipate that experimental improvements in elastic scattering will eventually permit detailed measurements of final polarizations, etc. We can show that final baryon polarization is most promising. Although observations of transverse nuclear polarization with respect to the p', k' plane would indicate a failure of T invariance, calculations show that observations of longitudinal polarization would provide a more sensitive test for the second-class form factor $g_2(q^2)$ than is possible from direct cross-section measurements (Pai 70).

13.4 INELASTIC NEUTRINO-NUCLEON COLLISIONS

13.4.1 Weak Pion Production

We now consider briefly reactions of the form

$$\nu_e + p \to l^- + p + \pi^+(\sigma_1) \qquad (13.21)$$

$$\nu_e + n \to l^- + n + \pi^+(\sigma_2) \qquad (13.22)$$

$$\nu_e + n \to l^- + p + \pi^0(\sigma_3) \qquad (13.23)$$

Let us first discuss the relative strengths of these transitions from the point of view of isotopic spin. Here we may think in each case of the initial nucleon absorbing a *spurion* h with ($I = 1$, $I_3 = +1$):

$$h + p \to \pi^+ + p$$

$$h + n \to \pi^+ + n$$

$$h + n \to \pi^0 + p$$

The transition operator leading to the final pion-nucleon state is then assumed to be isospin invariant. It is a simple matter to work out the Clebsch–Gordan coefficients for the three reactions and show that if the amplitudes to go to $I = \frac{3}{2}, \frac{1}{2}$ final states are denoted by A_3, A_1 respectively, we have for the ratios of the cross sections for (13.21) to (13.23)

$$\sigma_1 : \sigma_2 : \sigma_3 = |A_3|^2 : \tfrac{1}{9}|A_3 + 2A_1|^2 : \tfrac{2}{9}|A_3 - A_1|^2$$

If the final state is pure $I = \frac{3}{2}$, this becomes

$$\sigma_1 : \sigma_2 : \sigma_3 = 9 : 1 : 2 \qquad (13.24)$$

In practice, virtually all of the events actually observed are

$$\nu + p \to l^- + p + \pi^+(\sigma_1)$$

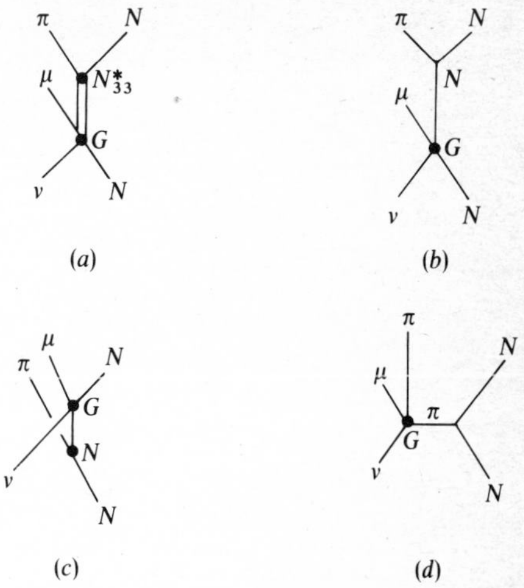

FIGURE 13.4
Feynman diagrams for $\nu + p \to l^- + p + \pi^+$. The weak vertex is labeled G.

How is this reaction to be described theoretically? Analyses have been carried out by Adler (Adl 68), Salin (Sal 67), Berman and Veltman (Ber 65), and others. One finds that the most important contribution to the amplitude comes from $N^*(1238)$ exchange (Fig. 13.4*a*), although substantial contributions also arise from diagrams of Fig. 13.4*b* and *c*, which represent nucleon exchange, and Fig. 13.4*d*, which represents pion exchange. In order to compute the amplitudes, the vector weak coupling is obtained by using the analogous electromagnetic form factors for photo- and electroproduction (CVC), while the axial coupling is derived from the static pion-nucleon theory of Chew, Goldberger, Low, and Nambu, plus the Goldberger–Treiman relation. The predictions of Salin and Adler are based on somewhat similar arguments, but their results differ by a factor of 2.

Experimentally we find that the invariant mass for $\nu + p \to p + \pi^+ + \mu^-$ peaks around $N^*(1238)$, as expected. However, the data show nonresonant contributions as well (Fra 69). The experimental results (CERN, Fra 69) are compared with theoretical predictions in Fig. 13.5. It can be seen that the observations are still much too crude to allow a choice between the various theoretical models.

13.4.2 Adler's Relation for Forward Scattering

The reaction

$$\nu_\mu + p \to \mu^- + N^*(1238)$$

is a specific example of the general case

$$\nu + B \to l + B' \qquad (\Delta S = 0) \qquad (13.25)$$

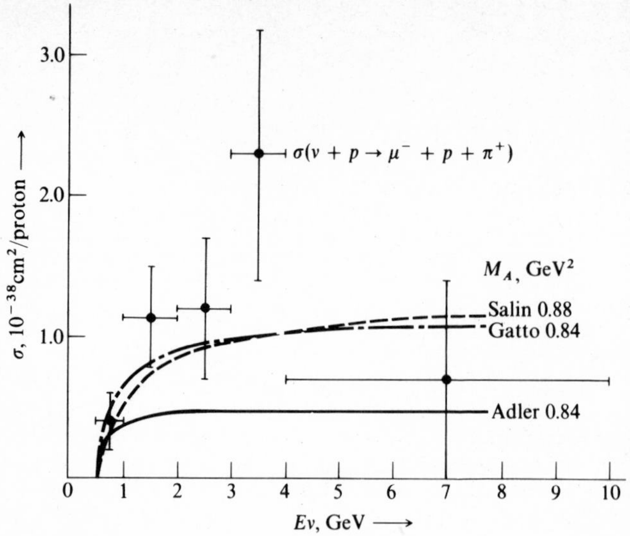

FIGURE 13.5
Cross section for $\nu_\mu + p \rightarrow \mu^- + p + \pi^+$. Experimental points are compared with theoretical predictions (Per 69).

where $m_{B'} \neq m_B$. The matrix element is

$$\mathcal{M} = \frac{G}{\sqrt{2}} \cos \Theta \langle B' | \mathfrak{J}_0{}^\lambda | B \rangle \bar{u}_\mu \gamma_\lambda (1 + \gamma^5) u_\nu$$

and we have for $|\mathcal{M}|^2$ summed over final spins and averaged over initial baryon spin

$$|\overline{\mathcal{M}}|^2 = G^2 \cos \Theta \langle B' | \mathfrak{J}_0{}^\lambda | B \rangle \langle B | \mathfrak{J}_0{}^{\dagger\sigma} | B' \rangle$$
$$(k_\lambda k'_\sigma + k'_\lambda k_\sigma - 2k \cdot k' g_{\lambda\sigma} - i\varepsilon_{\alpha\beta\lambda\sigma} k^\alpha k^\beta) \qquad (13.26)$$

Suppose we denote the incoming neutrino direction by z and consider forward scattering (Adl 64). In this case, $q^2 = 0 = k \cdot k'$. Thus

$$k_0 k_0 = k_z k_z$$
$$k'_0 k'_0 = k'_z k'_z$$

and

$$k_0 k'_0 = k_z k'_z$$

It follows that $k = \beta k'$, where β is some constant. Furthermore, $q_0 \neq 0$ since $m_{B'} \neq m_B$. It follows that since $q_0 \neq 0$ in the laboratory frame and

$q^2 = 0$, q_0 must be nonzero in all frames. Thus we can write for forward scattering

$$k = \frac{k \cdot p}{q \cdot p} q \qquad k' = \frac{k' \cdot p}{q \cdot p} q$$

and the lepton factor in (13.26) becomes

$$k_\lambda k'_\sigma + k'_\lambda k_\sigma - 2k \cdot k' g_{\lambda\sigma} - i\varepsilon_{\alpha\beta\lambda\sigma} k^\alpha k^{\beta'} = \frac{2k \cdot p}{q \cdot p} \frac{k' \cdot p}{q \cdot p} q_\lambda q_\sigma \tag{13.27}$$

Also we have

$$\frac{\partial}{\partial_x{}^\lambda} \langle B' | \mathfrak{J}_0{}^\lambda | B \rangle = - iq_\lambda \langle B' | \mathfrak{J}_0{}^\lambda | B \rangle$$

and so we obtain from (13.26) and (13.27)

$$\langle B' | \frac{\partial \mathfrak{J}_0{}^\lambda}{\partial x^\lambda} | B \rangle \langle B | \frac{\partial \mathfrak{J}_0{}^{\dagger\sigma}}{\partial x^\sigma} | B \rangle = q_\lambda q_\sigma \langle B' | \mathfrak{J}_0{}^\lambda | B \rangle \langle B | \mathfrak{J}_0{}^{\dagger\sigma} | B' \rangle$$

$$= \frac{1}{G^2 \cos^2 \Theta} \frac{(q \cdot p)^2}{k \cdot pk' \cdot p} |\bar{\mathcal{M}}|^2 \tag{13.28}$$

However, the left-hand side of (13.28) may be reduced by noting that

$$\frac{\partial \mathfrak{J}_0{}^\lambda}{\partial x^\lambda} = \frac{\partial \mathcal{V}_0{}^\lambda}{\partial x^\lambda} + \frac{\partial \mathcal{A}_0{}^\lambda}{\partial x^\lambda}$$

The CVC hypothesis requires that $\partial \mathcal{V}_0{}^\lambda / \partial x^\lambda = 0$, and PCAC states that $\partial \mathcal{A}_0{}^\lambda / \partial x^\lambda = \text{const}\ \phi_\pi$. It follows that

$$|\langle B' | \phi_\pi | B \rangle|^2 = \text{kinematic factor} \times |\mathcal{M}|^2 \tag{13.29}$$

Therefore, for forward scattering ($q^2 = 0$) the cross sections for $\nu + B \to \mu + B'$ and $\pi + B \to B'$ are proportional, with the proportionality factor known. It follows that the distribution of final states B' should be the same for the weak and strong reactions. Actually one must observe *forward* scattering in a small but finite cone $\theta \neq 0$; and also $m_\mu \neq 0$. Thus, corrections must be introduced for these effects. The experimental data are so far quite meager, but they do not contradict Adler's relation (Bon 68).

13.4.3 Deep Inelastic Scattering

As we go to the regime of large-energy and high-momentum transfer, deep inelastic scattering becomes dominant. We mentioned earlier that neutrino-nucleon collisions are quite analogous to electron-proton collisions in the deep

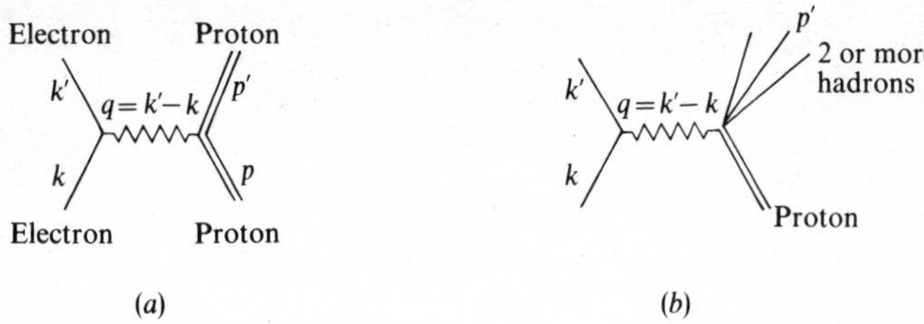

FIGURE 13.6
Feynman diagrams for (*a*) elastic and (*b*) inelastic *ep* scattering by one-photon exchange.

inelastic limit, and so it seems appropriate to summarize the latter at this juncture. Figure 13.6*a* and *b* represent *ep* elastic and inelastic scattering, respectively, in the simplest approximation of single-photon exchange.

The cross section for unpolarized *ep* scattering is given by the basic formula

$$d\sigma = \frac{(2\pi)^4\delta^4(k+p-k'-p')}{4\sqrt{(k\cdot p)^2 - m_p{}^2 m_e{}^2}}\,\frac{1}{2\omega'}\,\frac{1}{2E'}\,\frac{d^3\mathbf{p}'\,d^3\mathbf{k}'}{(2\pi)^6}\,|\overline{\mathcal{M}}|^2 \tag{13.30}$$

where $\mathcal{M}$ is the amplitude for elastic or inelastic scattering and $|\overline{\mathcal{M}}|^2$ indicates a sum over final spin states and an average over initial spin states. In the laboratory frame, if we take the limit $m_e \to 0$, we obtain

$$d\sigma = \frac{1}{16(2\pi)^2}\,\frac{\delta^4(k+p-k'-p')}{\omega\omega' m_p E'}\,d^3\mathbf{p}'\,\omega'^2\,d\omega'\,d\Omega'\,|\overline{\mathcal{M}}|^2$$

where $d\Omega'$ is the differential solid angle into which the final electron is emitted. Integrating over the unobserved $\mathbf{p}'$, we obtain

$$d\sigma = \frac{1}{16(2\pi)^2}\,\frac{\delta(\omega + m_p - \omega' - E')\omega'\,d\omega'\,d\Omega}{\omega E' m_p}\,|\overline{\mathcal{M}}|^2 \tag{13.31}$$

Now we consider the amplitude $\mathcal{M}_{\text{el}}$ for elastic scattering:

$$\mathcal{M}_{\text{el}} = -\,L_\lambda \frac{1}{q^2} H^\lambda \tag{13.32}$$

where $-1/q^2$ is the photon propagator and L_λ and H^λ are factors representing the lepton and hadron vertices, respectively:

$$L_\lambda = \sqrt{4\pi\alpha}\,\bar{u}_e(k')\gamma_\lambda u_e(k) \tag{13.33}$$

$$H^\lambda = \sqrt{4\pi\alpha}\,\bar{u}_p(p')[C_p(q^2)\gamma^\lambda + i\mathrm{M}_p(q^2)\sigma^{\lambda\nu}q_\nu]u_p(p) \tag{13.34}$$

and where $$\alpha = \frac{e^2}{\hbar c} \approx \frac{1}{137}$$

Here $C_p(q^2)$ and $M_p(q^2)$ are the electric and magnetic form factors, respectively, for the proton (real by T invariance). We have

$$|\bar{\mathcal{M}}_{\text{el}}|^2 = \frac{1}{4} \sum_{\substack{e \\ \text{spin}}} \sum_{\substack{p \\ \text{spin}}} L_\lambda L_\sigma^* \frac{1}{q^4} H^\lambda H^{\sigma *} \tag{13.35}$$

It is easy to show by the usual methods that

$$\sum_{\substack{e \\ \text{spin}}} L_\lambda L_\sigma^* = 16\pi\alpha[k'_\lambda k_\sigma + k'_\sigma k_\lambda + (m_e{}^2 - k' \cdot k)g_{\lambda\sigma}] \tag{13.36}$$

and that

$$H^\lambda = \bar{u}(p')[C_p + 2M_p m_p]\gamma^\lambda u(p) - M_p(p + p')^\lambda \bar{u}(p')u(p) \tag{13.37}$$

Thus, neglecting the electron mass and choosing the laboratory frame, we find

$$|\bar{\mathcal{M}}_{\text{el}}|^2 = \frac{2^8\pi^2 m_p{}^2 \omega\omega'\alpha^2}{q^4} \left[(C_p{}^2 - M_p{}^2 q^2)\cos^2\frac{\theta}{2} - \frac{q^2}{2m_p{}^2} \right.$$
$$\left. \times (C_p + 2M_p m_p)^2 \sin^2\frac{\theta}{2}\right] \tag{13.38}$$

In the inelastic case it is no longer possible to write the hadronic portion of the matrix element in the simple form (13.34). Instead we have

$$|\bar{\mathcal{M}}_{\text{incl}}|^2 = \frac{1}{4} \sum_{\substack{e \\ \text{spin}}} \sum_{\substack{p \\ \text{spin}}} L_\lambda L_\sigma^* \frac{1}{q^4} W^{\lambda\sigma} \tag{13.39}$$

where $W^{\lambda\sigma}$ is a tensor which depends on the kinematic invariants q^2 and $q \cdot p$ and is expressed most generally as follows (Dre 64):

$$\sum_{\substack{p \\ \text{spin}}} W^{\lambda\sigma} = 16\pi\alpha[Ag^{\lambda\sigma} + Bq^\lambda q^\sigma + Cp^\lambda p^\sigma$$
$$+ D(q^\lambda p^\sigma + p^\lambda q^\sigma) + E(q^\lambda p^\sigma - p^\lambda q^\sigma)] \tag{13.40}$$

where $A, \ldots, E$ are functions of p^2 and $p \cdot q$. However, the electromagnetic current is conserved. Therefore, it follows that $q_\lambda W^{\lambda\sigma} = q_\sigma W^{\lambda\sigma} = 0$, and this implies that $E = 0$ and

$$C = \frac{q^2}{(q \cdot p)^2}(A + Bq^2)$$

$$D = -\frac{1}{q \cdot p}(A + Bq^2)$$

Therefore we obtain

$$\sum_{\substack{p \\ \text{spin}}} W^{\lambda\sigma} = 32\pi\alpha\left[a\left(p^\lambda - \frac{q \cdot pq^\lambda}{q^2}\right)\left(p^\sigma - \frac{q \cdot pq^\sigma}{q^2}\right) - bm_p{}^2\left(g^{\lambda\sigma} - \frac{q^\lambda q^\sigma}{q^2}\right)\right]$$

where a and b are two new coefficients depending on $q \cdot p$ and q^2. Thus, in the limit $m_e \to 0$,

$$|\bar{\mathcal{M}}_{\text{incl}}|^2 = \frac{2^7\pi^2\alpha^2}{q^4}\left[a\left(p^\lambda - q \cdot p\frac{q^\lambda}{q^2}\right)\left(p^\sigma - \frac{q \cdot pq^\sigma}{q^2}\right) - bm_p{}^2\left(g^{\lambda\sigma} - \frac{q^\lambda q^\sigma}{q^2}\right)\right]$$
$$\times (k'_\lambda k_\sigma + k'_\sigma k_\lambda - k' \cdot kg_{\lambda\sigma})$$

which reduces in the laboratory frame to

$$|\bar{\mathcal{M}}_{\text{incl}}|^2 = \frac{2^8\pi^2 m_p{}^2\omega\omega'\alpha^2}{q^4}\left(a\cos^2\frac{\theta}{2} + 2b\sin^2\frac{\theta}{2}\right) \tag{13.41}$$

Comparing (13.38) with (13.41), we see that the latter reduces to the former when

$$\begin{aligned} a &= (C_p{}^2 - M_p{}^2 q^2) \\ b &= -\frac{q^2}{4m_p}(C_p + 2M_p m_p)^2 \end{aligned} \tag{13.42}$$

We now insert (13.41) into (13.21) and make the change of variable

$$d(-q^2) = \frac{\omega\omega'}{\pi}\,d\Omega'$$

to obtain

$$\frac{\partial^2\sigma_{\text{incl}}(e, p)}{\partial q^2\,\partial\nu} = \frac{4\pi\alpha^2\omega'}{q^4\omega}\cos^2\frac{\theta}{2}\left[W_2(q^2, \nu) + 2\tan^2\frac{\theta}{2}W_1(q^2, \nu)\right] \tag{13.43}$$

where

$$W_2(q^2, \nu) = \frac{am_p}{E'}\delta(\omega + m_p - \omega' - E') \tag{13.44}$$

and

$$W_1(q^2, \nu) = \frac{bm_p}{E'}\delta(\omega + m_p - \omega' - E') \tag{13.45}$$

Formula (13.43) is the basic result we need for electron-proton inelastic scattering.

We now return to weak interactions and consider the inelastic neutrino-nucleon cross section. By entirely analogous arguments (Adl 66), we find

$$\frac{\partial^2\sigma_{\text{incl}}(\nu, N)}{\partial q^2\,\partial\nu} = \frac{\omega'}{\omega}\frac{G^2}{2\pi}\left[\mathbf{W}_2\cos^2\frac{\theta}{2} + 2\mathbf{W}_1\sin^2\frac{\theta}{2} + (\omega + \omega')\mathbf{W}_3\sin^2\frac{\theta}{2}\right] \tag{13.46}$$

where $\mathbf{W}_1$ and $\mathbf{W}_2$ are analogous to the structure functions W_1, W_2, respectively, of the electromagnetic case [Eq. (13.43)], while $\mathbf{W}_3$ is a new function which arises because of the existence of axial vector as well as vector coupling in the weak interaction.

Bjorken (Bjo 69a) has demonstrated from a sum rule of current algebra at infinite momentum that the following behavior is to be expected for the functions W_1, W_2, $\mathbf{W}_1$, $\mathbf{W}_2$, $\mathbf{W}_3$, as $|q|^2 \to \infty$, $\nu \to \infty$, and $q^2/\nu \equiv \rho$ is maintained constant:

$$\nu W_2 \to F_2(\rho) \qquad (13.47)$$

$$W_1 \to F_1(\rho) \qquad (13.48)$$

$$\nu \mathbf{W}_2 \to \mathbf{F}_2(\rho) \qquad (13.49)$$

$$\mathbf{W}_1 \to \mathbf{F}_1(\rho) \qquad (13.50)$$

$$\nu \mathbf{W}_3 \to \mathbf{F}_3(\rho) \qquad (13.51)$$

where the F's and $\mathbf{F}$'s are certain universal functions. The asymptotic dependence of these quantities on ρ alone is indicative of the absence of a natural finite scattering length and would suggest that the scattering takes place as though the nucleon consisted of point particles in the asymptotic limit (deep inelastic continuum) (Bjo 69b, Fey 69).

The experimental results for electron-proton scattering (at SLAC) and also for neutrino-nucleon scattering (at CERN) are in agreement with this picture. In the electron-proton case (Blo 69), the scattering angle is rather small ($\theta \approx 6^0$). Therefore, the first term on the right-hand side of (13.43) is dominant. It is found that the experimental values for W_2 do, in fact, conform to (13.47) for sufficiently large $|q^2|$ and ν, and for $|\rho| > 1.5$.

In the neutrino-nucleon case, the data are, of course, of poorer quality, but we find essentially the same dependence of $\mathbf{W}_2$ on q^2 for given ν and the same variation with ν (Bud 69b). These experiments show that

$$\left.\begin{aligned} \mathbf{F}_2(\rho) &\approx 0.9 \pm 0.3 \\ \frac{1}{\rho}\mathbf{F}_1(\rho) &\approx 0.9 \pm 0.3 \\ \frac{1}{\rho}\mathbf{F}_3(\rho) &\approx 0 \end{aligned}\right\} \qquad (13.52)$$

The above results are consistent with certain theoretical speculations (Har 69, Aba, 69, Dre 69a), which suggest that $\mathbf{F}_2(\rho)$ and $\mathbf{F}_1(\rho)/\rho$ should be constant and $\mathbf{F}_3(\rho)/\rho$ should be zero.[1]

Another prediction arising from Bjorken's sum-rule argument (Bjo 69a) is that the *total* cross section for neutrino-nucleon collisions should rise linearly

[1] A more detailed discussion of these questions may be found, e.g., in the review by Pais (Pai 71).

with energy in the asymptotic limit. This has also been verified by the CERN experiments (Bud 69b):

$$\sigma = (0.6 \pm 0.15)\frac{G^2}{\pi} m_p \omega \qquad (13.53)$$

where σ is in units 10^{-38} cm^2/nucleon and ω is in gigaelectronvolts.

REFERENCES

Aba 69 Abarbanel, H., M. L. Goldberger, and S. Treiman: *Phys. Rev. Letters,* **22**:500 (1969).

Adl 63 Adler, S. L.: *Nuovo Cimento,* **30**:1020 (1963)

Adl 64 Adler, S. L.: *Phys. Rev.,* **135**:B963 (1964).

Adl 66 Adler, S. L.: *Phys. Rev.,* **143**:1144 (1966).

Adl 68 Adler, S. L.: *Ann. Phys. (N.Y.),* **50**:189 (1968).

Ber 65 Berman, S. M., and M. Veltman: *Nuovo Cimento,* **38**:993 (1965).

Bie 64 Bienlein, J., et al.: *Phys. Letters,* **13**:80 (1964).

Bjo 69a Bjorken, J. D.: *Phys. Rev.,* **179**:1547 (1969).

Bjo 69b Bjorken, J. D., and E. A. Paschos: *Phys. Rev.,* **185**:1975 (1969).

Blo 69 Bloom, E. D., et al.: *Phys. Rev. Letters,* **23**:930, 935 (1969).

Bon 68 Bonetti, S., et al.: *Proc. Vienna Conf. Elementary Particles,* 1968.

Bor 69 Borer K., et al.: *Phys. Letters,* **29B**:614 (1969).

Bud 69a Budagov, I., et al.: *Lett. Nuovo Cimento,* **2**:689 (1969).

Bud 69b Budagov, I., et al.: *Phys. Letters,* **30B**:364 (1969).

Bur 70 Burnstein, R. A.: *1970 Summer Study Rept.* 180, Natl. Accelerator Lab., Batavia, Ill.

Cab 65 Cabibbo, N., and F. Chilton: *Phys. Rev.,* **137**:1628 (1965).

Czy 64 Czyz, W., G. C. Sheppey, and J. D. Walecka: *Nuovo Cimento,* **34**:404 (1964).

Dan 62 Danby, G., J. M. Gaillard, K. Goulianos, L. M. Lederman, N. Mistry, M. Schwartz, and J. Steinberger: *Phys. Rev. Letters,* **9**:36 (1962).

Dre 64 Drell, S. D., and J. D. Walecka: *Ann. Phys.,* **28**:18 (1964).

Dre 69a Drell, S. D., and D. Levy: *Phys. Rev.,* **187**:2159 (1969).

Dre 69b Drell, S., D. Levy, and T. Yan: *Phys. Rev. Letters,* **22**:744 (1969).

Fey 69 Feynman, R. P.: *Phys. Rev. Letters,* **23**:1416 (1969).

Fra 69 Franzinetti, C.: *Proc. Topical Conf. Weak Interactions, CERN, Geneva,* 43 (1969).

Hag 68 Hagedorn, R., and J. Ranft: *Nuovo Cimento Suppl.,* **6**:169 (1968).

Har 69 Harari, H.: *Phys. Rev. Letters,* **22**:1078 (1969).

Led 67 Lederman, L.: Neutrino Physics, in E. H. S. Burhop (ed.), "High Energy Physics," vol. 2, Academic, New York, London, 1967.

Lee 60 Lee, T. D., and C. N. Yang: *Phys. Rev. Letters,* **4**:307 (1960).

Lee 62 Lee, T. D., and C. N. Yang: *Phys, Rev.*, **126**:2239 (1962).

Pai 70 Pais, A.: "Symposium on Modern Accelerator Experiment," The University of Wisconsin Press, Madison, 1970.

Pai 71 Pais, A.: *Ann. Phys.* (*N.Y.*), **63**:361 (1971).

Per 69 Perkins, D. H.: *Topical Conf. Weak Interactions, CERN, Geneva* (1969).

Pon 59 Pontecorvo, B.: *Zh. Eksperim. Teor. Fiz.*, **36**:1615 (1959); *Soviet Phys. JETP*, **9**:1148 (1959) (English translation).

Sal 67 Salin, Ph.: *Nuovo Cimento*, **48**:506 (1967).

Sch 60 Schwartz, M.: *Phys. Rev. Letters*, **4**:306 (1960).

Sir 65 Sirlin, A.: *Nuovo Cimento*, **37**:137 (1965).

Wac 69 Wachsmuth, H.: *Proc. Neutrino Meeting, CERN 69–28, Geneva* (1969).

14

NEUTRINO ASTROPHYSICS

14.1 INTRODUCTION

Astrophysical neutrinos bombard the earth from a variety of sources. Closest at hand are the secondary cosmic-ray (K, π, μ) decays in the atmosphere, which yield neutrinos (chiefly ν_μ, $\bar{\nu}_\mu$) over a broad energy spectrum which extends far beyond the range of planned accelerators. These neutrinos are investigated in the hope that some valuable insights may be found into the nature of weak processes at very high energies.

Neutrinos (ν_e) are also generated deep within the sun as a by-product of the nuclear reactions responsible for the solar luminosity. Reasonably quantitative estimates of the flux and energy spectrum of solar neutrinos are now available, and preliminary observations have been reported, which are expected to furnish important direct information about the sun's central temperature and chemical composition.

It seems almost impossible to observe neutrinos from other stars since the total neutrino flux from the stars divided by that from the sun ought to be roughly equal to the ratio of starlight to sunlight. Nevertheless, in order to draw a reasonably accurate picture of the structure and evolution of the stars, we have to take neutrino emission into account. In particular, although only 3 percent of the solar energy is radiated away in neutrinos, the emission of $\nu_e \bar{\nu}_e$ pairs by a hot dense star in an advanced stage of evolution may be enormous, and it can and probably does affect the final stages of evolution of most stars very drastically. Neutrinos of cosmological origin are more speculative. The "big bang" theory of the origin of the universe predicts the existence of a homogeneous isotropic sea of *relict* neutrinos and antineutrinos. Although these are unobservable for all practical purposes, their existence may have a profound effect on the Hubble expansion and other cosmological phenomena.

Neutrino astrophysics is still in its infancy, and hardly any definitive observations have yet been made. The theoretical obstacles to progress are immense, and the near-vanishing counting rates of terrestrial experiments aimed at detecting astrophysical neutrinos have prompted one pioneer to quip that longevity is the most important attribute for any worker in the field. Still, with all its difficulties, the study of astrophysical neutrinos raises deep and fascinating questions, and we may expect that their investigation will yield results of fundamental importance in the coming decades.

14.2 NEUTRINOS FROM THE SUN

14.2.1 The Total Flux of Solar Neutrinos

It seems firmly, if indirectly, established that the sun shines because of energy supplied from exothermic nuclear reactions deep in the interior. The main reactions are those of the proton-proton chain, although a few percent of the energy is contributed by the C—N—O cycle (see Table 14.1). In either case the net effect is the conversion of four protons into an alpha particle

$$2e^- + 4p \rightarrow \mathrm{He}^4 + 2\nu_e + 26.7\ \mathrm{MeV} \qquad (14.1)$$

where all but about 3 percent of the energy appearing on the right-hand side of Eq. (14.1) is delivered in the form of charged-particle and photon energy, which ultimately becomes part of the sun's thermal energy store. As to the 3 percent balance (the exact amount depends on which cycle we consider), it is carried off by the neutrinos, which depart from the sun with velocity c and are lost.

Therefore, 1 ν_e is created for about 13 MeV of thermal energy generated. Since the sun is at present quite stable in size and temperature, the rate of thermal energy generation must be equal to the rate at which energy is radiated from the surface. Since we know the *solar constant* S, that is, the amount of solar flux arriving per second at the top of the earth's atmosphere $S = 1.37 \times 10^6$ erg/cm^2-s, we easily calculate the solar neutrino flux at the earth:

$$\phi_\nu = \frac{S}{13 \text{ MeV}} \approx 6.7 \times 10^{10} \text{ neutrinos/cm}^2\text{-s} \qquad (14.2)$$

Table 14.1 NUCLEAR REACTIONS OF IMPORTANCE FOR SOLAR-ENERGY PRODUCTION

	Neutrino energy, MeV	
	Average	Maximum
Proton-proton chain		
$p + p \rightarrow d + e + \nu_e$	0.26	0.42
$p + p + e^- \rightarrow d + \nu_e$	1.44	1.44
$p + d \rightarrow He^3 + \gamma$		
$He^3 + He^3 \rightarrow He^4 + 2p$		
or		
$He^3 + He^4 \rightarrow Be^7 + \gamma$		
$Be^7 + e^- \rightarrow Li^7 + \nu_e$	0.86	0.86
$Li^7 + p \rightarrow 2He^4$		
or		
$Be^7 + p \rightarrow B^8 + \gamma$		
$B^8 \rightarrow Be^8 + e^+ + \nu_e$	7.2	14‡
$Be^8 \rightarrow 2He^4$		
C—N—O cycle		
$C^{12} + p \rightarrow N^{13} + \gamma$		
$N^{13} \rightarrow C^{13} + e^+ + \nu_e$	0.71	1.19§
$C^{13} + p \rightarrow N^{14} + \gamma$		
$N^{14} + p \rightarrow O^{15} + \gamma$		
$O^{15} \rightarrow N^{15} + e^+ + \nu_e$	1.00	1.70§
$N^{15} + p \rightarrow C^{12} + He^4$‡		

‡ Negaton capture accompanied by monoenergetic neutrino emission always competes to a degree with positron emission. However, the monoenergetic neutrino flux in these cases is quite negligible.

§ The last reaction competes with

$$\begin{aligned} N^{15} + p &\rightarrow O^{16} + \gamma \\ O^{16} + p &\rightarrow F^{17} + \gamma \\ F^{17} &\rightarrow O^{17} + e^+ + \nu_e \\ O^{17} + p &\rightarrow N^{14} + He^4 \end{aligned}$$

but the neutrino flux from F^{17} decay is very small and may be neglected.

Unfortunately, it is very difficult to detect the solar neutrinos. It will be recalled that identification of *antineutrinos* (available in profusion from powerful nuclear reactors) is *relatively* practicable via the reaction

$$\bar{\nu}_e + p \rightarrow n + e^+ \qquad (14.3)$$

An unambiguous identification of such an event may be made by observing the delayed coincidence between the characteristic annihilation x-ray from the positron and the γ ray following neutron capture in some suitable surrounding material (cadmium, for example). However, there is only one known way to detect low-energy electron neutrinos, and that is by radio-chemical identification of the final nucleus in the endothermic reaction

$$\nu_e + (Z, N) \rightarrow (Z + 1, N - 1) + e^- \qquad (14.4)$$

The one practical case for which an actual experiment has been realized so far is

$$\nu_e + \mathrm{Cl}^{37} \rightarrow \mathrm{Ar}^{37} + e^- \qquad (14.5)$$

This reaction has a threshold of 0.81 MeV, and the cross section for neutrino capture is a steeply rising function of neutrino energy above 0.81 MeV. It therefore follows (see Table 14.1) that the great bulk of the neutrino flux from the sun (that from the $pp \rightarrow de^+\nu$ reaction) is *unobservable*, using Cl^{37} as target material. Moreover, although less than 10^{-4} of the total neutrino flux originates in B^8 decay, it turns out that these neutrinos, by virtue of their high energy, should account for about 80 percent of the estimated Cl^{37} capture rate!

Now, the total rate of B^8 production in the solar interior depends sensitively on the central temperature and composition. Therefore, even though it is trivial to calculate the total solar neutrino flux, calculation of the expected Cl^{37} capture rate requires a detailed estimate of central temperature and composition and thus construction of a precise solar model.

14.2.2 Theoretical Solar Models

The equations governing the structure and evolution of the sun[1] relate the pressure $P(r)$, the mass $M(r)$ interior to radial distance r from the center, the density $\rho(r)$, the temperature $T(r)$, the energy production by nuclear reactions per gram $\varepsilon(r)$, the luminosity $L(r)$, and the *Rosseland mean* opacity $\overline{\kappa(r)}$, as follows:

Hydrostatic equilibrium:

$$\frac{dP}{dr} = -\frac{G_0 M(r)\rho(r)}{r^2} \qquad (14.6)$$

[1] The equations apply to any model star in mechanical equilibrium, where rotations and magnetic fields can be neglected. See, for example, Cla 68.

Mass:

$$\frac{dM(r)}{dr} = 4\pi r^2 \rho \tag{14.7}$$

Energy balance:

$$\frac{dL(r)}{dr} = 4\pi r^2 \rho \varepsilon \tag{14.8}$$

Energy transport:

$$\frac{dT}{dr} = -\frac{3}{4ac}\frac{\bar{\kappa}}{T^3}\frac{L(r)}{4\pi r^2} \qquad \text{(radiative)} \tag{14.9}$$

$$\frac{dT}{dr} = \left(1 - \frac{1}{\gamma}\right)\frac{T}{P}\frac{dP}{dr} \qquad \text{(convective)} \tag{14.10}$$

In these equations, G_0 is the newtonian constant of gravitation, a is the Stefan–Boltzmann constant, c is the velocity of light, and γ is the adiabatic index $\gamma = C_p/C_V$. Equation (14.9) describes the temperature gradient in the core of the sun, where energy transport is by radiative transfer, whereas (14.10) is appropriate for the outer, convective zone which reaches to the photosphere and throughout which there is probably considerable turbulence and mixing. In addition to these four differential equations, we have the *constitutive* relations.

Equation of state:

$$P = P(\rho, T, X, Y, Z) \tag{14.11}$$

Opacity:

$$\bar{\kappa} = \bar{\kappa}(\rho, T, X, Y, Z) \tag{14.12}$$

Energy production:

$$\varepsilon = \varepsilon(\rho, T, X, Y, Z) \tag{14.13}$$

where X, Y, Z are the fractional abundances by mass of hydrogen, helium, and heavier elements. (By definition, $X + Y + Z = 1$.)

For the sun the equation of state is, to a very good approximation, that of a perfect gas throughout, although there are small but important corrections for electron degeneracy and coulomb effects (Debye–Hückel factor). The opacity is an exceedingly complicated function of density, temperature, and composition (especially the heavy element content Z). It can be estimated only from a detailed knowledge of atomic transition probabilities for free-free, free-bound, and bound-bound transitions, as well as electron scattering effects in which corrections due to many-electron correlations are important (Wat 69). Calculation of ε depends on detailed knowledge of cross sections and decay probabilities for the reactions listed in Table 14.1. Of these, the cross section for the primary weak reaction

$$p + p \rightarrow de^{+}\nu_{e} \tag{14.14}$$

is far too small to be measured in the laboratory. It must be calculated from first principles of nuclear beta decay theory (Bet 38, Bah 69a).[1]

The cross sections for the other reactions have been measured. However, some uncertainty has existed in the past for

$$\sigma(\mathrm{He}^{3} + \mathrm{He}^{3} \rightarrow \mathrm{He}^{4} + 2p)\ddagger \qquad \text{and} \qquad \sigma(p + \mathrm{Be}^{7} \rightarrow \mathrm{B}^{8} + \gamma)$$

To the equations (14.6) to (14.13) we add the boundary conditions

$$\left.\begin{array}{lll} r = 0: & M(r) = 0 & L(r) = 0 \\ r = R: & M(R) = M_{\odot} & L(R) = L_{\odot} \end{array}\right\} \tag{14.15}$$

In addition, the pressure near the surface depends on the temperature in a definite but rather complicated way which is characteristic of convective stellar envelopes.

In principle, if we are given the mass of a star and its initial composition X, Y, Z (the latter assumed homogeneously distributed at $t = 0$, the start of hydrogen nuclear burning), then it is possible to solve the equations of stellar structure to find the radius, the surface luminosity $L(t)$ and temperature $T(r, t)$, and the composition $X(r, t)$, $Y(r, t)$, $Z(r, t)$ at any time $t > 0$. Of course, the composition changes with time as the star evolves and becomes inhomogeneous as the result of nuclear reactions.

[1] The transition $pp \rightarrow de^{+}\nu_{e}$ involves only the axial vector coupling. For, while the spin of the deuteron is unity, the initial protons have very low energy and interact only in a relative s state, to a very good approximation. From the Pauli principle, the initial proton wave function must be antisymmetric, and so it is a ${}^{1}S_{0}$ state; therefore $\Delta J = 1$. The cross section for $pp \rightarrow de^{+}\nu_{e}$ is thus proportional to $|C_{A}|^{2}$. The latter is determined from measurements of the neutron half-life: $|C_{A}| = 1.23 \pm 0.01$. The 1967 measurements of the neutron half-life revised $|C_{A}|^{2}$ upward and thus increased $\sigma(pp \rightarrow de^{+}\nu_{e})$. The rate for $pp \rightarrow de^{+}\nu_{e}$ is proportional to σ and to a positive power of the central solar temperature. Therefore to obtain the observed solar luminosity, revision of the neutron half-life led to a downward revision of the expected central solar temperature.

‡ Cla 68, p. 373.

Actually, in the case of the sun, we know the mass $M_\odot = 1.989 \pm 0.002 \times 10^{33}$ g and present luminosity $L_\odot = 3.90 \pm 0.04 \times 10^{33}$ erg-s very well, and we also know the probable *main sequence* age, or time since the start of hydrogen burning: $t \approx 4.7 \times 10^9$ years. (We also know the radius accurately, but this number is not useful for computing the neutrino flux because of the complications associated with the convective nature of the outer envelope.) However, we do *not* know the initial composition, or even the present composition, of the sun. The fractional abundance of helium Y, cannot be estimated from photospheric absorption spectroscopy since there are no helium lines in the visible spectrum. The heavy element abundance Z can be estimated from spectroscopy (current values put $Z \approx 0.015$), but this determination is fraught with difficulties.

To give only one example, the abundance of iron in the solar photosphere is determined from observed Fe absorption line intensities together with laboratory measurements of Fe transition oscillator strengths. The solar iron abundances thus obtained were in disagreement with iron abundances in meteorites by about a factor of 10 for a decade. Recently, new laboratory determinations of iron oscillator strengths have eliminated the discrepancy (Gar 69) and increased the estimated solar Fe abundance by a factor of about 10. However, these measurements are viewed with skepticism by some (Ros 70). Since the opacity depends sensitively on the iron/hydrogen abundance ratio, calculations of the central solar temperature and thus the neutrino energy spectrum are strongly affected by this recent possible change.

In light of the foregoing remarks, the procedure for calculating a solar model is as follows. We take $M_\odot$ and $L_\odot$ as input data which the model must satisfy. The photospheric value $Z = 0.015$ is taken as representative of Z_{initial} throughout the sun since Z has not been altered by the light-nucleus thermonuclear reactions of Table 14.1. The opacity is then calculable from knowledge of heavy element abundances. Trial initial values of $Y/X)_{\text{initial}}$ are chosen and various solar models are calculated as a function of Y/X until the correct present luminosity is found at $t = 4.7 \times 10^9$ years. This fixes the initial value of Y and also the central temperature. Since the thermonuclear reaction rates depend on composition and temperature, the neutrino flux from each of the contributing reactions of Table 14.1 is also an output of the calculation.

Estimation of the Cl^{37} capture rate requires, not only the neutrino spectrum, but also knowledge of the cross section for absorption of ν_e by Cl^{37} as a function of incoming ν_e energy; this has been computed by Bahcall (Bah 64, 66). It is important to note (see Fig. 14.1) that for energetic neutrinos (such as those which come from B^8 decay) the major contribution to the cross section comes from

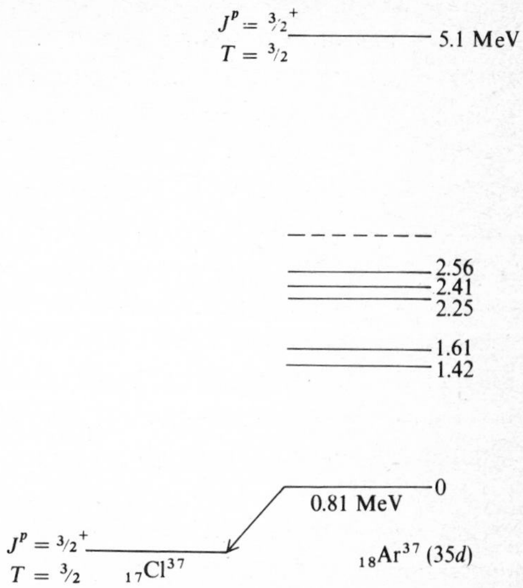

FIGURE 14.1
Energy-level diagram of Cl^{37} and Ar^{37}. The cross section for absorption of energetic neutrinos by $Cl^{37} \rightarrow Ar^{37*}$ (5.1 MeV) is very large since the matrix element is "superallowed" in this case.

transitions to the 5.1-MeV ($J = \frac{3}{2}^+$, $T = \frac{3}{2}$) excited level of Ar^{37}. The latter is the isobaric analogue of the Cl^{37} ground state, and the transition amplitude is therefore superallowed in this case.

With the calculated cross sections (estimated uncertainty no more than 25 percent) and the solar neutrino fluxes for each of the neutrino-emitting reactions of Table 14.1, we can estimate the total conversion rate per Cl^{37} atom per second. The latter is conveniently expressed in terms of

$$\begin{aligned} 1 \text{ SNU} &= 1 \textit{ solar neutrino unit} \\ &= 10^{-36} \text{ capture per target particle per second} \end{aligned}$$

The main conclusions are summarized in Table 14.2, taken from a recent paper by Bahcall and Ulrich (Bah 70). Solar models 1 and 2 are identical except that model 1 utilizes the higher iron photospheric abundance advocated by Garz et al. (Gar 69), whereas model 2 uses the iron abundance accepted earlier (Ros 70). Both models incorporate the most up-to-date methods of computing the opacity and corrections to the ideal gas equation of state. In Table 14.2 the quantities T_c, ρ_c, X_c, and Y_c are the present-day values for the central temperature, density, and H and He abundances by mass, while $X_{\text{primordial}}$ and $Y_{\text{primordial}}$ are the original hydrogen and helium abundances.

It is to be noted from Table 14.2, as we have already remarked, that less than 10^{-4} of the total solar neutrino flux comes from B^8 decays, yet these neutrinos, by virtue of their high energy, account for about 80 percent of the estimated capture rate in the Cl^{37} experiment. Another conclusion of importance is the estimated primordial helium abundance. Model 1 gives $Y_{\text{primordial}} \approx 0.26$, in good agreement with that calculated by Peebles from the "big bang" model of the origin of the universe (Pee 66a,b), while the prediction of model 2 is reasonably close. However, estimates of $Y_{\text{primordial}}$ from solar model calculations are subject to large uncertainties and possibly should not be taken too seriously (see, e.g., Ibe 68). Finally, it is to be noted that the estimated Cl^{37} capture rates in models 1 and 2 differ by almost a factor of 2, while the estimated central temperatures and densities differ by less than 3 percent. This may be taken as an indication of the extreme difficulty of calculating the expected Cl^{37} capture rate with any real confidence in the absence of more precise data concerning the solar composition and in view of the hazards and uncertainties involved in computations of the opacity. One may be skeptical, in short, of the possibility of calculating the Cl^{37} capture rate to better than a factor of 2 at present.

Table 14.2 ESTIMATED NEUTRINO FLUXES, Cl^{37} CAPTURE RATES, AND OTHER CHARACTERISTICS OF RECENT SOLAR MODELS

	Model 1	Model 2
Neutrino fluxes at the earth (10^{10} cm^{-2}/s^{-1})		
$p + p \rightarrow d + e^+ + \nu$	6.0	6.2
$p + e^- + p \rightarrow d + \nu$	1.55×10^{-2}	1.6×10^{-2}
$Be^7 + e^- \rightarrow Li^7 + \nu$	4.0×10^{-1}	2.9×10^{-1}
$B^8 \rightarrow Be^8 + e^+ + \nu$	4.5×10^{-4}	2.4×10^{-4}
$N^{13} \rightarrow C^{13} + e^+ + \nu$	3.7×10^{-2}	2.6×10^{-2}
$O^{15} \rightarrow N^{15} + e^+ + \nu$	2.6×10^{-2}	1.6×10^{-2}
$F^{17} \rightarrow O^{17} + e^+ + \nu$	3.0×10^{-4}	1.7×10^{-4}
Cl^{37} capture rates (SNU)		
R (all neutrinos)	7.8	4.5
R (B^8 neutrinos)	6.1	3.2
T_c, °K	1.52×10^7	1.48×10^7
ρ_c, g/cm^3	139	142
X_c	0.383	0.414
Y_c	0.602	0.571
$X_{\text{primordial}}$	0.726	0.764
$Y_{\text{primordial}}$	0.259	0.221

14.2.3 The Cl^{37} Experiment

We now turn to a brief description of the Cl^{37} experiment undertaken by Davis, Harmer, and Hoffman (Dav 68, 71). The apparatus is located 4,850 ft underground in an abandoned gold mine at Lead, South Dakota. The massive earth shielding is necessary to reduce the muon cosmic-ray background to a tolerable level. (Muons would produce fast protons, which, in turn, would generate Ar^{37} via the $Cl^{37}(p, n)Ar^{37}$ reaction.)

The target is a tank with 390,000 l (about 600 tons) of C_2Cl_4 (ordinary cleaning fluid), containing the normal isotopic abundance of Cl^{37} (about 25 percent). If the estimates of Table 14.2 are correct, approximately 1.0 Ar^{37} atom/48 h should be produced in this huge target from solar neutrino reactions! However, this value is larger than the total production rate from all background sources, including radioactivity in the surrounding rock as well as cosmic-ray muons (estimated total background <0.2 Ar^{37}/day).[1] Moreover, Davis and his collaborators have carefully demonstrated that it is possible to extract the minute quantity of Ar^{37} from the C_2Cl_4 with better than 90 percent efficiency. The method is to expose the C_2Cl_4 to the solar neutrino flux for a month or so [$\tau_{1/2}(Ar^{37}) = 35$ days] and then circulate large quantities of He^4 gas through the liquid target. The helium flushes out the Ar^{37} in about 24 h. The latter is subsequently separated from the helium, purified by a series of traps, and pumped into a small proportional counter (volume about 0.5 cm^3) in which the 2.8 keV Auger electrons accompanying Ar^{37} electron-capture decay are detected with about 50 percent efficiency. The entire system is carefully calibrated by means of auxiliary experiments.

Preliminary results of this difficult and important experiment yield an upper limit on the capture rate:

$$R_{\text{exp}}(Cl^{37}) \leq 1.0 \text{ SNU} \qquad (14.16a)$$

This value is to be compared with the estimated capture rates given in Table 14.2 and more recent estimates which are all in the range

$$5 \text{ SNU} \leq R_{\text{theo}}(Cl^{37}) \leq 10 \text{ SNU} \qquad (14.16b)$$

Evidently a serious discrepancy exists, which may indicate a fundamental deficiency in our understanding of solar structure or even that our understanding of the basic physical processes is incomplete.

[1] Neutrinos from other stars, even from a nearby supernova, would produce at least an order of magnitude fewer Ar^{37} atoms per day.

14.2.4 Other Proposed Solar Neutrino Experiments

Since the $Cl^{37} + \nu \rightarrow Ar^{37} + e^-$ reaction is sensitive mainly to B^8 neutrinos, it is very desirable to observe the neutrino flux by some alternative scheme. When the requirements of convenient half-life, low threshold, ease of chemical separation, detectability of final nucleus decay, etc., are imposed, we find that among the various possible target nuclei only a few candidates are acceptable, such as Rb^{87} (Sun 60), H^2 (Kel 66), and Ga^{71} (Kuz 66). However, when all practical considerations are taken into account, the transition

$$Li^7 + \nu_e \rightarrow Be^7 + e^- \qquad (14.17)$$

seems to be the most favorable case of all. The level scheme is shown in Fig. 14.2.

Even though the neutrino threshold energies for the Li^7 and Cl^{37} reactions are comparable, the cross section for capture of low-energy neutrinos (such as the 1.4-MeV ν_e from the pe^-p reaction) is much larger for Li^7 than for Cl^{37}. We find (Bah 69b) that, for a Li^7 detector, models 1 and 2 give

	Model 1	Model 2
R (all neutrinos)	39.4 SNU	≈26 SNU
R (B^8 neutrinos)	22.1 SNU	≈12 SNU

which are substantially larger than the values for Cl^{37} given in Table 14.2. Perhaps even more important, the pe^-p neutrinos contribute substantially to the Li^7 capture rate in a way which is relatively *insensitive* to central temperature and composition. Thus, experimental observation of solar neutrino capture by Cl^{37} *and* Li^7 would undoubtedly bring about a dramatic advance in our understanding of the solar interior.

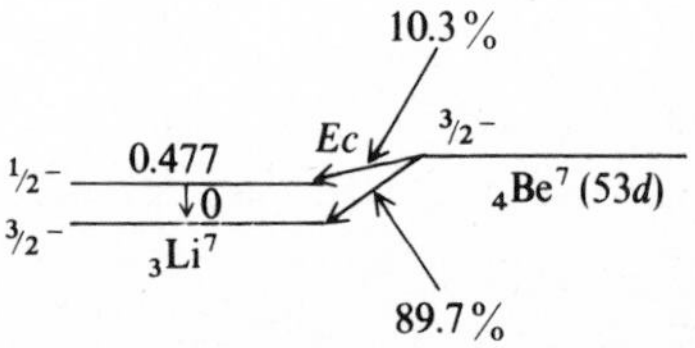

FIGURE 14.2
Energy-level diagram of $_3Li^7$ and $_4Be^7$.

14.3 EMISSION OF $\nu_e \bar{\nu}_e$ PAIRS FROM HOT STARS

14.3.1 General Remarks

In this section we shall consider the various hypothetical mechanisms by which $\nu_e \bar{\nu}_e$ pairs are radiated from the hot, dense cores of highly evolved stars. It may be helpful to the reader if we first sketch very briefly the events which lead to the existence of these stars, from which intense $\nu_e \bar{\nu}_e$ radiation should occur.

It is generally believed that stars originate from tenuous clouds of gas which contract, slowly at first, because of self-gravitation. The first stars are thought to have been composed of primordial matter, which we have reason to believe was exclusively hydrogen plus helium ($X = 0.74$, $Y = 0.26$). Later generations of stars, including the sun, contain small but significant fractions of heavy elements, which were probably synthesized in earlier, highly evolved stars and then deposited in space by various mass-loss mechanisms, including violent events such as supernova explosions.

As a young protostar contracts, gravitational energy is converted into kinetic energy of mass motion and then into thermal energy. Therefore, pressure builds up which slows the contraction, until ultimately the star reaches a quasi-static configuration which may be described to a good approximation at any instant by the equation of hydrostatic equilibrium [Eq. (14.6)]. Application of the virial theorem then shows that if the gas is nonrelativistic and monatomic, half of the released gravitational energy must go into thermal energy and half into energy lost by radiation. Thus the temperature of the core continues to rise as the star contracts quasi-statically, until hydrogen-burning reactions (*pp* chain, see Table 14.1) are ignited at a temperature $T \approx 10^7$ °K. At this point, gravitational contraction is halted, the temperature is stabilized, and the rate of energy generation by nuclear reactions is just sufficient to account for radiation loss from the surface. The star has thus arrived at its position on the main sequence of the Hertzsprung–Russell (HR) diagram, and there it remains, more or less stationary, during the entire period of main sequence hydrogen burning. (See Fig. 14.3.) This time interval is greater for light stars than for massive ones and is about 10^{10} years for the sun. (The long time scale on the main sequence is set by the slow rate of the weak reaction $pp \to de^+ \nu$.)

Once the hydrogen in the core is exhausted, gravitational contraction resumes and the core temperature and density once more increase. Meanwhile, hydrogen burning may continue in a spherical shell around the core while the envelope expands and cools. Thus the star moves relatively rapidly from its main sequence position to the *red giant* region (Fig. 14.3). Once the core

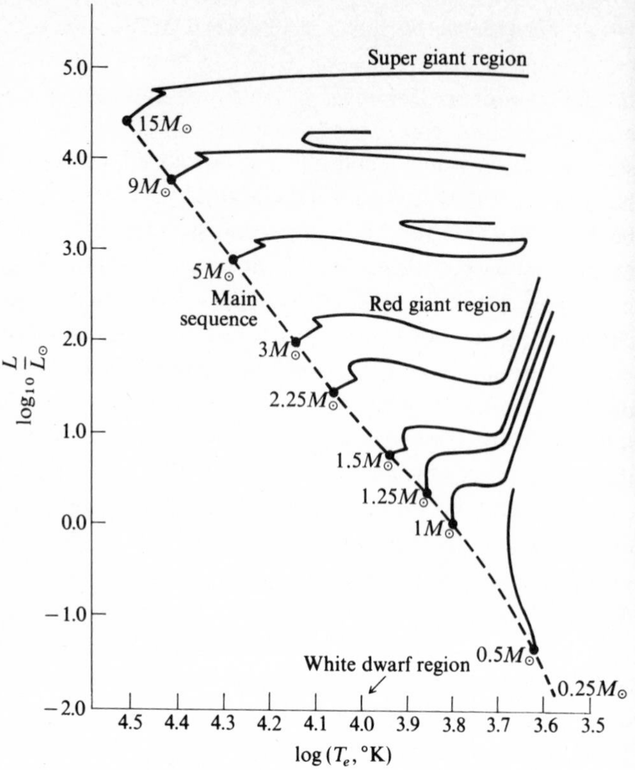

FIGURE 14.3
Schematic Hertzsprung–Russell diagram. Post-main-sequence evolutionary tracks of typical stars are indicated by solid curves. T_e = effective surface temperature. [*Adapted from I. Iben, Jr., Ann. Rev. Astron. Astrophys.*, **5**:*583* (*1967*).]

reaches temperatures $T \approx 10^8$ °K (and typical densities $\rho \approx 10^4$ g/cm^3), helium burning commences:

$$3\text{He}^4 \to \text{C}^{12*} \to \text{C}^{12} + \gamma \qquad (14.18)$$

This reaction supplies sufficient energy to halt, temporarily, the gravitational contraction. Other reactions also begin to occur at this stage, although they are inhibited to an increasing degree by high coulomb barriers:

$$\begin{aligned} \text{He}^4 + \text{C}^{12} &\to \text{O}^{16} + \gamma \\ \text{He}^4 + \text{O}^{16} &\to \text{Ne}^{20} + \gamma \\ \text{He}^4 + \text{Ne}^{20} &\to \text{Mg}^{24} + \gamma \\ \text{He}^4 + \text{Mg}^{24} &\to \text{Si}^{28} + \gamma \end{aligned} \qquad (14.19)$$

However, once the helium in the core is exhausted, gravitational contraction resumes again, accompanied as always by heating. What happens next depends on the mass of the star. It can be shown that for a given central temperature and composition, the central density of a light star is greater than that of a heavy star. If the density is sufficiently great, pressure due to electron degeneracy halts the gravitational contraction before further thermonuclear reactions can begin. Thus for light stars the core may be stabilized at some temperature below 6×10^8 °K with a composition primarily of carbon and/or oxygen. By some mechanisms which are not clearly understood, the envelope of the star may be shed, and the remainder, a light white dwarf, cools by radiation.[1]

If the star is more massive, gravitational contraction continues unabated until at $T = 6$ to 7×10^8 °K a whole new series of thermonuclear reactions is initiated, involving the burning of carbon and oxygen. At each stage, then, as thermonuclear fuel is exhausted, gravitational contraction sets in and continues either until halted by electron degeneracy pressure or until a temperature is reached where the next thermonuclear fuel can be ignited. [However, it is well known that if the star is sufficiently massive ($M \geq 1.2M_{\odot}$, the *Chandrasekhar limit*), *no* amount of electron degeneracy pressure can halt the contraction.]

As T approaches 2×10^9 °K, carbon and oxygen have burned mainly to Si^{28} and the thermal radiation becomes so intense and energetic that photo-disintegration rearrangement of the nuclei begins to proceed very rapidly. Therefore, in this temperature range, one rapidly arrives at a situation where the core nuclear abundances are described by equations of statistical equilibrium and where the most stable, tightly bound nuclei (these are in the vicinity of $A = 56$—Fe^{56} has the maximum binding energy per nucleon) are produced in profusion. At this point, which is reached in the temperature range 3 to 5×10^9 °K and at densities in the range 10^6 to 10^9 g/cm^3, no further energy can be extracted from thermonuclear reactions.

Meanwhile, as T has increased past 10^8 °K, $\nu_e \bar{\nu}_e$ production processes have begun to make their appearance and the neutrino-pair luminosity mounts very rapidly with temperature. In fact, the very weakness of neutrino interactions with matter allows $\nu_e \bar{\nu}_e$ pairs to escape from stellar cores directly and thus carry off what amounts to colossal energies at $T > 10^9$ °K. The neutrino energy losses can be made up for only by further gravitational contraction and heating. Thus the whole tempo of evolution in the late stages is drastically affected and greatly accelerated bv $\nu_e \bar{\nu}_e$ emission. We now consider the chief mechanisms

[1] A planetary nebula consists of an expanding shell of gas, mainly hydrogen, which surrounds a very hot star of fairly low mass ($M < 1.2\ M_{\odot}$). It is conjectured that we are witnessing the formation of a hot white dwarf when we observe a planetary nebula.

responsible for this emission. Each depends on the existence of the hypothesized direct $(e\nu)(e\nu)$ weak interaction, discussed in Chap. 3.

14.3.2 The Pair Process

At high temperatures, the intense thermal radiation will produce e^+e^- pairs, some of which decay by $\nu_e \bar{\nu}_e$ pair production:

$$e^+ + e^- \rightarrow \nu_e + \bar{\nu}_e \qquad (14.20)$$

According to the standard theory elaborated in Chap. 3, the square of the invariant amplitude, averaged over e^+, e^- spins, is

$$|M|^2 = 32G^2 p \cdot q' p' \cdot q \qquad (14.21)$$

where p, p', q, q' are the momentum four-vectors of e^-, e^+, ν_e, and $\bar{\nu}_e$, respectively. Let us calculate the cross section in the CM frame. We have

$$d\sigma = \frac{(2\pi)^4 \delta^4(p + p' - q - q')}{(2\pi)^6 4\sqrt{(p \cdot p')^2 - m_e^{\,4}}} \frac{d^3\mathbf{q}\, d^3\mathbf{q}'}{2\omega 2\omega'} 32G^2 p \cdot q' p' \cdot q \qquad (14.22)$$

where, as usual, $\mathbf{q}$, $\mathbf{q}'$ are the three-momenta of ν_e and $\bar{\nu}_e$, and ω, ω' are their energies. In the CM frame, the relative velocity of e^+ and e^- is $v = 2v_0$. Each charged particle has energy E, and the three-momenta are

$$\mathbf{p} = +\mathbf{v}_0 E$$

$$\mathbf{p}' = -\mathbf{v}_0 E$$

Thus

$$p \cdot p' = E^2 + |\mathbf{p}|^2 = 2E^2 - m_e^{\,2}$$

$$[(p \cdot p')^2 - m_e^{\,4}]^{1/2} = \sqrt{(2E^2 - m_e^{\,2})^2 - m_e^{\,4}} = 2E|\mathbf{p}|$$

And thus,

$$d\sigma = \frac{G^2}{(2\pi)^2} \frac{\delta^4(p + p' - q - q')}{E|\mathbf{p}|\omega\omega'} p \cdot q' p' \cdot q \, d^3\mathbf{q}\, d^3\mathbf{q}'$$

Integrating over $d^3\mathbf{q}'$, we have

$$d\sigma = \frac{G^2}{(2\pi)^2} \frac{\delta(2E - 2\omega)\omega^2 \, d\omega \, d\Omega}{E|\mathbf{p}|\omega\omega'} (E\omega + \mathbf{p} \cdot \mathbf{q})^2$$

$$\sigma = \frac{G^2}{(2\pi)^2} \int \omega^2 d\omega \, d\Omega \, \delta(2E - 2\omega) \frac{E^2\omega^2}{E|\mathbf{p}|\omega^2} (1 + \mathbf{v}_0 \cdot \hat{\mathbf{n}})^2 \qquad (14.23)$$

where $\hat{\mathbf{n}}$ is a unit vector in the direction of $\mathbf{q}$. After an elementary manipulation, (14.23) becomes

$$\sigma = \frac{G^2}{(2\pi)^2 v_0} \frac{E^2}{2} \left(4\pi + \frac{4\pi v_0^{\,2}}{3}\right)$$

Thus,

$$\sigma = \frac{G^2 m_e{}^2}{3\pi v}\left[\frac{4E^2}{m_e{}^2} - 1\right] \tag{14.24a}$$

or, in cgs units,

$$\sigma\frac{v}{c} = 1.4 \times 10^{-45}\left(\frac{4E^2}{m_e{}^2 c^4} - 1\right)\ \text{cm}^2 \tag{14.24b}$$

Equation (14.24a) may be generalized quite easily to the laboratory frame, in which case we find, in cgs units,

$$\sigma = \frac{G^2 m_e{}^4}{6\pi v E E'}\left[1 + 3\left(\frac{p \cdot p'}{m_e{}^2 c^2}\right) + 2\left(\frac{p \cdot p'}{m_e{}^2 c^2}\right)^2\right] \tag{14.25}$$

where, in this equation, v is the relative velocity in the laboratory frame and E, E', $\mathbf{p}$, $\mathbf{p}'$ also refer to the laboratory frame. The total energy loss per cubic centimeter per second Q arising from the pair process depends not only on σ but also on the available densities of electrons and positrons:

$$Q_{\text{pair}} = \int (E + E')\sigma v \, dn_- \, dn_+ \tag{14.26}$$

In order to calculate the electron and positron densities we make use of Fermi–Dirac statistics. Let the average number density of protons in the stellar material be $n_0 = N\rho/\mu_e$, where N is Avogadro's number, ρ is the mass density, and

$$\frac{1}{\mu_e} \equiv \sum_i X_i \frac{Z_i}{A_i} \tag{14.27}$$

where X_i, Z_i, A_i are the mass fraction, atomic number, and atomic weight, respectively, of the ith nuclear species. Assuming overall charge neutrality, we have

$$n_- = n_0 + n_+ = \frac{1}{\pi^2\hbar^3}\int_0^\infty \frac{p^2\,dp}{e^{(E-\mu_-)/kT} + 1} \tag{14.28}$$

and

$$n_+ = \frac{1}{\pi^2\hbar^3}\int_0^\infty \frac{p^2\,dp}{e^{(E-\mu_+)/kT} + 1} \tag{14.29}$$

where μ_- and μ_+ are the chemical potentials for electrons and positrons, respectively. The physical mechanism whereby electron-positron pairs exist is, of course, pair production from the thermal radiation bath, and at any given temperature we may safely assume equilibrium between electromagnetic pair production and annihilation:

$$n\gamma \leftrightarrow e^+ + e^- \tag{14.30}$$

Now, for any reaction in equilibrium $A \leftrightarrow B + C$, the chemical potentials satisfy $\mu_A = \mu_B + \mu_C$; and since the chemical potential of thermal radiation is zero, it follows that $\mu_+ = -\mu_-$. Therefore, defining the new quantities

$$\nu = \frac{\mu_-}{kT} \qquad \lambda = \frac{kT}{m_e c^2} \qquad \text{and} \qquad x = \frac{E}{kT} = \frac{1}{kT}(p^2c^2 + m_e^2c^4)^{1/2}$$

we find that Eqs. (14.28) and (14.29) may be solved for ρ/μ_e:

$$\frac{\rho}{\mu_e} = K[G_0^- - G_0^+] \tag{14.31}$$

where

$$G_n^{\pm}(\lambda, \nu) = \lambda^{3+2n} \int_{\lambda^{-1}}^{\infty} \frac{x^{2n+1}(x^2 - \lambda^{-2})^{1/2}\, dx}{1 + e^{x \pm \nu}} \tag{14.32}$$

and

$$K = \frac{1}{\pi^2}\left(\frac{m_e c}{\hbar}\right)^3 \frac{1}{N_0} = 2.922 \times 10^6 \text{ g/cm}^3 \tag{14.33}$$

It can be seen that

1 ν is never negative.
2 If $(\nu - \lambda^{-1}) \ll 0$, the electrons and positrons are very nondegenerate.
3 If $(\nu - \lambda^{-1}) \gg 0$, the electrons are very degenerate.

Equation (14.31) amounts to a functional relationship between ν, λ, and ρ/μ_e. The integrals $G_n^{\pm}(\lambda, \nu)$ have been evaluated numerically for useful ranges of λ and ν. We can then regard ρ/μ_e as the independent variable and plot $\nu = \mu_-/kT$ for various values of the temperature $\lambda = kT/m_e c^2$. Moreover, we can also express Eq. (14.26) in terms of the functions $G_n^{\pm}$ with the aid of (14.25), (14.28), (14.29), and (14.32). Thus we find

$$\begin{aligned} Q_{\text{pair}} = {} & \frac{G^2 m_e^4}{18\pi^5}(m_e c^2)\left(\frac{m_e c}{\hbar}\right)^3\left(\frac{m_e c^2}{\hbar}\right) \\ & \times \{G_0^-[7G_{1/2}^+ + 5G_{-1/2}^+] + G_0^+[7G_{1/2}^- + 5G_{-1/2}^-] \\ & + G_1^-[8G_{1/2}^+ - 2G_{-1/2}^+] + G_1^+[8G_{1/2}^- - 2G_{-1/2}^-]\} \end{aligned} \tag{14.34}$$

which is an exact result for all values of λ and ν, whether or not the electrons are degenerate or relativistic.

Expression (14.34) has been evaluated numerically by Beaudet, Petrosian, and Salpeter (Bea 67) for the range of temperatures and densities of interest for stellar interiors, and some of their results are shown in Fig. 14.5. Unfortunately, expression (14.34) is complicated, and so it is of some interest to

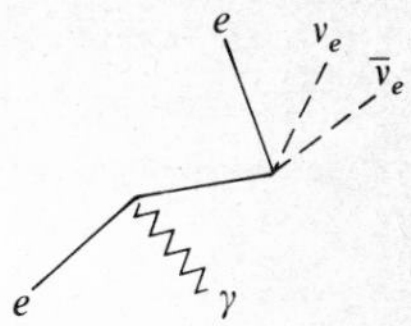

FIGURE 14.4
Feynman diagram for the photoneutrino process.

consider a simple and rather crude approximation for Q_{pair} which will reveal some of its main features. Thus we write

$$Q_{pair} \approx n_+ n_-(\sigma v)(E_+ + E_-) \tag{14.35}$$

with

$$n_+ \approx n_- = \frac{1}{\pi^2 \hbar^3} \int_0^\infty \frac{p^2\,dp}{e^x + 1} \tag{14.36}$$

Equation (14.36) ignores the density of nuclear charges. If $kT \ll mc^2$, then we may approximate $(e^x + 1)^{-1}$ by $e^{-m_e c^2/kT} e^{-p^2/2m_e kT}$. Thus we obtain

$$n_+ \approx n_- \approx \left(\frac{2m_e kT}{\pi \hbar^2}\right)^{3/2} e^{-m_e c^2/kT} \qquad kT \ll m_e c^2 \tag{14.37}$$

For $T = 10^9\,°\mathrm{K}$, for example, this would give $n_+ \approx n_- = 2 \times 10^{27}/\mathrm{cm}^3$. Writing $\sigma v \approx 5 \times 10^{-45}\,c\,\mathrm{cm}^3/\mathrm{s}$ and $E_+ + E_- \approx mc^2$, we have

$$Q_{pair} \approx 5 \times 10^{14}\ \mathrm{erg/cm^3\text{-}s} \qquad \text{at } 10^9\,°\mathrm{K} \tag{14.38}$$

Although this crude calculation contains many rough (and somewhat inconsistent) approximations, it is within an order of magnitude of the correct result for the pair process at low and moderate densities at $10^9\,°\mathrm{K}$.

The pair energy-loss rate increases very rapidly with temperature. For $kT \gg m_e c^2$ the integral in (14.36) may be approximated by

$$\frac{1}{\pi^2 \hbar^3 c^3} \int_0^\infty \frac{E^2\,dE}{e^{E/kT} + 1} = \frac{(kT)^3}{\pi^2 \hbar^3 c^3} \int_0^\infty \frac{x^2\,dx}{1 + e^x} \approx 1.8\,\frac{(kT)^3}{\pi^2 \hbar^3 c^3}$$

Thus, since $\sigma v \propto E^2 \approx (kT)^2$ for large energies, we have

$$Q_{pair} \approx n_+ n_-(\sigma v)(E_+ + E_-) \propto (kT)^3 (kT)^3 (kT)^2 (kT) \propto T^9 \tag{14.39}$$

In fact, as Fig. 14.5 shows, Q_{pair} increases by something like 10 orders of magnitude between $T = 10^9\,°\mathrm{K}$ and $T = 10^{10}\,°\mathrm{K}$.

It is interesting to note just how large the energy-loss rates are for $\nu_e \bar{\nu}_e$ emission at these high temperatures. Let us compare, for example, $Q_{pair}(10^{10}\,°\mathrm{K})$ with U, the total thermal energy stored in radiation per cubic centimeter. Whereas $Q_{pair}(10^{10}\,°\mathrm{K}) \approx 10^{25}\,\mathrm{erg/cm^3\cdot s}$, $U = aT^4 \approx 7 \times 10^{-15}\,\times$

$10^{40} = 7 \times 10^{25}$. The ratio $U/Q \approx 10$ s indicates approximately the kind of time scale for major changes in stellar structure brought on by neutrino losses at these temperatures.

14.3.3 The Plasma Process

Consider a nondegenerate, nonrelativistic plasma such as exists in the solar core. Ordinary (transverse) photons are described by a dispersion relation for this plasma which differs from that for vacuum. In the plasma

$$\omega^2 = k^2c^2 + \omega_0{}^2 \tag{14.40}$$

where k is the photon wave number and ω_0 is the plasma frequency. Multiplying both sides of Eq. (14.40) by $\hbar^2$, we obtain

$$E^2 = p^2c^2 + m_{pl}{}^2c^4$$

where m_{pl} is the effective mass of the transverse photon (*plasmon*):

$$m_{pl}c^2 = \hbar\omega_0 = \sqrt{\frac{4\pi n_e e^2\hbar^2}{m_e{}^2}}$$

where the last equality holds for a nondegenerate, nonrelativistic plasma. A photon in vacuum cannot decay spontaneously to an e^+e^- pair or to a $\nu_e\bar{\nu}_e$ pair via a virtual e^+e^- pair, etc., because of energy-momentum conservation and the helicity properties of ν_e, $\bar{\nu}_e$. However, a transverse plasmon with an effective mass m_{pl} can so decay, and the transition probability is a rapidly increasing function of m_{pl}.

More precisely, generalizing to a dense relativistic plasma, we can show that the dispersion relations for transverse and longitudinal plasmons are valid in general:

$$\text{Transverse} \qquad \omega^2 = \omega_0{}^2 + \cdots + k^2c^2 + \tfrac{1}{5}\left(\frac{\omega_1}{\omega}\right)^2 k^2c^2 \tag{14.41}$$

$$\text{Longitudinal} \qquad \omega^2 = \omega_0{}^2 + \cdots + \tfrac{3}{5}\left(\frac{\omega_1}{\omega}\right)^2 k^2c^2 \tag{14.42}$$

where ω_0 is the plasma frequency and ω_1 is a small relativistic correction. In terms of the integrals $G_n{}^{\pm}$ previously defined,

$$\frac{(\hbar\omega_0)^2}{(m_e c^2)^2} = \frac{4\alpha}{3\pi}\left[2G_{-1/2}{}^+ + 2G_{-1/2}{}^- + G_{-3/2}{}^+ + G_{-3/2}{}^-\right] \tag{14.43}$$

which reduces to $\omega_0{}^2 = 4\pi n_e e^2/m_e{}^2$ in the nonrelativistic, nondegenerate limit, and

$$\frac{(\hbar\omega_1)^2}{(m_e c^2)^2} = \frac{4\alpha}{3\pi}\left[2G_{-1/2}{}^+ + 2G_{-1/2}{}^- + G_{-3/2}{}^+ + G_{-3/2}{}^- - 3G_{-5/2}{}^+ - 3G_{-5/2}{}^-\right] \tag{14.44}$$

By means of a rather complicated calculation, it can be shown that the energy-loss rates for transverse and longitudinal plasmons are

$$Q_T = A_0 \gamma^6 \lambda^9 \int_\gamma^\infty \frac{x(x^2 - \gamma^2)^{1/2}}{e^x - 1}\, dx \tag{14.45}$$

and

$$Q_L = A_0 \gamma^9 \lambda^9 (\tfrac{5}{3})^{7/2} \tfrac{1}{2} \left(\frac{\omega_0}{\omega_1}\right)^7 \int_1^a \frac{y^{10}(y^2 - a^2)^2 (y^2 - 1)^{1/2}}{e^{\gamma y} - 1}\, dy \tag{14.46}$$

respectively. Here,

$$A_0 = \frac{G^2}{48\pi^2 \alpha} (m_e c^2) \left(\frac{m_e c^2}{\hbar}\right) \left(\frac{m_e c}{\hbar}\right)^3 = 2.912 \times 10^{21} \text{ erg-cm}^{-3}/\text{s}^{-1}$$

and

$$\gamma = \frac{\hbar \omega_0}{kT} \qquad a^2 = 1 + \tfrac{3}{5} \left(\frac{\omega_1}{\omega_0}\right)^2$$

The total plasma-loss rate $Q_{\text{plasma}} = Q_T + Q_L$ has been evaluated numerically (Bea 67) and some of the results are plotted in Fig. 14.5. Earlier calculations of this process were reported by Adams, Ruderman, and Woo (Ada 63), Zaidi (Zai 65), and Tsytovich (Tsy 64).

14.3.4 The Photoneutrino Process

A very substantial contribution to $\nu_e \bar{\nu}_e$ emission in hot stars comes from the photoneutrino process in which a $\nu_e \bar{\nu}_e$ pair replaces the outgoing photon in the Feynman diagram for Compton scattering. (See Fig. 14.4.) It is easy enough to construct the invariant matrix element for the photoneutrino process, but computation of the cross section involves evaluation of a clumsy multidimensional integral over outgoing neutrino and electron momenta (see Pet 67). An accurate calculation of the energy-loss rate for the photoneutrino process is therefore very complicated, and we merely present some of the results (Bea 67) in Fig. 14.5. It is important to note that account must be taken of plasma effects when estimating Q_{photo} at high densities.

Finally we note that *neutrino bremsstrahlung* can also occur. This is the analogue of ordinary *bremsstrahlung* in a coulomb field, except that a $\nu\bar{\nu}$ pair is emitted instead of a photon. It turns out that this contribution is rather minor, and it is not discussed further here.[1]

[1] See, for example: M. Ruderman, Neutrinos in Astrophysics, *CERN Topical Conf. Weak Interactions*, 1969.

14.3.5 Total Emission from Pair, Plasma, and Photo Processes

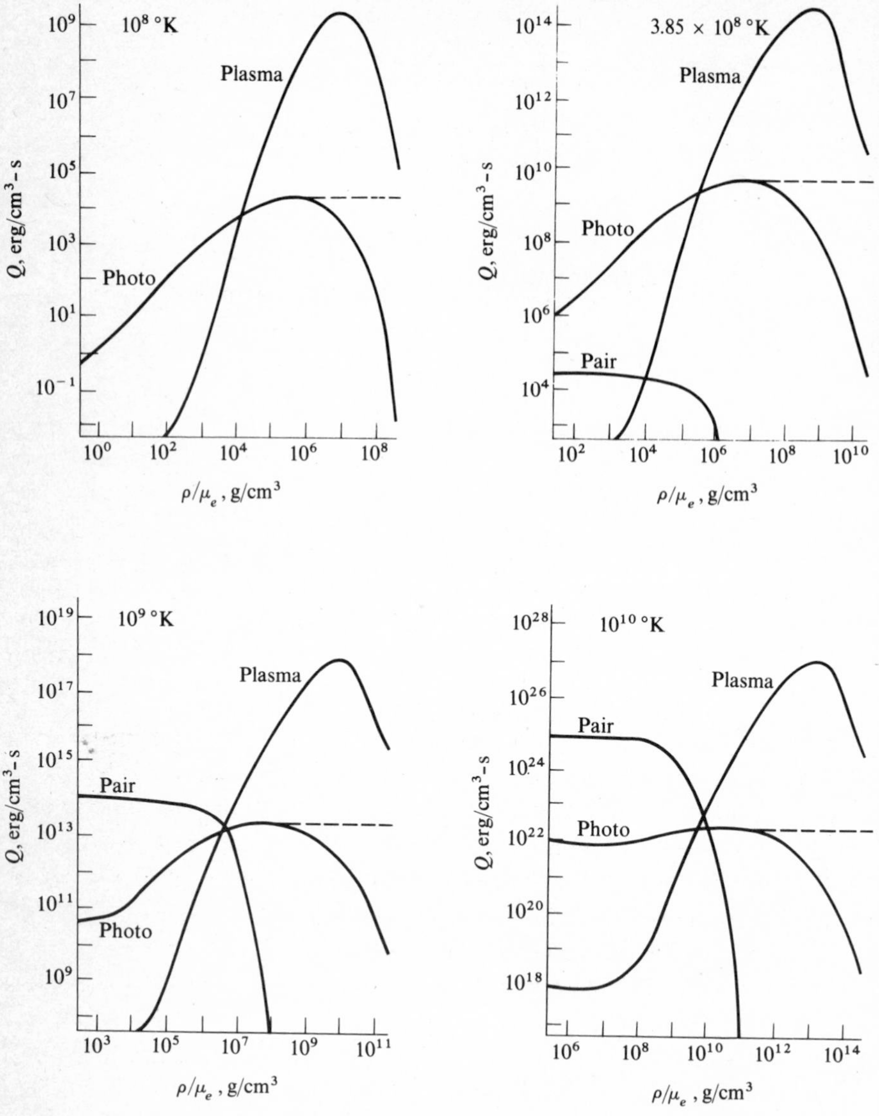

FIGURE 14.5
Neutrino-pair emission rates are plotted versus ρ/μ_e for four representative temperatures. Note that the pair process dominates at high temperatures and moderate densities, whereas the plasma process dominates at high densities at any temperature. The dashed extension of the "photo" curve represents what would occur if plasma corrections were not taken into account for the photo process. [*From G. Beaudet et al., Astrophys. J.*, ***150*:***979* (*1967*).]

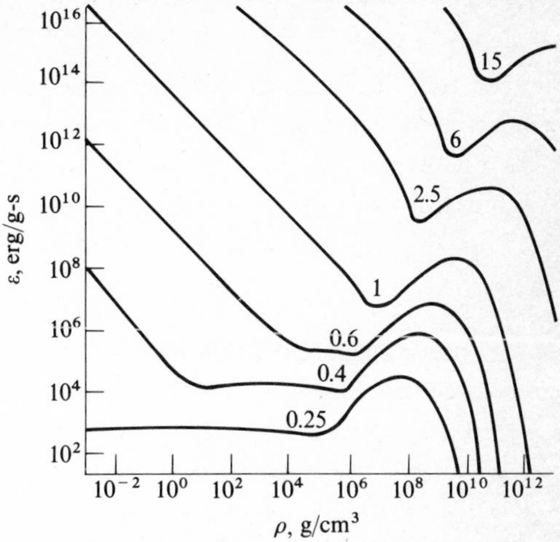

FIGURE 14.6
The total rate of energy loss ε into $\nu\bar{\nu}$ pairs in erg/g-s is plotted versus ρ for the three processes (pair + plasma + photo) combined, for $\mu_e = 2$. Each curve represents a constant temperature, labeled by the value of T in units of 10^9°K. The horizontal portions of the curves at low temperatures are due to the photoneutrinos. The pair-annihilation process gives a density-independent rate per unit volume and thus contributes to the negative slope of the curves at low densities and high temperatures. The hump in each curve at high densities arises from the plasma process. [*From G. Beaudet et al., Astrophys. J.,* ***150****:979 (1967)*.]

The main qualitative features of $\nu_e \bar{\nu}_e$ emission are made apparent from inspection of Figs. 14.5 and 14.6. For the lowest temperatures at which $Q_{\text{total}} = Q_{\text{pair}} + Q_{\text{plasma}} + Q_{\text{photo}}$ is at all significant ($T \approx 10^8$ °K), the pair process may in fact be neglected. However, at low to moderate densities and extremely high temperatures ($T > 10^9$ °K), Q_{pair} dominates. In the region of highest densities Q_{plasma} dominates at any temperature (but for $T \approx 10^{10}$ °K, $\rho > 10^{11}$ or 10^{12} g/cm^3 it may be that the Urca process, to be described below, is even more important). In Fig. 14.6 the total emission from pair, plasma, and photo processes is summarized.

14.3.6 The Urca Process

There is still another mechanism which can play a substantial role in principle, although detailed evaluation is difficult and uncertain. This is the so-called *Urca* process, originally suggested by Gamow and Schoenberg (Gam 41).[1]

[1] The name *Urca* refers to a casino in Rio de Janeiro where the theory is said to have been worked out on a tablecloth. As neutrino pairs inexorably depart from a star, so, it is said, did Gamow's money depart from his pocket.

Consider a beta decay

$$(Z, N) \rightarrow (Z+1, N-1)e^- \bar{\nu}_e \qquad (14.47)$$

and let the difference in rest energies of initial and final nuclei be $m_e c^2 + \delta$. Neglecting recoil of the final nucleus, δ appears as kinetic energy shared by the final leptons. Typically $0.1 < \delta < 10$ MeV, and in the roughest approximation the decay probability is proportional to δ^5.

In ordinary matter at low temperatures the reaction

$$e^- + (Z+1, N-1) \rightarrow \nu_e + (Z, N) \qquad (14.48)$$

cannot occur because the thermal energy of ambient electrons is not sufficient to overcome the threshold δ. However, in a hot stellar core ($T \geq 10^9$ °K, $kT \geq 100$ keV), there will be appreciable numbers of electrons in the high-energy tail of the thermal velocity distribution (ordinarily maxwellian), and (14.48) can proceed. The net effect of (14.47) and (14.48) is that the nucleus catalyzes the conversion of electron thermal energy to energy of the $\nu\bar{\nu}$ pair. Although the principle is simple enough (and, unlike the previous mechanisms, depends only on well-understood features of the weak interaction), the following complexities must be taken into account if one wishes to arrive at a reasonably quantitative estimate of the Urca process:

1 At the high temperatures required, the nuclei will be evolved at least to Si^{28} and probably beyond (into the Fe^{56} region).

2 The intense thermal radiation implies a good deal of nuclear photo-excitation, and many beta decays and electron captures will occur from excited nuclear states. Thus energy may also be converted from thermal radiation to $\nu_e \bar{\nu}_e$ pairs via the Urca process.

3 For high densities the electron degeneracy will give a relatively large number of electrons at high energies, and thus the e^- capture probability will be greatly enhanced.

4 If positrons can exist in large numbers via pair production, the reaction

$$e^+ + (Z, N) \rightarrow (Z+1, N-1) + \bar{\nu}_e$$

can also contribute to the Urca process. However, this is thought to be much less important than the pair process of Sec. 14.3.2.

Detailed calculations of Urca loss rates are very complex and thus somewhat uncertain. (See, for example, Tsu 65.) Qualitatively the conclusion is that the Urca process becomes competitive with the other three mechanisms only in regions of extremely high densities and temperatures, where the plasma process must be most important among those already mentioned. (See Fig. 14.7.)

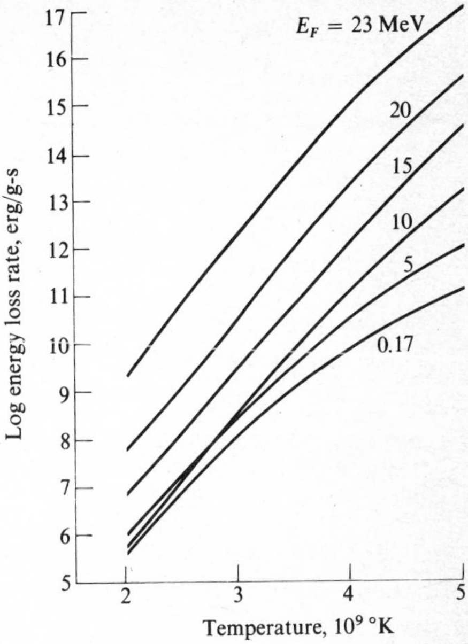

FIGURE 14.7
Urca process $\nu\bar{\nu}$ losses versus temperature. The rate of energy generation per gram of stellar material is plotted for matter in nuclear statistical equilibrium as a function of temperature. The density dependence is contained in the Fermi energy E_F of the degenerate electron gas, $E_f \approx 0.5 \times 10^{-2}\rho^{1/3}$ (ρ in g/cc, E_F in MeV), the latter formula being valid for $\rho > 10^7$ g/cc. [*From S. Tsuruta and A. G. W. Cameron, Canad. J. Phys., 43:2056 (1965).*]

14.3.7 Observational Evidence for $\nu_e \bar{\nu}_e$ Pairs from Hot Stars

What observations bear on the hypothesis that $\nu_e \bar{\nu}_e$ emission occurs? The main information, although quite fragmentary and uncertain, leans in favor of the existence of the interaction. Such evidence involves stellar statistics of blue and red supergiants, evolution rates of planetary nebulae, cooling of very hot white dwarfs, and the existence of supernovae. In the following paragraphs we consider each of these items very briefly.

1 Stars with very large mass ($M \geq 15M_\odot$) are called *supergiants*. Detailed calculations show that their evolutionary paths are nearly horizontal after leaving the upper main sequence of the HR diagram (see Fig. 14.8). A study of the statistics of blue supergiants (beginning of core helium burning) and red supergiants (C/O burning) in clusters reveals a relatively large ratio of blues to reds (Sto 69a,b). In other words, the distribution of points along a track of given mass is quite sparse toward the red region, which implies that evolution times become quite short when the C/O burning stage is reached. Such short times cannot be accommodated in the theory of evolution of supergiants unless $\nu_e \bar{\nu}_e$ losses are included.

2 Approximately seven hundred planetary nebulae are known in our galaxy, and about thirty have been studied in great detail. A nebula consists of a very hot central star (whose total mass is probably less than $1.2M_\odot$) and an

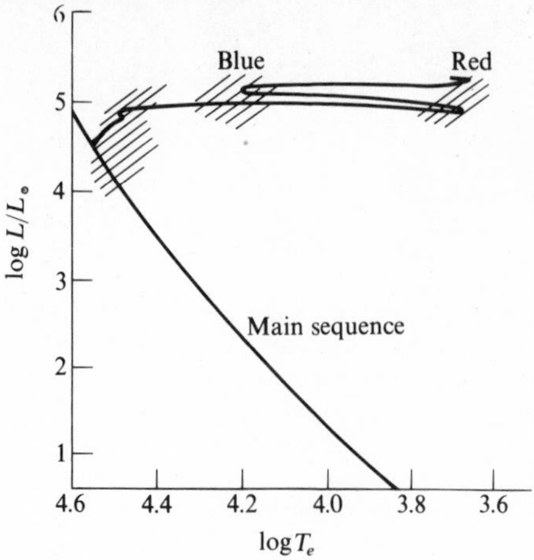

FIGURE 14.8
Diagram (schematic only) of the evolutionary path of a massive star ($M \approx 15M_\odot$) after leaving the main sequence. The relatively small number of points found in the diagram for a galactic cluster in the red region implies rapid evolution and hence $\nu\bar{\nu}$ losses.

outer, roughly spherical shell of expanding gas, the radial velocity of which is typically about 20 km/s. (The latter is determined from doppler shifts of visible emission lines, mainly Balmer lines of hydrogen.) By observing planetary nebulae in various stages of development and noticing that the shell becomes fainter as it expands outward until it becomes quite invisible beyond $R \approx 2 \times 10^{18}$ cm, we conclude that all the planetary nebulae we now observe were formed within a time of the order of 10^5 years. Between the beginning and end of this 10^5-year period the central star changes rapidly. Its luminosity goes down—from, typically, $L = 10^4 L_\odot$ to $L = 10^2 L_\odot$—and the surface temperature goes up—from $T \approx 4 \times 10^4$ to $T \approx 10^5$ °K. Since the luminosity for an object of radius R and blackbody temperature T is $L = 4\pi R^2 \sigma T^4$, the radius must decrease rapidly from $R \approx R_\odot$ to $R \approx 10^{-2} R_\odot$. Such behavior leads many observers to conclude that we are witnessing a light- to moderate-mass star in the post-main-sequence phase, and that the star is in the act of shedding its hydrogen-rich envelope. A hot, dense, at least partially degenerate core is left behind. For very small masses the composition could be mainly helium, while for slightly larger masses it is probably carbon and oxygen.

The evolutionary track of the central star of a planetary nebula can be represented schematically as in Fig. 14.9 ($A \rightarrow B$). Because of transient effects, it is extremely difficult to calculate the evolutionary path from a basic theory with great reliability. However, preliminary indications do exist that unless neutrino losses are included, the resulting time scales are much too long compared with 10^5 years (Vil 67, Ros 67).

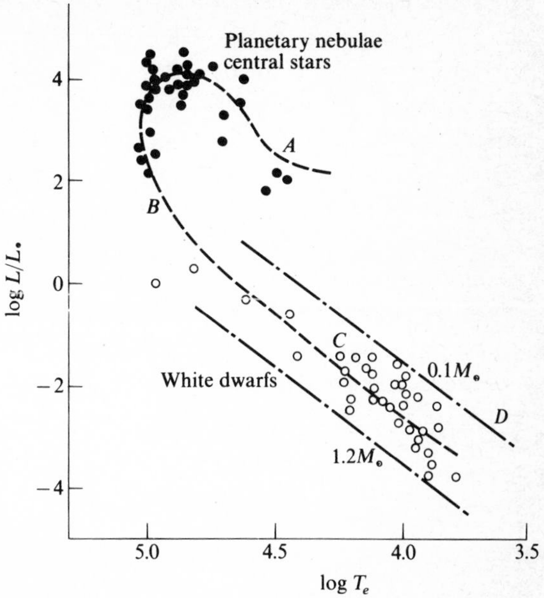

FIGURE 14.9
Schematic diagram of that portion of the HR diagram which contains central stars of planetary nebulae and white dwarfs. The evolution time from *A* to *B* is too short to be explained without $\nu\bar{\nu}$ losses. Also the low density of points from *B* to *C* indicates that cooling times of hot white dwarfs in this region are very short. Such rapid cooling occurs because of $\nu\bar{\nu}$ losses also.

3 In Fig. 14.9 we have also indicated an area of the HR diagram which is heavily populated by white dwarfs (*C* to *D*). As previously noted, these are light stars ($M < 1.2M_\odot$ = Chandrasekhar limit), where the inward force of gravity is balanced by electron degeneracy pressure. To a good approximation this pressure is independent of the temperature and depends only on ρ/μ_e, where ρ is the density and $\mu_e \approx 2/1 + X$ (X is the mass fraction of hydrogen). A simple argument then shows that the total mass and radius are increasing and decreasing functions of central density, respectively. It also follows that *M* and *R* are uniquely related for such a star. Since $\log L = \text{const} + 2\log R + 4\log T$, a given value of *R* corresponds to a given value of *M*, which is in turn represented by a straight line in the HR diagram, as shown in Fig. 14.9.

The fact that almost all observed white dwarfs fall into a narrow band between $M = 1.2M_\odot$ and $M \approx 0.1M_\odot$ confirms the basic idea that the pressure is due to electron degeneracy and that a Chandrasekhar limit exists. From our point of view another observational result, possibly of great significance, is the low density of white dwarfs in the high-temperature region between B and C in Fig. 14.9, as opposed to the high density of observed white dwarfs in the low-temperature region (C to D). Although only tentative conclusions can be drawn at this point (Sto 66, 69c, 70), one possible explanation for these statistics is that plasma neutrino losses from a young hot white dwarf (with a central temperature $T \approx 3$ to 5×10^8 °K) cool it very rapidly, so that the star evolves quickly from B to C. Subsequently the neutrino emission becomes insignificant, and cooling proceeds very slowly from C to D via thermal radiation from the surface.

A statistical analysis of white dwarf stars, when combined with the other evidence mentioned so far, leads to the conclusion that the electron-neutrino weak interaction can be described by the conventional theory of weak interactions, and limits are placed on the coupling constant of

$$g^2 = 10^{(0 \pm 2)} G^2$$

4 Finally, we mention supernova explosions. According to the description presented earlier in Sec. 14.3.1, we may crudely sketch a highly evolved star of, say, 10 solar masses, at a certain instant of time as follows: There is a dense hot core ($\rho \approx 10^8$ g/cm^3, $T \approx 4 \times 10^9$ °K) in which thermonuclear reactions have been carried to completion and the material is almost exclusively Fe^{56}.‡ This core is surrounded by a mantle at somewhat lower temperature and pressure consisting mainly of Si^{28} and, in addition, protons, neutrons, alphas, other α-particle nuclei, and, as we proceed further outward from the center, carbon and oxygen. Surrounding the mantle at still lower temperatures and densities is a large envelope of helium and hydrogen. Let us now consider what happens as time elapses. Emission of $\nu_e \bar{\nu}_e$ pairs from the core is enormous. This energy loss cannot be made up for by thermonuclear reactions since these have been carried to completion. The only source of energy is gravitation, and so the core contracts, the temperature rises (according to the virial theorem), and the neutrino losses become larger. The tempo for gravitational contraction and temperature rise thus becomes accelerated until the matter of the core is essentially in free fall (time scale $\approx 1/G_0^{1/2}\rho^{1/2}$).

A further drain on the gravitational energy is imposed by the photodisintegration of iron nuclei, which occurs at extremely high temperatures ($\approx 5 \times 10^9$ °K):

$$Fe^{56} \rightarrow 13\alpha + 4n$$

‡ The relative amounts of Fe^{56}, Fe^{54}, and Ni^{56} (which decays to Fe^{56}) depend on ρ, T, and the time available for establishing statistical equilibrium. The latter may be determined by $\nu\bar{\nu}$ losses as well as beta decay. Therefore the relative abundances of nuclei in the iron peak give a clue about $\nu\bar{\nu}$ losses, which is, unfortunately, not unambiguous.

Qualitatively this is somewhat similar to the thermal ionization of hydrogen atoms at approximately 10^4 °K, but the breakup of Fe^{56} is much more endothermic; it costs approximately 100 MeV of energy per iron nucleus. What ensues is not thoroughly clear, but the outlines can be seen. The core collapses with enormous rapidity as iron is transformed to helium and $\nu\bar{\nu}$ pairs are given off. At still higher densities the helium is broken down to hydrogen in a dissociation which is also highly endothermic. Meanwhile the electron Fermi energy rises so high that electron capture reactions become energetically favorable:

$$e^- + p \rightarrow n + \nu_e$$
$$e^- + (Z, N) \rightarrow (Z - 1, N + 1) + \nu_e$$

resulting in the *neutronization* of the core with further emission of neutrinos in the energy range 10 to 100 MeV. The collapse of the core cannot be halted by any known mechanism until it approaches nuclear densities, $\rho \geq 10^{14}$ g/cm^3, at which point the degeneracy pressure of the neutron gas (and repulsive nuclear forces at extremely short range) may be sufficient to balance the gravitational crush.

Meanwhile, collapse of the core apparently has a catastrophic effect on the mantle, although that effect is by no means clearly understood. Since support of the mantle is rather suddenly removed, it will also have a tendency to compress and heat quickly, with the consequent, possibly explosive, initiation of thermonuclear reactions. Intense γ-ray fluxes from the core collapse may also heat the mantle quickly enough to start an explosion. From our point of view the most interesting hypothetical mechanism for transferring energy from the core to the mantle is neutrino emission and scattering. Ordinarily one thinks that mean free paths for interaction of neutrinos of stellar thermal energies with stellar material are many orders of magnitude larger than the star diameter. However, a number of factors may conspire to make the mantle of a supernova partially opaque to neutrinos generated in the core. All of these depend on the fact that the cross section for $\nu_e e$ scattering increases as the square of the $\nu_e e$ or $\bar{\nu}_e e$ relative energy in the CM frame.

(*a*) Neutrinos emitted from the core during neutronization have very high energy: 10 to 100 MeV.

(*b*) If the mantle temperature is high enough ($T \gg 10^9$ °K), electron-positron pairs will be created in great profusion from thermal radiation. As we noted earlier, this number density varies as T^3 for sufficiently high temperatures.

(*c*) If electron thermal energies are sufficiently high ($kT \gg mc^2$), the total relativistic energy of ν and e in the CM frame is much greater than that of either ν or e alone in the laboratory frame.

Naturally, very little is yet known about the exceedingly complex phenomena accompanying supernova explosions, and we may view *any* detailed model with some skepticism at this stage. However, the transfer of energy from core to mantle via $\nu_e e$ and $\bar{\nu}_e e$ scattering does seem to be a reasonable hypothetical mechanism, and preliminary calculations (Col 66) yield approximately the right orders of magnitude.

14.4 NEUTRINOS OF COSMOLOGICAL ORIGIN

The simplest plausible assumption one can make regarding the large-scale distribution of mass energy in the universe is that it is homogeneous and isotropic. This so-called *cosmological principle*, together with *general relativity*, lies at the basis of the well-known and widely accepted "big bang" cosmological model.

The roots of the big bang hypothesis go back to 1922, when the Soviet mathematician A. Friedmann discovered solutions to Einstein's general relativistic field equations for a homogeneous isotropic fluid of energy density $\varepsilon = \rho c^2$ and pressure p. One of these solutions corresponds to expansion of the fluid with time. Thus, in a sense, Friedmann anticipated the important discovery by the American astronomer E. Hubble in 1929 of the cosmological red shift. As is well known, the latter is interpreted as a doppler shift of light emitted by distant galaxies which recede from us with a velocity v proportional in the first approximation to their distance from us:

$$v = H_0 r$$

where H_0 is Hubble's constant; $H_0^{-1} \approx 1.3 \times 10^{10}$ years. If the universal fluid contains radiation (photons, neutrinos) as well as material particles, then in the course of the expansion the matter density ρ_m falls off as R^{-3}, where R is a distance scale factor (called the *radius of the universe*), whereas the radiation density ρ_r falls off as R^{-4}, as long as matter and radiation are decoupled.

During the present epoch the average matter density, although uncertain, appears to be about $\bar{\rho}_{m0} \approx 10^{-30}$ g/cm^3, while the universal blackbody photon temperature is $T_{r0} \approx 3°$, corresponding to a mass density of only

$$\rho_{r0} = \frac{a T_{r0}^{\ 4}}{c^2} \approx 5 \times 10^{-35} \text{ g/cm}^3$$

Thus, at present, the matter density is much more important than the radiation. However, as was first pointed out by Gamow (Gam 46), at much earlier times R was much smaller and the ratio of radiation density to matter density, proportional to $R^{-4/3}$, must have been much larger than it is now. Indeed, at sufficiently early times, the radiation dominated the expansion. Assuming it possessed a blackbody spectrum with temperature $T(t)$, we can easily show from the field equations that in the early stages of the universe, T was enormously high:

$$T = \left(\frac{3c^2}{32\pi G_0 a}\right)^{1/4} t^{-1/2} \approx 10^{10} t^{-1/2}\,{}^\circ\mathrm{K}$$

where G_0 is the newtonian constant of gravitation and t is the age of the universe in seconds, as measured with a comoving clock.

At the extremely large temperatures which prevailed during the first few seconds of evolution ($T \geq 10^9\,{}^\circ$K), electron-positron pairs must have existed in thermal equilibrium with the photons.[1] From previous remarks in this chapter it is clear that some of these pairs must have decayed to $\nu_e \bar{\nu}_e$ pairs. Indeed, since for sufficiently high temperatures the rate of formation of $\nu_e \bar{\nu}_e$ pairs by $e^+e^- \to \nu_e \bar{\nu}_e$ goes like T^8, at early enough times the $\nu_e \bar{\nu}_e$ pairs were created (and destroyed) with sufficient rapidity to establish thermal equilibrium between them and the thermal photon-plus-e^+e^- bath. However, as time elapsed and T decreased, the $\nu_e \bar{\nu}_e$ pairs became decoupled (at $kT \approx 30$ MeV); subsequently the $\nu_e \bar{\nu}_e$ radiation expanded and cooled adiabatically until at present, as detailed calculations show (Alp 53), the temperature of the *relict* $\nu_e \bar{\nu}_e$ sea is $T_\nu = (\frac{4}{11})^{1/3} T_r$. These neutrinos are described by a Fermi–Dirac distribution with zero chemical potential. It is thus easy to show that the energy density of the relict $\nu_e \bar{\nu}_e$ sea is of the same order of magnitude as that of the cosmic blackbody radiation:

$$\rho_{\nu\bar{\nu}} \approx 5 \times 10^{-35}\ \mathrm{g/cm^3}$$

Even if $\rho_{\nu\bar{\nu}}$ were 8 to 10 orders of magnitude larger, these neutrinos would still escape detection. Nevertheless, they could have a profound effect on the Hubble expansion if this were the case. For, as Friedmann discovered in the 1920s, if the total density of matter plus radiation plus neutrinos is sufficiently great,

$$\rho_{\text{total}} > \frac{3H_0{}^2}{8\pi G_0} \approx 2 \times 10^{-29}\ \mathrm{g/cm^3}$$

then gravitational attraction will ultimately halt and reverse the expansion. Thus the relict neutrinos may play a decisive role in the evolution of the universe even if, for all practical purposes, they are unobservable.

[1] At still higher temperatures and earlier times $\mu^+\mu^-$ pairs also must have existed. The description which follows ignores the $\nu_\mu \bar{\nu}_\mu$ neutrinos but could be enlarged in an obvious way to include them.

REFERENCES

Ada 63	Adams, J., M. Ruderman, and C. Woo: *Phys. Rev.*, **129**:1383 (1963).
Alp 53	Alpher, R. A., J. W. Follin, and R. C. Herman: *Phys. Rev.*, **92**:1347 (1953).
Bah 64	Bahcall, J. N.: *Phys. Rev.*, **135**:B137 (1964).
Bah 66	Bahcall, J. N.: *Phys. Rev. Letters*, **17**:398 (1966).
Bah 68	Bahcall, J. N., N. A. Bahcall, and R. Ulrich: *Astrophys. Letters*, **2**:91 (1968).
Bah 69a	Bahcall, J. N., and R. M. May: *Astrophys. J.*, **155**:501 (1969).
Bah 69b	Bahcall, J. N.: *Phys. Rev. Letters*, **23**:251 (1969).
Bah 70	Bahcall, J. N., and R. Ulrich: *Astrophys. J.*, **160**:L57 (1970).
Bea 67	Beaudet, G., V. Petrosian, and E. Salpeter: *Astrophys. J.*, **150**:979 (1967).
Bet 38	Bethe, H. A., and C. L. Critchfield: *Phys. Rev.*, **54**:248 (1938).
Cla 68	Clayton, D. R.: "Principles of Stellar Evolution and Nucleosynthesis," McGraw-Hill, New York, 1968.
Col 66	Colgate, S. A., and R. H. White: *Astrophys. J.*, **143**:626 (1966).
Dav 68	Davis, R., Jr., D. S. Harmer, and K. C. Hoffman: *Phys. Rev. Letters*, **20**:1205 (1968).
Dav 71	Davis, R., Jr., L. C. Rogers, and V. Radeka: *Amer. Phys. Soc. Bul.*, **16**:631 (1971)
Gam 41	Gamow, G., and M. Schoenberg: *Phys. Rev.*, **109**:193 (1941).
Gam 46	Gamow, G.: *Phys. Rev.*, **70**:572 (1946).
Gar 69	Garz, T., et al.: *Nature*, **223**:1254 (1969).
Ibe 68	Iben, I., Jr.: *Phys. Rev. Letters*, **21**:1208 (1968).
Ibe 69	Iben, I., Jr.: *Phys. Rev. Letters*, **22**:100 (1969).
Kel 66	Kelly, F. J., and H. Überall: *Phys. Rev. Letters*, **16**:145 (1966).
Kuz 66	Kuzmin, V. A.: *Soviet Phys. JETP*, **22**:1051 (1966).
Pee 66a	Peebles, P. J. E.: *Phys. Rev. Letters*, **16**:410 (1966).
Pee 66b	Peebles, P. J. E.: *Astrophys. J.*, **146**:542 (1966).
Pet 67	Petrosian, V., G. Beaudet, and E. Salpeter: *Phys. Rev.*, **154**:1445 (1967).
Ros 67	Rose, W. K.: *Astrophys. J.*, **150**:193 (1967).
Ros 70	Ross, J. E.: *Nature*, **225**:610 (1970).
Sto 66	Stothers, R.: *Astron. J.*, **71**:943 (1966).
Sto 69a	Stothers, R.: *Astrophys. J.*, **155**:935 (1969).
Sto 69b	Stothers, R., and C. W. Chin: *Astrophys. J.*, **158**:1039 (1969).
Sto 69c	Stothers, R.: *Astrophys. Letters*, **4**:187 (1969).
Sto 70	Stothers, R.: *Phys. Rev. Letters*, **24**:538 (1970).
Sun 60	Sunyar, A. W., and M. Goldhaber: *Phys. Rev.*, **120**:871 (1960).
Tsu 65	Tsuruta, S., and A. G. W. Cameron: *Canad. J. Phys.*, **43**:2056 (1965).
Tsy 64	Tsytovich, V. N.: *Soviet Phys. JETP*, **18**:816 (1964).
Vil 67	Villa, S.: *Astrophys. J.*, **149**:613 (1967).
Wat 69	Watson, W. D.: *Astrophys. J.*, **158**:303 (1969).
Zai 65	Zaidi, M. H.: *Nuovo Cimento*, **40**:502 (1965).

GENERAL BIBLIOGRAPHY

Adler, S. L., and R. F. Dashen: "Current Algebras and Applications to Particle Physics," Benjamin, New York, 1968.

Bell, J. S. (ed.): *Proc. CERN Topical Conf. Weak Interactions, CERN, Geneva* (1969).

Björken, J. D., and M. Nauenberg: Current Algebra, *Ann. Rev. Nucl. Sci.*, **18**:229 (1968).

Blin-Stoyle, R., and S. C. K. Nair: Fundamentals of β Decay Theory, *Adv. Phys.*, **15**:493–545 (1966).

Cabibbo, N.: Weak Interactions, *Proc. Internatl. Conf. High-Energy Physics, Berkeley*, 29–50 (1966).

Chiu, H. Y.: Neutrinos in Astrophysics and Cosmology, *Ann. Rev. Nucl. Sci.*, **16**:591 (1966).

Feinberg, G., and L. M. Lederman: The Physics of Muons and Muon-Neutrinos, *Ann. Rev. Nucl. Sci.*, **13**:431 (1963).

Gasiorowicz, S.: "Elementary Particle Physics," Wiley, New York, 1966.

Gell–Mann, M., and Y. Ne'eman: "The Eightfold Way," Benjamin, New York, 1964.

Henley, E. M.: Parity and Time-Reversal Invariance in Nuclear Physics, *Ann. Rev. Nucl. Sci.*, **19**:367 (1969).

Jackson, J. D.: "Lecture Notes from Brandeis Summer School," pp. 263–425, 1962.

Kabir, P. K.: "The Development of Weak Interaction Theory," Gordon and Breach, New York, 1963.

Kabir, P. K.: "The CP Puzzle: Strange Decays of the Neutral Kaon," Academic, New York, 1968.

Källen, G.: "Elementary Particle Physics," Addison-Wesley, Reading, Mass., 1964.

Konopinski, E. J.: "The Theory of Beta Radioactivity," Clarendon Press, Oxford, 1966.

Lee, T. D.: (director of course,): Weak Interactions and High-Energy Neutrino Physics, *Proc. Internatl. School Phys., Course 32: "Enrico Fermi,"* Academic, New York, 1964.

Lee, T. D., and C. S. Wu: Weak Interactions I, II, *Ann. Rev. Nucl. Sci.*, **15**:381 (1965); *ibid.*, **16**:471 (1966).

Marshak, R. E.: Ten Years of the Universal (V A) Theory of Weak Interactions, in G. Puppi (ed.), "Old and New Problems in Elementary Particles," pp. 177–201, Academic, New York, 1968.

Marshak, R. E., Riazuddin, and C. P. Ryan: "Theory of Weak Interactions in Particle Physics," Wiley, New York, 1969.

Morpurgo, G.: A Short Guide to the Quark Model, *Ann. Rev. Nucl. Sci.*, **20**:79 (1970).

Okun', L. B.: "Weak Interactions of Elementary Particles," Addison-Wesley, Reading, Mass., 1965.

Okun', L. (ed).: *Proc. Symp. Probs. Related to the Violation of CP Invariance*, *Soviet Phys. Uspekhi*, **11**:461–602 (1969).

Primakoff, H.: Theory of Muon Capture, *Rev. Mod. Phys.*, **31**:802–822 (1959).

Renner, B.: "Current Algebra," Pergamon, New York, 1968.

Ruderman, M.: Neutrino Astrophysics, *Repts. Progr. Phys.*, **28**:411 (1965).

Schopper, H.: "Weak Interactions and Nuclear Beta Decay," North-Holland, Amsterdam, 1966.

Siegbahn, K. (ed).: "Alpha, Beta, and Gamma Ray Spectroscopy," North-Holland, Amsterdam, 1965.

Weissenberg, A. O.: "Muons," North-Holland, Amsterdam, 1966.

Wolfenstein, L.: Weak Interaction Theory, *Proc. Heidelberg Internatl. Conf. Elementary Particles*, North-Holland, Amsterdam, 1968.

Wu, C. S.: History of Beta Decay, in O. R. Frisch et al. (eds.), "Beiträge zur Physik und Chemie des 20. Jahrhunderts," Friedr. Vieweg & Sohn, Braunschweig, Germany, 1959.

Wu, C. S., and S. A. Moszkowski: "Beta Decay," Interscience, New York, 1966.

Zichichi, A. (ed.): "Symmetries in Elementary Particle Physics," Academic, New York, 1965.

Zichichi, A. (ed.): "Strong and Weak Interactions, Present Problems," Academic, New York, 1966.

APPENDIX 1

UNITS, CONSTANTS, AND NOTATION

A1.1

Throughout the text we employ "natural" units unless otherwise noted. These are defined by setting $\hbar = c = m_p = 1$. For convenience the table below lists values of fundamental quantities frequently encountered in the text.

Quantity	Value in natural units	Value in other units
Proton mass m_p	1	1.672614×10^{-24} g 938.256 MeV/c^2 $6.721 m_{\pi^\pm}$ $1{,}836.109 m_e$
Proton compton wavelength $\lambda\!\!\!^{-}_p = \hbar/m_p c$	1	2.103139×10^{-14} cm
Velocity of light c	1	2.9979250×10^{10} cm/s
Time $\lambda\!\!\!^{-}_p/c$	1	7.015×10^{-25} s
Proton rest energy $m_p c^2$	1	938.256 MeV
Action $\hbar$	1	$1.0545919 \times 10^{-27}$ erg-s
Weak coupling constant G	1.026×10^{-5}	$1.43506(26) \times 10^{-49}$ erg-cm^3

A1.2 OTHER USEFUL CONSTANTS AND CONVERSIONS

$$e = 4.803250(21) \times 10^{-10} \text{ esu}$$

$$1 \text{ eV} = 1.6021917(70) \times 10^{-12} \text{ erg}$$

$$\alpha^{-1} = 137.03602(21)$$

$$\hbar c = 1.9732891(66) \times 10^{-11} \text{ MeV-cm}$$

$$m_e = 0.5110041(16) \text{ MeV}$$

$$\lambda\!\!\!^{-}_e = 3.861592(12) \times 10^{-11} \text{ cm}$$

$$\mu_N = \hbar/2m_p c = 3.152526(21) \times 10^{-18} \text{ MeV-G}^{-1}$$

A1.3 RELATIVISTIC NOTATION

We employ essentially the same notation as Bjorken and Drell, in "Relativistic Quantum Mechanics."[1] Space-time coordinates $(t,x,y,z) = (t,\mathbf{x})$ are denoted by the contravariant four-vector

$$x^\mu \equiv (x^0, x^1, x^2, x^3) \equiv (t, x, y, z)$$

The covariant coordinate four-vector x_μ is

$$x_\mu \equiv (x_0, x_1, x_2, x_3) = (t, -x, -y, -z)$$

A covariant four-vector A_μ and a contravariant four-vector A_μ are related by

$$A_\mu = g_{\mu\nu} A^\nu$$

where

$$g_{\mu\nu} = \begin{bmatrix} 1 & 0 & 0 & 0 \\ 0 & -1 & 0 & 0 \\ 0 & 0 & -1 & 0 \\ 0 & 0 & 0 & -1 \end{bmatrix} = g^{\mu\nu}$$

The scalar product of two four-vectors A and B is

$$A \cdot B = A_\mu B^\mu = A^\mu B_\mu = A_t B_t - \mathbf{A} \cdot \mathbf{B}$$

The momentum operator in coordinate representation is

$$p^\mu = i \frac{\partial}{\partial x_\mu} = \left(i \frac{\partial}{\partial t}, -i\nabla \right)$$

Also,

$$p_\mu = i \frac{\partial}{\partial x^\mu} = \left(i \frac{\partial}{\partial t}, +i\nabla \right)$$

In coordinate representation,

$$p^2 = p \cdot p = p^\mu p_\mu = -\frac{\partial}{\partial x^\mu}\frac{\partial}{\partial x_\mu} = -\frac{\partial^2}{\partial t^2} + \frac{\partial^2}{\partial x^2} + \frac{\partial^2}{\partial y^2} + \frac{\partial^2}{\partial z^2} = -\Box$$

If A and B are four-vectors, the quantities $A_\mu B_\nu$, $A_\mu B^\nu$, and $A^\mu B^\nu$ are second-rank covariant, mixed, and contravariant tensors, respectively.

[1] McGraw-Hill, New York, 1964.

APPENDIX 2

THE DIRAC EQUATION, DIRAC ALGEBRA, AND DIRAC FIELDS

A2.1 THE DIRAC EQUATION

The Dirac equation for a particle of spin 1/2 and mass m, in the absence of external fields, is

$$i\gamma^\mu \frac{\partial \psi}{\partial x^\mu} - m\psi = 0$$

where ψ is a four-component spinor

$$\psi = \begin{pmatrix} \psi_1 \\ \psi_2 \\ \psi_3 \\ \psi_4 \end{pmatrix}$$

and the γ^μ's are 4×4 matrices satisfying the commutation relations

$$\gamma^\mu\gamma^\nu + \gamma^\nu\gamma^\mu = 2g^{\mu\nu}I$$

The sixteen 4×4 matrices

$$I$$

$$\gamma^\mu$$

$$\sigma^{\mu\nu} = \frac{i}{2}(\gamma^\mu\gamma^\nu - \gamma^\nu\gamma^\mu)$$

$$\gamma^\mu\gamma^5$$

$$\gamma^5 = -i\gamma^0\gamma^1\gamma^2\gamma^3$$

are linearly independent and form a basis for the vector space of all 4×4 matrices.

The Dirac conjugate, or *adjoint* spinor, $\bar{\psi} = \psi^\dagger\gamma^0$, satisfies the equation

$$-i\frac{\partial\bar{\psi}}{\partial x^\mu}\gamma^\mu - m\bar{\psi} = 0$$

Let ψ be a plane-wave solution of the Dirac equation, representing a fermion (e.g., an electron) with three-momentum $\mathbf{p}$ and energy $E = +\sqrt{|\mathbf{p}|^2 + m^2}$:

$$\psi = \frac{1}{\sqrt{V}} u(p, s)e^{-iEt}e^{i\mathbf{p}\cdot\mathbf{x}}$$

where V is a large, arbitrary volume. Then u is a four-component spinor satisfying the equation

$$(E\gamma^0 - \mathbf{p}\cdot\boldsymbol{\gamma} - m)u = 0$$

or in covariant notation

$$(p_\mu\gamma^\mu - m)u \equiv (\not{p} - m)u = 0$$

Similarly,

$$\bar{u}(\not{p} - m) = 0$$

where

$$\bar{u} \equiv u^\dagger\gamma^0$$

The corresponding Dirac wave function for an antifermion (e.g., a positron) with three-momentum $\mathbf{p}$ and energy $E = +\sqrt{|\mathbf{p}|^2 + m^2}$ is

$$\psi' = \frac{1}{\sqrt{V}} v(\mathbf{p}, s)e^{iEt}e^{-i\mathbf{p}\cdot\mathbf{x}}$$

where v satisfies

$$(\not{p} + m)v = 0$$

and

$$\bar{v}(\not{p} + m) = 0$$

A2.2 BILINEAR COVARIANTS

Consider two arbitrary Dirac wave functions ψ_1 and ψ_2. Under homogeneous Lorentz transformations, including reflections,

	$\bar{\psi}_1\psi_2$	transforms as a scalar
	$\bar{\psi}_1\gamma^\mu\psi_2$	transforms as a polar vector
	$\bar{\psi}_1\sigma^{\mu\nu}\psi_2$	transforms as an antisymmetric tensor
	$\bar{\psi}_1\gamma^\mu\gamma^5\psi_2$	transforms as an axial vector
and	$\bar{\psi}_1\gamma^5\psi_2$	transforms as a pseudoscalar

A2.3 REPRESENTATIONS OF γ MATRICES

A2.3.1 Weyl Representation

Here,

$$I=\begin{pmatrix} I & 0 \\ 0 & I \end{pmatrix} \qquad \gamma^0=\begin{pmatrix} 0 & I \\ I & 0 \end{pmatrix} \qquad \boldsymbol{\gamma}=\begin{pmatrix} 0 & -\boldsymbol{\sigma} \\ \boldsymbol{\sigma} & 0 \end{pmatrix} \qquad \gamma^5=\begin{pmatrix} -I & 0 \\ 0 & I \end{pmatrix}$$

where the σ's are 2×2 Pauli spin matrices satisfying

$$\sigma_r\sigma_s = \delta_{rs}I + i\varepsilon_{rst}\,\sigma_t$$

We define the helicity projection operators

$$a = \tfrac{1}{2}(I+\gamma^5)$$

$$\bar{a} = \tfrac{1}{2}(I-\gamma^5)$$

which satisfy

$$a^2 = \bar{a}^2 = I$$

$$a\bar{a} = \bar{a}a = 0$$

In the Weyl representation, these take the form

$$a=\begin{pmatrix} 0 & 0 \\ 0 & I \end{pmatrix}$$

$$\bar{a}=\begin{pmatrix} I & 0 \\ 0 & 0 \end{pmatrix}$$

We write

$$\psi=\begin{pmatrix} \chi \\ \phi \end{pmatrix}$$

where $\quad \chi=\begin{pmatrix} \psi_1 \\ \psi_2 \end{pmatrix} \quad$ and $\quad \phi=\begin{pmatrix} \psi_3 \\ \psi_4 \end{pmatrix}$

Then, in the present representation,

$$a\psi = \begin{pmatrix} 0 \\ \phi \end{pmatrix} \qquad \text{and} \qquad \bar{a}\psi = \begin{pmatrix} \chi \\ 0 \end{pmatrix}$$

In the Weyl representation the Dirac equation becomes

$$i\frac{\partial \chi}{\partial t} + i\boldsymbol{\sigma}\cdot\nabla\chi = m\phi$$

$$i\frac{\partial \phi}{\partial t} - i\boldsymbol{\sigma}\cdot\nabla\phi = m\chi$$

For plane waves of energy E and three-momentum $\mathbf{p}$ these equations reduce to

$$E\chi - \mathbf{p}\cdot\boldsymbol{\sigma}\chi = m\phi$$
$$E\phi + \mathbf{p}\cdot\boldsymbol{\sigma}\phi = m\chi$$

If $m = 0$, these equations are decoupled and

$$h\chi \equiv \frac{\boldsymbol{\sigma}\cdot\mathbf{p}}{|\mathbf{p}|}\chi = \chi$$

$$h\phi \equiv \frac{\boldsymbol{\sigma}\cdot\mathbf{p}}{|\mathbf{p}|}\phi = -\phi$$

where h is the helicity operator. Thus χ and ϕ are eigenstates of helicity with eigenvalues ± 1, respectively. We interpret χ as a (two-component) antineutrino spinor and ϕ as a neutrino spinor.

A2.3.2 The Pauli–Dirac Representation

Here

$$I = \begin{pmatrix} I & 0 \\ 0 & I \end{pmatrix} \qquad \gamma^0 = \begin{pmatrix} I & 0 \\ 0 & -I \end{pmatrix} \qquad \boldsymbol{\gamma} = \begin{pmatrix} 0 & \boldsymbol{\sigma} \\ -\boldsymbol{\sigma} & 0 \end{pmatrix} \qquad \gamma^5 = \begin{pmatrix} 0 & -I \\ -I & 0 \end{pmatrix}$$

and

$$\sigma^{ij} = \begin{pmatrix} \sigma_k & 0 \\ 0 & \sigma_k \end{pmatrix} \qquad i, j, k = 1, 2, 3\text{(cyclic)}$$

Consider a fermion of finite rest mass m at rest in frame F with spin polarization unit three-vector $\hat{\mathbf{s}}$. It is described by the four-spinor

$$u(0, \hat{\mathbf{s}}) = \sqrt{2m}\begin{pmatrix} \eta \\ 0 \end{pmatrix}$$

where η is a two-component Pauli spinor satisfying

$$\boldsymbol{\sigma}\cdot\hat{\mathbf{s}}\eta = \eta$$

$$\eta\eta^\dagger = \frac{1 + \boldsymbol{\sigma}\cdot\hat{\mathbf{s}}}{2}$$

In another Lorentz frame F' in which the particle has three-momentum $\mathbf{p}$ and energy $E=+\sqrt{|\mathbf{p}|^2+m^2}$, u becomes

$$u(\mathbf{p}, s) = \sqrt{E+m}\begin{pmatrix} \eta \\ \dfrac{\boldsymbol{\sigma}\cdot\mathbf{p}}{E+m}\eta \end{pmatrix}$$

In particular, if $\hat{\mathbf{s}} = \hat{k}$, where $\hat{k}$ is the z-axis unit vector, we have

$$u(\mathbf{p}, s) = \sqrt{E+m}\begin{pmatrix} 1 \\ 0 \\ \dfrac{\boldsymbol{\sigma}\cdot\mathbf{p}}{E+m}\begin{pmatrix}1\\0\end{pmatrix} \end{pmatrix}$$

The corresponding spinor $v(\mathbf{p}, s)$ for an antiparticle with three-momentum $+\mathbf{p}$, energy $E = +\sqrt{|\mathbf{p}|^2+m^2}$, and $\hat{\mathbf{s}} = \hat{k}$ is

$$v(\mathbf{p}, s) = \sqrt{E+m}\begin{pmatrix} \dfrac{\boldsymbol{\sigma}\cdot\mathbf{p}}{E+m}\begin{pmatrix}0\\1\end{pmatrix} \\ \begin{pmatrix}0\\1\end{pmatrix} \end{pmatrix}$$

In general,

$$\bar{u}(\mathbf{p}, s)u(\mathbf{p}, s) = 2m$$

$$\bar{v}(\mathbf{p}, s)v(\mathbf{p}, s) = -2m$$

and

$$u^\dagger(\mathbf{p}, s)u(\mathbf{p}, s) = v^\dagger(\mathbf{p}, s)s(\mathbf{p}, s) = 2E$$

For particles with finite rest mass m, s is the spin polarization four-vector, which reduces to $\hat{\mathbf{s}}$ in the rest frame:

$$s^\mu \to (0, \hat{\mathbf{s}})$$

One has

$$s^0 = \frac{\mathbf{p}\cdot\hat{\mathbf{s}}}{m}$$

$$\mathbf{s} = \hat{\mathbf{s}} + \frac{\mathbf{p}\cdot\hat{\mathbf{s}}}{m(E+m)}\,\mathbf{p}$$

Also, for m finite,

$$s\cdot s = -1$$

and

$$s\cdot p = 0$$

A2.4 USEFUL RELATIONSHIPS INVOLVING γ MATRICES

1 $(\gamma^\mu)^\dagger = \gamma^0\gamma^\mu\gamma^0$

$\gamma^\mu\gamma^5 = -\gamma^5\gamma^\mu$

$\gamma^{0\dagger} = \gamma^0$

$\boldsymbol{\gamma}^\dagger = -\boldsymbol{\gamma}$ independent of representation

2 In the Pauli–Dirac, or *standard*, representation,

$$\tilde{\gamma}^0 = \gamma^0$$
$$\tilde{\gamma}^1 = -\gamma^1 \qquad \sim \text{ means transpose}$$
$$\tilde{\gamma}^2 = \gamma^2$$
$$\tilde{\gamma}^3 = -\gamma^3$$

and

$$\gamma^{0*} = \gamma^0$$
$$\gamma^{1*} = \gamma^1$$
$$\gamma^{2*} = -\gamma^2$$
$$\gamma^{3*} = \gamma^3$$

3 $\text{Tr}\ \gamma^\mu \gamma^\nu = 4g^{\mu\nu}$

$$\text{Tr}\ (\gamma^\mu \gamma^\nu \gamma^\rho \gamma^\sigma) = 4g^{\mu\nu}g^{\rho\sigma} - 4g^{\mu\rho}g^{\nu\sigma} + 4g^{\nu\rho}g^{\mu\sigma}$$
$$\text{Tr}\ (\gamma^\mu \gamma^\nu \gamma^\rho \gamma^\sigma \gamma^5) = -4i\varepsilon^{\mu\nu\rho\sigma}$$

where $\varepsilon^{\mu\nu\rho\sigma}$ is the completely antisymmetric unit four-tensor.

$$\tfrac{1}{4}\text{Tr}[\gamma^\rho \gamma^\alpha \gamma^\sigma \gamma^\beta (1 + \gamma^5)] = \chi^{\rho\sigma\alpha\beta} = g^{\rho\alpha}g^{\sigma\beta} + g^{\alpha\sigma}g^{\rho\beta} - g^{\rho\sigma}g^{\alpha\beta} - i\varepsilon^{\rho\alpha\sigma\beta}$$

A2.5 DIRAC FIELDS

The Dirac quantum field operator $\Psi(\mathbf{x}, t)$ is usually written as a superposition of single-particle plane-wave solutions to the Dirac equation:

$$\Psi(\mathbf{x}, t) = \frac{1}{\sqrt{V}} \sum_{\pm s} \int d^3\mathbf{p}\, \frac{1}{\sqrt{2E}} \times [b(p, s)u(p, s)e^{-ip\cdot x} + d^\dagger(p, s)\, v(p, s)e^{ip\cdot x}]$$

Here the one-particle solutions $u(p, s)e^{-ip\cdot x}$, $v(p, s)e^{ip\cdot x}$ have already been defined and correspond to particle and antiparticle (e.g., electron and positron), respectively. The Dirac conjugate field is

$$\overline{\Psi}(\mathbf{x}, t) = \Psi^\dagger(\mathbf{x}, t)\gamma^0$$
$$= \frac{1}{\sqrt{V}} \sum_{\pm s} \int d^3\mathbf{p}\, \frac{1}{\sqrt{2E}} [b^\dagger(p, s)\bar{u}(p, s)e^{ip\cdot x} + d(p, s)\bar{v}(p, s)e^{-ip\cdot x}]$$

The quantities b and d are, respectively, annihilation operators for fermions and antifermions (e.g., electrons and positrons), while $b^\dagger$ and $d^\dagger$ are the corresponding creation operators. Let the vacuum state containing no electrons or positrons be written as $|0\rangle$. Then

$$b(p, s)|0\rangle = d(p, s)|0\rangle = 0$$
$$b^\dagger(p, s)|0\rangle = |e^-(p, s)\rangle$$
$$d^\dagger(p, s)|0\rangle = |e^+(p, s)\rangle$$

The operators b, $b^\dagger$, d, and $d^\dagger$ satisfy

$$
\begin{aligned}
\{b^\dagger(p, s), b(p', s')\} &= \delta_{ss'}\,\delta^3(\mathbf{p} - \mathbf{p}') \\
\{d^\dagger(p, s), d(p', s')\} &= \delta_{ss'}\,\delta^3(\mathbf{p} - \mathbf{p}') \\
\{b^\dagger(p, s), b^\dagger(p', s')\} &= \{b(p, s), b(p', s')\} = 0 \\
\{d^\dagger(p, s), d^\dagger(p', s')\} &= \{d(p, s), d(p', s')\} = 0 \\
\{b(p, s), d^\dagger(p', s')\} &= \{b^\dagger(p, s), d(p', s')\} = 0 \\
\{b(p, s), d(p', s')\} &= \{b^\dagger(p, s), d^\dagger(p', s')\} = 0
\end{aligned}
$$

APPENDIX 3

PARITY, CHARGE CONJUGATION, AND TIME REVERSAL

A3.1 PARITY

Consider the Dirac equation for a fermion (e.g., electron) with rest mass m, momentum $\mathbf{p}$, and energy $E = +\sqrt{|\mathbf{p}|^2 + m^2}$:

$$\left(i\gamma^0 \frac{\partial}{\partial t} + i\boldsymbol{\gamma} \cdot \nabla - m\right) u e^{-iEt} e^{i\mathbf{p}\cdot\mathbf{x}} = 0 \qquad \text{(A3.1)}$$

Under a space inversion (parity transformation) $\mathbf{x} \to -\mathbf{x}$, $t \to t$, this becomes

$$\left(i\gamma^0 \frac{\partial}{\partial t} - i\boldsymbol{\gamma} \cdot \nabla - m\right) u e^{-iEt} e^{-i\mathbf{p}\cdot\mathbf{x}} = 0 \qquad \text{(A3.2)}$$

which is no longer a Dirac equation. However, multiplying on the left by $\eta_p \gamma^0$ where η_p is some constant and noting that

$$\gamma^0 \boldsymbol{\gamma} = -\boldsymbol{\gamma}\gamma^0$$

we obtain

$$\left(i\gamma^0 \frac{\partial}{\partial t} + i\boldsymbol{\gamma} \cdot \nabla - m\right) \eta_p \gamma^0 u(\mathbf{p}, s) e^{-iEt} e^{-i\mathbf{p}\cdot\mathbf{x}} = 0 \qquad \text{(A3.3)}$$

Since $$\gamma^0 u(\mathbf{p}, s) = u(-\mathbf{p}, s)$$

and $$\gamma^0 v(\mathbf{p}, s) = -v(-\mathbf{p}, s)$$

we obtain

$$\left(i\gamma^0 \frac{\partial}{\partial t} + i\boldsymbol{\gamma} \cdot \nabla - m\right) \eta_p u(-\mathbf{p}, s) e^{-iEt} e^{i(-\mathbf{p}) \cdot \mathbf{x}} = 0 \tag{A3.4}$$

which is a Dirac equation for a particle of momentum $-\mathbf{p}$.

In general, for some arbitrary one-particle wave function $\psi(x, t)$, the Dirac equation is invariant under the replacement $\mathbf{x} \to -\mathbf{x}$ if we also make the replacement $\psi(\mathbf{x}, t) \to \eta_p \gamma^0 \psi(-\mathbf{x}, t)$. The transformation is unitary if $|\eta_p| = 1$. As in the single-particle case, in order to preserve the invariance of the Dirac equation under spatial inversion, it is necessary to transform the Dirac field $\Psi(\mathbf{x}, t)$ to a new field $\eta_p \gamma^0 \Psi(-\mathbf{x}, t)$:

$$\hat{P} \Psi(\mathbf{x}, t) \hat{P}^{-1} = \eta_p \gamma^0 \Psi(-\mathbf{x}, t)$$

and similarly we must have

$$\hat{P} \overline{\Psi}(\mathbf{x}, t) \hat{P}^{-1} = \eta_p^* \overline{\Psi}(-\mathbf{x}, t) \gamma^0$$

Let F be a 4×4 matrix. Since

$$\gamma^0 u(\mathbf{p}, s) = u(-\mathbf{p}, s)$$

and $$\gamma^0 v(\mathbf{p}, s) = -v(-\mathbf{p}, s) \tag{A3.5}$$

we have

$$\bar{u}_a(\mathbf{p}, s) F u_b(\mathbf{p}', s') \to \bar{u}_a(\mathbf{p}, s) \gamma^0 F \gamma^0 u_b(\mathbf{p}', s')$$

$$\bar{v}_a(\mathbf{p}, s) F v_b(\mathbf{p}', s') \to \bar{v}_a(\mathbf{p}, s) \gamma^0 F \gamma^0 v_b(\mathbf{p}', s')$$

and $$\bar{u}_a(\mathbf{p}, s) F v_b(\mathbf{p}', s') \to -\bar{u}_a(\mathbf{p}, s) \gamma^0 F \gamma^0 v_b(\mathbf{p}', s') \tag{A3.6}$$

under spatial inversion.

A3.2 CHARGE CONJUGATION

The one-particle Dirac equation for an electron with charge $e (e < 0)$ in an external electromagnetic field A_μ is

$$i\gamma^\mu \frac{\partial \psi}{\partial x^\mu} - e\gamma^\mu A_\mu \psi - m\psi = 0 \tag{A3.7}$$

A positron has charge $-e$, and in the same field it is described by wave function $\psi'(x, t)$:

$$i\gamma^\mu \frac{\partial \psi'}{\partial x^\mu} + e\gamma^\mu A_\mu \psi' - m\psi' = 0 \tag{A3.8}$$

Taking the complex conjugate of Eq. (A3.7) and multiplying on the left by a nonsingular matrix Γ chosen so that $\Gamma\gamma^{\mu*}\Gamma^{-1} = -\gamma^{\mu}$, we obtain

$$i\gamma^{\mu}\frac{\partial}{\partial x^{\mu}}(\Gamma\psi^*) + e\gamma^{\mu}A_{\mu}(\Gamma\psi^*) - m(\Gamma\psi^*) = 0$$

which is the same equation as (A3.8) if we identify $\Gamma\psi^*$ with ψ'. Setting $A_{\mu} = 0$, ψ and ψ' reduce to free-particle solutions and we have

$$\Gamma[u(p, s)e^{-ip\cdot x}]^* = v(p, s)e^{ip\cdot x}$$

or

$$\Gamma u^*(p, s) = v(p, s)$$

A solution to the equations $\Gamma\gamma^{\mu*}\Gamma^{-1} = -\gamma^{\mu}$ is $\Gamma = i\gamma^2$ in the standard representation. Also,

$$\tilde{\bar{u}} = \gamma^0 u^*$$

Thus, under charge conjugation,

$$u(p, s) \to i\gamma^2\gamma^0\tilde{\bar{u}}(p, s) \equiv M\tilde{\bar{u}}(p, s) = v(p, s) \tag{A3.9}$$

and also

$$v(p, s) \to i\gamma^2\gamma^0\tilde{\bar{v}}(p, s) = M\tilde{\bar{v}}(p, s) = u(p, s) \tag{A3.10}$$

Similarly, we can show that the Dirac fields transform as follows under charge conjugation:

$$\hat{C}\Psi(\mathbf{x}, t)\hat{C}^{-1} = \eta_c{}^* M\tilde{\bar{\Psi}}(\mathbf{x}, t) \tag{A3.11}$$

and

$$\hat{C}\bar{\Psi}(\mathbf{x}, t)\hat{C}^{-1} = \eta_c\tilde{\Psi}(\mathbf{x}, t)M \tag{A3.12}$$

where η_c is some constant. If $|\eta_c| = 1$, $\hat{C}$ is unitary.

$M = i\gamma^2\gamma^0$ has the following properties:

$$M = M^* = -M^{\dagger} = -\tilde{M}$$

$$M^2 = -I$$

Under $\hat{C}$ a bilinear form $\bar{u}_a F u_b$ transforms as follows:

$$\bar{u}_a F u_b \to \tilde{u}_a \gamma^0\gamma^2 F\gamma^2\gamma^0\tilde{\bar{u}}_b$$

$$= \bar{u}_b(\gamma^0\gamma^2\tilde{F}\gamma^2\gamma^0)u_a \tag{A3.13}$$

Similarly,

$$\bar{u}_a F v_b \to \bar{v}_b(\gamma^0\gamma^2\tilde{F}\gamma^2\gamma^0)u_a \tag{A3.14}$$

and

$$\bar{v}_a F v_b \to \bar{v}_b(\gamma^0\gamma^2\tilde{F}\gamma^2\gamma^0)v_a \tag{A3.15}$$

A3.3 TIME REVERSAL

Whereas the parity and charge-conjugation transformations are unitary, the time-reversal ($\hat{T}$) transformation is antiunitary. This follows from the requirement that

$$\hat{T}\hat{\mathbf{J}}\hat{T}^{-1} = -\hat{\mathbf{J}}$$

and

$$\hat{T}\hat{\mathbf{p}}\hat{T}^{-1} = -\hat{\mathbf{p}}$$

where $\hat{\mathbf{J}}$ and $\hat{\mathbf{p}}$ are angular momentum and linear momentum operators, respectively, together with the requirement that $\hat{T}$ leave invariant the fundamental quantum mechanical commutation relations $[\hat{\mathbf{x}}, \hat{\mathbf{p}}] = i$ and $\hat{\mathbf{J}} \times \hat{\mathbf{J}} = i\hat{\mathbf{J}}$.

Consider the one-particle Dirac equation for an electron in an external electromagnetic field ($A^0 = V, A$):

$$i\gamma^0 \frac{\partial\psi(t)}{\partial t} - eV(t)\gamma^0\psi(t) = -i\boldsymbol{\gamma} \cdot \boldsymbol{\nabla}\psi(t) - e\mathbf{A}(t) \cdot \boldsymbol{\gamma}\psi(t) + m\psi(t) \qquad \text{(A3.16)}$$

Here, the dependence of V, A, and ψ on $\mathbf{x}$ is understood. Replacing t by $t' = -t$, we have $V(-t) = V(t)$ and $\mathbf{A}(-t) = -\mathbf{A}(t)$. Thus

$$-i\gamma^0 \frac{\partial\psi(-t')}{\partial t'} - eV(t')\gamma^0\psi(-t') = -i\boldsymbol{\gamma} \cdot \boldsymbol{\nabla}\psi(-t') + e\mathbf{A}(t') \cdot \boldsymbol{\gamma}\psi(-t') + m\psi(-t') \qquad \text{(A3.17)}$$

Dropping the prime and multiplying on the left bv the antiunitary operator $\hat{T}$, with $\hat{T}\psi(-t) = \psi'(t)$, we have

$$-\hat{T}(i\gamma^0)\hat{T}^{-1} \frac{\partial}{\partial t}\psi'(t) - eV(t)\hat{T}\gamma^0\hat{T}^{-1}\psi'(t)$$
$$= -\hat{T}i\boldsymbol{\gamma}\hat{T}^{-1} \cdot \boldsymbol{\nabla}\psi'(i) + e\mathbf{A}(t) \cdot \hat{T}\boldsymbol{\gamma}\hat{T}^{-1}\psi'(t) + m\psi'(t) \qquad \text{(A3.18)}$$

Now we choose

$$\begin{aligned} \hat{T}i\gamma^0\hat{T}^{-1} &= -i\gamma^0 \\ \hat{T}\gamma^0\hat{T}^{-1} &= \gamma^0 \\ \hat{T}i\boldsymbol{\gamma}\hat{T}^{-1} &= i\boldsymbol{\gamma} \\ \hat{T}\boldsymbol{\gamma}\hat{T}^{-1} &= -\boldsymbol{\gamma} \end{aligned} \qquad \text{(A3.19)}$$

recalling that $\hat{T}$ is antiunitary. Then (A3.18) reduces to the original (A3.16) with a time-reversed solution $\psi'(x, t) = \hat{T}\psi(x, -t)$. Equations (A3.19) define the properties of the antiunitary operator $\hat{T}$ as follows, in the *standard* representation: $\hat{T}$ is antiunitary: $\hat{T} = T_0 \hat{K}$ where T_0 is a unitary matrix, and $\hat{K}$ is the *complex conjugation* operator.

$$\begin{aligned} T_0\gamma^0 &= \gamma^0 T_0 \\ T_0\gamma^1 &= -\gamma^1 T_0 \\ T_0\gamma^2 &= \gamma^2 T_0 \\ T_0\gamma^3 &= -\gamma^3 T_0 \end{aligned}$$

Thus we find $T_0 = i\gamma^1\gamma^3$ and

$$\hat{T}\psi(\mathbf{x}, t) = \psi'(\mathbf{x}, -t) = T_0\psi^*(\mathbf{x}, t) = i\gamma'\gamma^3\psi^*(\mathbf{x}, t) \qquad \text{(A3.20)}$$

For the free-particle solutions,

$$\begin{aligned}
\hat{T}[u(\mathbf{p}, \hat{k})e^{-ip\cdot x}] &= -iu(-\mathbf{p}, -\hat{k})e^{-iE(-t)}e^{i(-\mathbf{p})\cdot\mathbf{x}} \\
\hat{T}[u(\mathbf{p}, -\hat{k})e^{-ip\cdot x}] &= +iu(-\mathbf{p}, +\hat{k})e^{-iE(-t)}e^{i(-\mathbf{p})\cdot\mathbf{x}} \\
\hat{T}[v(\mathbf{p}, -\hat{k})e^{ip\cdot x}] &= -iv(-\mathbf{p}, +\hat{k})e^{iE(-t)}e^{-i(-\mathbf{p})\cdot\mathbf{x}} \\
\hat{T}[v(\mathbf{p}, +\hat{k})e^{ip\cdot x}] &= +iv(-\mathbf{p}, -\hat{k})e^{iE(-t)}e^{-i(-\mathbf{p})\cdot\mathbf{x}}
\end{aligned} \tag{A3.21}$$

As we expect, the time-reversal transformation reverses the spins and momenta of the particles and changes the sign of the time. (Also it introduces certain phase factors $\pm i$ which are defined by our choice of the factor $+i$ in matrix T_0.) Under a time-reversal transformation $\hat{T}$, the bilinear form $\bar{u}_a(p, s)Fu_b(p, s)$ transforms as follows:

$$\bar{u}_a(p, s)Fu_b(p, s) \to \bar{u}_b(p, s)\gamma^0\gamma^1\gamma^3\tilde{F}\gamma^3\gamma^1\gamma^0 u_a(p, s) \tag{A3.22}$$

A3.4 TRANSFORMATIONS OF BILINEAR FORMS

Collecting Eqs. (A3.6), (A3.13) to (A3.15), and (A.3.22), we have the results presented in Table A3.1.

Table A3.1 TRANSFORMATION PROPERTIES OF BILINEAR FORMS

standard representation

$\bar{u}_a = u_a^\dagger\gamma^0$	$\gamma^5 = -i\gamma^0\gamma^1\gamma^2\gamma^3$	$\gamma^0 = \begin{pmatrix} I & 0 \\ 0 & -I \end{pmatrix}$
$\gamma^\mu = \gamma^0, \boldsymbol{\gamma}$	$\sigma^{\mu\nu} = \frac{i}{2}(\gamma^\mu\gamma^\nu - \gamma^\nu\gamma^\mu)$	$\boldsymbol{\gamma} = \begin{pmatrix} 0 & \boldsymbol{\sigma} \\ -\boldsymbol{\sigma} & 0 \end{pmatrix}$
$\gamma_\mu = \gamma^0, -\boldsymbol{\gamma}$	$\sigma_{\mu\nu} = \frac{i}{2}(\gamma_\mu\gamma_\nu - \gamma_\nu\gamma_\mu)$	$\gamma^5 = \begin{pmatrix} 0 & -I \\ -I & 0 \end{pmatrix}$

	P	C	T	Hermitian conjugate
$\bar{u}_b F u_a \to$	$\bar{u}_b\gamma^0F\gamma^0u_a$	$\bar{u}_a M\tilde{F}Mu_b$	$\bar{u}_a\gamma^0TFT\gamma^0u_b$	$\bar{u}_a(\gamma^0F^\dagger\gamma^0)u_b$
$\bar{u}_b u_a$	$\bar{u}_b u_a$	$-\bar{u}_a u_b$	$\bar{u}_a u_b$	$\bar{u}_a u_b$
$\bar{u}_b\gamma^\mu u_a$	$\bar{u}_b\gamma_\mu u_a$	$\bar{u}_a\gamma^\mu u_b$	$\bar{u}_a\gamma_\mu u_b$	$\bar{u}_a\gamma^\mu u_b$
$\bar{u}_b\sigma^{\mu\nu}u_a$	$\bar{u}_b\sigma_{\mu\nu}u_a$	$\bar{u}_a\sigma^{\mu\nu}u_b$	$-\bar{u}_a\sigma_{\mu\nu}u_b$	$\bar{u}_a\sigma u^{\mu\nu}u_b$
$\bar{u}_b\gamma^5u_a$	$-\bar{u}_b\gamma^5u_a$	$-\bar{u}_a\gamma^5u_b$	$-\bar{u}_a\gamma^5u_b$	$-\bar{u}_a\gamma^5u_b$
$\bar{u}_b\gamma^\mu\gamma^5u_a$	$-\bar{u}_b\gamma_\mu\gamma^5u_a$	$-\bar{u}_a\gamma^\mu\gamma^5u_b$	$\bar{u}_a\gamma_\mu\gamma^5u_b$	$\bar{u}_a\gamma^\mu\gamma^5u_b$
$\bar{u}_b\sigma^{\mu\nu}\gamma^5u_a$	$-\bar{u}_b\sigma_{\mu\nu}\gamma^5u_a$	$\bar{u}_a\sigma^{\mu\nu}\gamma^5u_b$	$\bar{u}_a\sigma_{\mu\nu}\gamma^5u_b$	$-\bar{u}_a\sigma^{\mu\nu}\gamma^5u_b$

INDEX